TRUST THEORY

Wiley Series in Agent Technology

Series Editor: Michael Wooldridge, *University of Liverpool, UK*

The 'Wiley Series in Agent Technology' is a series of comprehensive practical guides and cutting-edge research titles on new developments in agent technologies. The series focuses on all aspects of developing agent-based applications, drawing from the Internet, telecommunications, and Artificial Intelligence communities with a strong applications/technologies focus.

The books will provide timely, accurate and reliable information about the state of the art to researchers and developers in the Telecommunications and Computing sectors.

Titles in the series:

Padgham/Winikoff: Developing Intelligent Agent Systems 0-470-86120-7 (June 2004)

Bellifemine/Caire/Greenwood: Developing Multi-Agent Systems with JADE 0-470-05747-5 (February 2007)

Bordini/Hübner/Wooldrige: Programming Multi-Agent Systems in AgentSpeak using Jason 0-470-02900-5 (October 2007)

Nishida: Conversational Informatics: An Engineering Approach 0-470-02699-5 (November 2007)

Jokinen: Constructive Dialogue Modelling: Speech Interaction and Rational Agents 0-470-06026-3 (April 2009)

TRUST THEORY

A SOCIO-COGNITIVE AND COMPUTATIONAL MODEL

Cristiano Castelfranchi

Institute of Cognitive Sciences and Technologies (ISTC) of the Italian National Research Council (CNR), Italy

Rino Falcone

Institute of Cognitive Sciences and Technologies (ISTC) of the Italian National Research Council (CNR), Italy

A John Wiley and Sons, Ltd., Publication

This edition first published 2010
© 2010 John Wiley & Sons Ltd.,

Registered office
John Wiley & Sons Ltd, The Atrium, Southern Gate, Chichester, West Sussex, PO19 8SQ, United Kingdom

For details of our global editorial offices, for customer services and for information about how to apply for permission to reuse the copyright material in this book please see our website at www.wiley.com.

Library of Congress Cataloging-in-Publication Data

Castelfranchi, Cristiano.
    Trust theory : a socio-cognitive and computational model / Cristiano Castelfranchi, Rino Falcone.
      p. cm.
    Includes index.
    ISBN 978-0-470-02875-9 (cloth)
    1. Trust.   2. Trust–Simulation methods.   3. Artificial intelligence–Psychological aspects.   4. Cognitive science.
I. Falcone, Rino.   II. Title.
    BF575.T7C37 2010
    302'.1–dc22

                                                                                        2009040166

A catalogue record for this book is available from the British Library.

ISBN 9780470028759 (H/B)

Typeset in 10/12pt Times by Aptara Inc., New Delhi, India
Printed and Bound in Singapore by Markono Print Media Pte Ltd.

This book is dedicated
to Mario, Ketty, Eugenio, and Maria, our roots,
to Rosanna and Ivana, our life companions;
to Yurij, Vania and Giulio, our realized dreams;
to the many collegues who consciously or unconsciuously contributed to the
ideas included in it;
to our Country, better to the Idea of Country, of Collective, of Institutional
Entity, in which as such it is possible develop socially centred hopes,
ambitions, dreams, so contributing to the dignity and future for any individual.

Contents

For a schematic view of the main terms introduced and analyzed in this book see the Trust, Theory and Technology site at http://www.istc.cnr.it/T3/.

Foreword

I turn up to give a lecture at 9 am on a Monday morning, trusting that my students will attend; and they in turn reluctantly drag themselves out of bed to attend, trusting that I will be there to give the lecture. When my wife tells me that she will collect our children from school, I expect to see the children at home that night safe and sound. Every month, I spend money, trusting that, on the last Thursday of the month, my employer will deposit my salary in my bank account; and I trust my bank to safeguard this money, investing my savings prudently. Sometimes, of course, my trust is misplaced. Students don't turn up to lectures; my bank makes loans to people who have no chance of repaying them, and as a consequence they go bankrupt, taking my savings with them. But despite such disappointments, our lives revolve around trust: we could hardly imagine society functioning without it.

The rise of autonomous, computer-based agents as a technology gives trust an interesting new dimension. Of course, one issue is that we may not be comfortable trusting a computer program to handle our precious savings. But when software agents interact with people an entirely new concern arises: why or how should a computer program trust 'us'? How can we design computer programs that are safe from exploitation by un-trustworthy people? How can we design software agents that can understand how trust works in human societies, and live up to human expectations of trust? And what kind of models of trust make sense when software agents interact with 'other' software agents?

These considerations have led to attempts by cognitive scientists, computer scientists, psychologists, and others, to develop models of trust, and to implement these tentative models of trust in computer programs. The present book is the first comprehensive overview of the nascent field of modeling trust and computational models of trust. It discusses trust and the allied concept of reputation from a range of different backgrounds. It will be essential reading for anybody who wants to understand the issues associated with building computer systems that work with people in sensitive situations, and in particular for researchers in multi-agent systems, who will deploy and build on the techniques and concepts presented herein. The journey to understand trust from a scientific, technological, and computational perspective may only just have begun, but this book represents a critical milestone on that journey.

Michael Wooldridge

Introduction

The aim of this book, carried out in quite a user-friendly way, is clear from its title: to systematize a general *theory* of 'trust'; to provide an organic *model* of this very complex and dynamic phenomenon on cognitive, affective, social (interactive and collective) levels.

Why approach such a scientific project, not only from the point of view of Cognitive and Behavioral Sciences, but also from Artificial Intelligence (AI) and in particular 'Agent' theory domains? Actually, trust for Information and Communication Technologies (ICT) is for us just an application, a technological domain. In particular, we have been working (with many other scholars)[1] in promoting and developing a tradition of studies about trust with Autonomous Agents and in Multi-Agent Systems (MAS). The reason is that we believe that an AI oriented approach can provide – without reductionisms – good systematic and operational instruments for the explicit and well-defined representation of goals, beliefs, complex mental states (like expectations), and their dynamics, and also for modeling social action, mind, interaction, and networks. An AI approach with its programmatic 'naiveté' (but being careful to avoid simplistic assumptions and reductions of trust to technical tricks – see Chapter 12) is also useful for revising the biasing and distorting 'traditions' that we find in specific literature (philosophy, psychology, sociology, economics, etc.), which is one of the causes of the recognized 'babel' of trust notions and definitions (see below, Section 0.2).

However, our 'tradition' of research at ISTC-CNR (Castelfranchi, Falcone, Conte, Lorini, Miceli, Paglieri, Paolucci, Pezzulo, Tummolini, and many collaborators like Poggi, De Rosis, Giardini, Piunti, Marzo, Calvi, Ulivieri, and several others) is a broader and Cognitive

---

[1] We are grateful to our collegues and friends in the AI Agent community discussing these issues with us for the last 10 years: Munindar Singh, Yao-Hua Tan, Suzanne Barber, Jordi Sabater, Olivier Boissier, Robert Demolombe, Andreas Herzig, Andrew Jones, Catholijn Jonker, Audun Josang, Stephen Marsh, Carles Sierra. And also to other colleagues from different communities, like Michael Bacharach, Sandro Castaldo, Michele Costabile, Roderick Kramer, Vittorio Pelligra, Raimo and May Tuomela. The following articles have been reproduced in this book: Cristiano Castelfranchi and Rino Falcone, *Principles of trust for MAS: Cognitive Anatomy, Social Importance, and Quantification,* Proceedings of the International Conference on Multi-Agent Systems (ICMAS'98), Paris, July, pp.72–79 (1998). Reproduced by Permission of ©1998 IEEE. Cristiano Castelfranchi and Rino Falcone, *The Human in the Loop of a Delegated Agent: The Theory of Adjustable Social Autonomy,* IEEE Transactions on Systems, Man, and Cybernetics, Part A: Systems and Humans, Special Issue on "Socially Intelligent Agents - the Human in the Loop, 31(5): 406–418, September 2001. Reproduced by Permission of ©2001 IEEE

Trust Theory   Cristiano Castelfranchi and Rino Falcone
© 2010 John Wiley & Sons, Ltd

Science-oriented tradition: to systematically study the *'cognitive mediators of social action'*: that is, the mental representations supporting social behaviors and collective and institutional phenomena; like: cooperation, social functions, norms, power, social emotions (admiration, envy, pity, shame, guilt, etc.).

Thus, trust was an unavoidable and perfect subject: on the one hand, it is absolutely crucial for social interaction and for collective and institutional phenomena (and one should explain 'why'); on the other hand, it is a perfect example of a necessary cognitive 'mediator' of sociality, and of integration of mind and interaction, of epistemic and motivational representations, of reasoning and affects. Our effort is in this tradition and frame (see below, Section 0.3).

Respecting and Analyzing Concepts

Quite frequently in science (especially in the behavioral and social sciences, which are still in search of their paradigmatic status and recognition) *'Assimilation'* (in Piaget's terms)[2] prevails on *'Accommodation'*.

That is, the *simplification* of factual data, the *reduction* of real phenomena in order they fit within the previously defined 'schemes', and in order to confirm the existing theories with their conceptual apparatus, strongly prevails on the adjustment of the concepts and schemes to the complexity and richness of the phenomenon in object.

In such a way, well-defined (and possibly formalized) schemes become blinkers, a too rigid and arbitrary filter of reality. Paradoxically reality must conform to theory, which becomes not just – as needed – abstract, parsimonious, 'ideal-type', and 'normative', but becomes 'prescriptive'. Scholars no longer try to develop a good general theory of 'trust' as conceived, used, perceived in 'natural' (cultural) contexts; they prescribe what 'trust' *should* be, in order to fit with their intangible theoretical apparatuses and previous defined basic notions. They deform their object by (i) pruning what is not interesting for their discipline (in its consolidated current asset), and by (ii) forcing the rest in its categories.

In this book, we try to assume an 'Accommodation' attitude.[3] For three reasons:

First, because the current trust 'ontology' is really a recognized mess, not only with a lot of domain-specific definitions and models, but with a lot of strongly contradictory notions and claims.

Second, because the separation from the current (and useful) notion of trust (in common sense and languages) is too strong, and loses too many interesting aspects and properties of the social/psychological phenomenon.

Third, because we try to show that all those ill-treated aspects not only deserve some attention, but are much more coherent than supposed, and can be unified and grounded in a principled way.

---

[2] See, for simplicity: http://www.learningandteaching.info/learning/assimacc.htm; http://projects.coe.uga.edu/epltt/index.php?title=Piaget%27s_Constructivism.

[3] Notice that both attitudes are absolutely natural and necessary for good cognitive development and adaptation; stabilizing categories and schemes; adjusting them when they become too deforming or selective.

The Characteristics of Our Trust Model

What we propose (or try to develop) is in fact a model with the following main features:

1) An *integrated model*
   - A definition/concept and a pre-formal 'model' that is composite or better layered: with various constituents and a core or kernel. A model that is able to assemble in a principled way 'parts' that are usually separated or lost for mere disciplinary and reductive interests.
   - Not a summation of features and aspects, but a 'gestalt'; a complex *structure* with specific components, relations, and functions.
   - A model apt to explain and justify in a coherent and non *ad hoc* way the various properties, roles, functions, and definitions of 'trust'.
2) *A socio-cognitive model*
   Where 'cognitive' does not mean 'epistemic' (knowledge), but means 'mental' (explicit mental representations); including motivational representations (various goal families).
   Trust should not be reduced to epistemic representations, 'beliefs' (like in many definitions that we will discuss: grounded prevision; subjective probability of the event; strength of the belief; statistic datum; etc.). Our 'integration' is architectural and 'pragmatic' where beliefs are integrated with *motivation* (goals and resultant affectivity, which is goal-based) and with *action*, and the consequential social effects and relations.
3) An *analytic* and *explicit* model
   Where the various components or 'ingredients' (epistemic, motivational, of action, and relational) are represented in an explicit format, ready to be formalized, and so on. And based on a 'normative' ('ideal-typical') frame in terms of those explicit mental constituents. However, there is a clear claim that this is just the prototypical model, the 'ideal' reference for analytical reasons. But *there are implicit and basic forms of trust*, either routine-based, mindless, and automated, or merely 'felt' and affect-based forms. In these 'implicit' forms, the same 'constituents' are just potentially present or are present in a tacit, procedural way; just primitive forerunners of the explicit advanced representations, but with the same functions: equifinal. This is, for example, the distinction between the true 'cognitive evaluation' and the 'affective appraisal' (see Chapter 8).
4) A *multi-factor and multi-dimensional* model of trustworthiness and of trust, and a *recursive* one.
   Where trust in agent Y is based on beliefs about its powers, qualities, capacities; which actually are the basis for the global trust in Y, but also are sub-forms of trust: trust in specific virtues of Y (like 'persistence', 'loyalty', 'expertise', etc.).
5) A *dynamic* model
   Where trust is not just a fixed attitude, or a context independent disposition, or the stable result of our beliefs and expectations about Y. But is context dependent, reasoning dependent, self-feed; and also reactive and interactive. There are two kinds of dynamics: one is 'internal' (mind-decision-action); the other is 'external': the dynamics of interactive, relational, and network trust links. And, not forgetting, they are intertwined.
6) A *structurally related notion*
   On such a basis, one should provide an explicit, justified, and systematic theory of the relationships between the notion/phenomenon of trust and other strongly related notions/phenomena: previsions, expectations, positive evaluations, trustworthiness,

uncertainty and risk, reliance, delegation, regularities and norms, cooperation, reputation, safety and security, and so on; and correlated emotions: relaxation and feeling safe, surprise, disappointment, betrayal, and so on.

7) *A non-prescriptive* model

We do not want to claim *'This is the "real", right, meaning; the correct use. The other current uses are mistaken or inappropriate uses of common languages'*. For example: *'Trust is just based on regularities and rules; even when the prevision is something that I do not like/want, that I worry about'*, or: *'The only true trust is the moral and personal one; the one that can be betrayed'*; or again: *'There is no trust when there are contracts, laws, authorities involved'*; *'Trust is there only in reciprocal and symmetric situations'*; and so on.

Our aim is not to abuse the concept, but at the same time to be able to situate in a precise and justified way a given special condition or property (like: 'grounded prediction') as a possible (and frequent) sub-component and usual basis of trust; or to categorize the moral-trust or the purely personal trust as (important) *types* of interpersonal trust.

The Structure of the Book

We start with a 'landscape' of the definitional debate and confusion; and with the discussion of some important definitions containing crucial ingredients. We try to show how and why some unification and abstraction is possible.

In Chapter 2 we present in a systematic way our basic model. How trust is not only a disposition or a set of beliefs (evaluation or prediction), but also a decision to rely on, and the following 'act' of, and the consequential social relation; and how these layers are embedded one into the other. How trust is not only about 'reliability' but also about 'competence', and about feeling safe, not being exposed to harms. How trust presupposes specific mental representations: evaluations, expectations, goals, beliefs of 'dependence', etc. How trust implies an 'internal' attribution to the trustee, based on external cues. How there are broader notions and more strict (but coherent) ones: like 'genuine' trust, relying on the other's goal-adoption (help), or trust relying on his 'morality'.

In Chapter 3 we present the quantification of trust. How trust has various strengths and degrees (precisely on the basis of its constituents: beliefs, goals). How trust enters the decision to delegate or not to delegate a task. How it copes with perceived risk and uncertainty. How we can say that trust is too great or too little.

In Chapter 4 we try to better understand the trust concept analyzing strictly related notions like: lack of trust, mistrust, diffidence. We also consider and develop the role of implicit trust: so relevant in many social actions.

In Chapter 5 we consider the affective trust. Even if in this book the emotional trust is (deliberately) a bit neglected, we briefly analyze this aspect and evaluate its relevance and show its interactions and influences with the more rational (reason-based) part.

In Chapter 6 trust dynamics is presented in its different aspects: how trust changes on the basis of the trustor's experiences; how trust is influenced by trust; how diffuse trust diffuses trust; how trust can change using generalization reasoning.

In Chapter 7 we consider the very interesting relationships between trust, control and autonomy, also with respect to the potential autonomy adjustments. In particular we show how very often, in the relationships between trust and control, some relevant aspects are neglected.

In Chapter 8 we present our deep disagreement with that economical point of view which reduces trust to a trivial quantification and measure of the costs/benefits ratio and risks, sacrificing a large part of the psychological and social aspects.

In Chapter 9 we underline the role of trust in Social Order, both as institutional, systemic glue producing shared rules, and as spontaneous, informal social relationships. In fact, we present trust as the basis of sociality.

In Chapter 10 we change the point of view in the trust relationship, moving to the trustee's side and analyzing how its own trustworthiness can be exploited as a relational capital. We consider in general terms the differences between simple dependence networks and trust networks.

In Chapter 11 we show a fuzzy implementation of our socio-cognitive model of trust. Although very simple and reduced, the results of the implementations present an interesting picture of the trust phenomenon and the relevance of a socio-cognitive analysis of it.

In Chapter 12 we present the main technological approaches to trust with their merits and limits. The growth of studies, models, experiments, research groups, and applications show how much relevance trust is gaining in this domain. How the bottleneck of technology can be measured by the capacity of integrating effective social mediators in it.

In Chapter 13 we draw conclusions and also present a potential challenge field: the interactions between neuro-trust (referring to the studies on the neurobiological evidence of trust) and the theoretical (socio-cognitive) model of trust: without this interaction the description of the phenomenon is quite poor, incomplete and with no prediction power.

For a schematic view of the main terms introduced and analyzed in this book see the Trust, Theory and Technology site at http://www.istc.cnr.it/T3/.

1

Definitions of Trust: From Conceptual Components to the General Core

In this chapter we will present a thorough review of the predominant definitions of trust in the literature, with the purpose of showing that, in cognitive and social sciences, there is not yet a shared or prevailing, and clear and convincing notion of *trust*. Not surprisingly, this appalling situation has engendered frequent and diffuse complaints.[1] However, the fact that the use of the term *trust* and its analytical definition are confused and often inaccurate should not become an unconscious alibi, a justification for abusing this notion, applying it in any *ad hoc* way, without trying to understand if, beyond the various specific uses and limited definitions, *there is some common deep meaning, a conceptual core to be enlightened.*

On the contrary, most authors working on trust provide their own definition, which frequently is not really general but rather tailored for a specific domain (commerce, politics, technology, organization, security, etc.). Moreover, even definitions aimed at being general and endowed with some cross-domain validity are usually incomplete or redundant: either they miss or leave implicit and give for presupposed some important components of trust, or they attribute to the general notion something that is just accidental and domain-specific.

The consequence is that there is very little overlapping among the numerous definitions of trust, while a strong common conceptual kernel for characterizing the general notion has yet to emerge. So far the literature offers only partial convergences and 'family resemblances' among different definitions, i.e. some features and terms may be common to a subset of definitions but not to other subsets.

This book aims to counteract such a pernicious tendency, and tries to provide *a general, abstract, and domain-independent notion and model of trust.*

---

[1] See for example Mutti (1987: 224): 'the number of meanings attributed to the idea of trust in social analysis is disconcerting. Certainly this deplorable state of things is the product of a general *theoretical negligence*. It is almost as if, due to some strange self-reflecting mechanism, social science has ended up losing its own trust in the possibility of considering trust in a significant way'.

Trust Theory   Cristiano Castelfranchi and Rino Falcone
© 2010 John Wiley & Sons, Ltd

This theoretical framework:

- should take inspiration from and further analyze the common-sense notion of trust (as captured by natural languages), as well as the intuitive notions frequently used in the social sciences, but
- should also define a technical scientific construct for a precise characterization of trust in cognitive and social theory, while at the same time
- accounting for precise relationships with the most important current definitions of trust, in order to show what they all have in common, regardless of their different terminological formulations.

We believe this generalization and systematization to be both possible and necessary. In this chapter, we will start identifying the most recurrent and important features in trust definitions, to describe them and explain their hidden connections and gaps. This will be instrumental to a twofold purpose: on the one hand, we will show how our layered definition and quite sophisticated model can account for those features of trust that appear to be most fundamental; on the other hand, we will discuss why other aspects of current definitions of trust are just local, i.e. relevant only for a very specific problem or within a restricted domain. In this analysis, we will take as initial inspiration Castaldo's content analysis of trust definitions (Castaldo, 2002).

This critical effort will serve both to clarify the distinctive features of our own perspective on trust, and to highlight the most serious limitations of dominant current approaches.

1.1 A Content Analysis

In dealing with the current 'theoretical negligence' and conceptual confusion in trust definitions, Castaldo (Castaldo, 2002) applied a more descriptive and empirical approach, rather different but partially complementary to our own. Castaldo performed a content analysis of 72 definitions of trust (818 terms; 273 different terms), as employed in the following domains: Management (46%), Marketing (24%), Psychology (18%), and Sociology (12%). The survey covered the period from the 1960s to the 1990s, as described in Table 1.1:

Table 1.1 Number of trust definitions in different periods

Year	Definitions	Fraction
1960–69	4	(5.6%)
1970–79	5	(7.0%)
1980–89	19	(26.4%)
1990–99	44	(51.0%)
Total	72	(100.0%)

This table is from Castaldo. For more sophisticated data and comments, based on cluster analysis, see (Castaldo, 2002).

Source: Reproduced with kind permission of © 2002 Società editrice il Mulino.

This analysis is indeed quite useful, since it immediately reveals the degree of confusion and ambiguity that plagues current definitions of trust. Moreover, it also provides a concrete framework to identify empirically different 'families' of definitions, important conceptual nucleuses, necessary components, and recurring terms. Thus we will use these precious results as a first basis for comparison and a source of inspiration, and only later will we discuss in detail specific definitions and models of trust.

Castaldo summarizes the results of his analysis underlining how the trust definitions are based on five inter-related categories. They are:

- The *construct*, where trust is conceived 'as an *expectation*, a *belief*, *willingness*, and an *attitude*' (Castaldo, 2002).
- The *trustee*, 'usually individuals, groups, firms, organizations, sellers, and so on' (Castaldo, 2002). Given the different nature of the trustee (individuals, organizations, and social institutions), there are different types of trust (personal, inter-organizational and institutional). These trustees 'are often described by reference to different characteristics in the definitions being analyzed – specific competencies, capacities, non-opportunistic motivations, personal values, the propensity to trust others, and so on' (Castaldo, 2002).
- *Actions* and *behaviors*, as underlined also from other authors (e.g. (Moorman Zaltman and Desphande, 1992)) the behavioral aspect of trust is fundamental for 'recognizing the concept of trust itself' (Castaldo, 2002); both trustor and trustee behaviors have to take into account the consistence of the trust relationship. Behavioral aspects of trust have been studied also showing its multi-dimensional nature (e.g (Cummings and Bromiley, 1996)).
- *Results* and *outputs* of behavior, trustee's actions are presumed to be both predictable and positive for the trustor. 'The predictability of the other person's behavior and the fact that the behavior produces outcomes that are favorable to the trustor's objectives, are two typical results of trust. This has been particularly studied in works which suggest models designed to identify the consequences of trust (e.g. (Busacca and Castaldo, 2002)) (Castaldo, 2002).
- The *risk*, without uncertainty and risk there is no trust. The trustor has to believe this. They have to willingly put themselves into a 'position of vulnerability with regard to the trustee'. Risk, uncertainty and ambiguity (e.g. (Johannisson, 2001)) are the fundamental analytic presuppositions of trust, or rather the elements that describe the situations where trust has some importance for predictive purposes. (. . .).

[There is some sort of] logical sequence (. . .) [which has] often been suggested in the definitions. This sequence often regards trust as the *expectation, belief* (and so on) that a subject with specific characteristics (honesty, benevolence, competencies, and so on) *will perform* actions designed to produce *positive results* in the future for the trustor, in situations of consistent *perceived risk* (Castaldo, 2002).

Notwithstanding its merits, the main limit of Castaldo's analysis is that it fails to provide a stronger account of the *relationships* among these recurrent terms in trust definitions, i.e. indicating when they are partial synonyms, rather than necessary interdependent parts of a larger notion, or consequences of each other, and so on. Just an empirical, descriptive and co-relational account remains highly unsatisfactory. For example, it is true that 'Trust has been predominantly conceived as an *expectation*, a *belief*, *willingness*, and an *attitude*'.

However, it remains to be understood what are the conceptual ties between *belief* and *expectation*, or between *belief* and *willingness* (is one a species of the other? Does one concept contain the other?). What are their exact roles in the processing of trust: For instance, what is the procedural relationship (e.g. sequential) between *belief* and *willingness*, which certainly is not a kind of belief? And why do some authors define trust *only* as a *belief*, while other authors only consider it as *willingness* and as a *decision* or *action*? Statistical relations do not even begin to address these questions.

An in-depth analysis of the *conceptual interconnections* among different facets of trust is also instrumental to achieve *a more adequate characterization of this notion*, since a good definition should be able to cover these different aspects and account for their relevance and their mutual relationships, or motivate their exclusion.

In particular, any theoretical model should take into account that trust is a *relational* construct, involving at the same time:

- A subject X (*the trustor*) which necessarily is an 'intentional entity', i.e. a system that we interpret according to Dennett's intentional stance (Dennett, 1989), and that is thus considered a cognitive agent.
- An addressee Y (*the trustee*) that is an *agent* in the broader sense of this term (Castelfranchi, 1998), i.e. an entity capable of causing some effect as the outcome of its behavior.
- The causal process itself (*the act,* or *performance*) and its result; that is, an act α of Y possibly producing the desired outcome O.

Moreover, we should also never forget that trust is a *layered notion*, used to refer to several different (although interrelated) meanings (see Chapter 2):

- in its basic sense, trust is just a mental and affective *attitude* or *disposition* towards Y, involving two basic types of *beliefs*: *evaluations* and *expectations*;
- in its richer use, trust is a *decision* and *intention* based on that disposition;
- as well as the *act* of *relying* upon Y's expected behavior;
- and the consequent social *relation* established between X and Y.

If we now apply this analysis to the results summarized in Table 1.2, we can make the following observations:

- As for the terms **Will, Expect, Belief, Outcome, Attitude**, they match the relation we postulate quite closely: *will* refers to the future (as Castaldo emphasizes), thus it is also included in the notion of *expectation*, which in turn involves a specific kind of *belief*: in its minimal sense, an expectation is indeed a belief about the future (Miceli and Castelfranchi, 2002; Castelfranchi and Lorini, 2003; Castelfranchi, 2003). Moreover, the term *belief* implies a mental attitude, and we can say that trust as evaluation and expectation is an *attitude* towards the trustee and his action: the *outcome*, the events, the situation, the environment.
- As for the terms **Action** and **Decision**, they refer to trust as the deciding process of X and the subsequent Y's course of action; hence they are general, but only with reference to the second and richer meaning of trust discussed above (see also below and Chapter 2).
- As for the terms **Expect, Outcome, Rely, Positive, Exploit**, and **Fulfill**, again they are tightly intertwined according to our relational view of trust: the positive outcome of the trustee's action is expected, relied upon, and exploited to fulfill the trustor's objective. In short: X *has*

Table 1.2 Most frequently used terms in trust definitions[2]

Terms	Frequency
Subject (Actor, Agent, Another, Company, Customer, Firm, Group, Individual, It, One, Other, Party, People, Person, Salesperson, Somebody, Trustee, Trustor)	180
Action (Action, Act, Behavior, Behave, Behavioral)	42
Will	29
Expect, Expectation, Expected, Expectancy	24
Belief, Believe	23
Outcome, Result, Performance, Perform	19
Rely, Reliable, Reliance, Reied, Reliability, Relying	18
Trust, Trusting, Trustworthy	17
Confident, Confidence	16
Willingness, Willing	14
Take, Taken, Taking, Accept, Accepted, Acceptable	11
Risk, Risky, Risking	11
Vulnerable, Vulnerability	11
Relationship	10
Exchange	9
Based	8
Competent, Competence, Capabilities	7
Positive	7
Cooperate, Cooperation, Coordination	6
Exploit, Exploitation	6
Situation	6
Attitude	5
Decide, Decision	5
Fulfill, Fulfilled, Fulfillment	5
Held	5
Intention, Intentionally, Intend	5
Involve, Involved, Involvement, Involving	5
Mutual, Mutually	5
Word	5
Would	5

Source: Reproduced with kind permission of © 2002 Società editrice il Mulino.

a goal (a desire or need) that is expected to be fulfilled thanks to Y's act; X intends to exploit the positive outcome of Y's act, and relies upon Y for fulfilling the goal.

- As for the terms **Taken**, **Accept**, **Risk**, and **Vulnerable**, their relationship is that while deciding to count on Y, to trust Y (according to trust as decision), X is necessarily accepting the risk of becoming *vulnerable* by Y, since there is uncertainty both in X's knowledge (incomplete, wrong, static) and in the (unpredictable, unknown) dynamics of the world.

---

[2]This table is from Castaldo. For more sophisticated data and comments, based on cluster analysis, see (Castaldo, 2002).

Whenever deciding to depend on Y for achieving O, X is exposed both to failure (not fulfilling O) and to additional harms, since there are intrinsic costs in the act of reliance, as well as retreats to possible alternatives, potential damages inflicted by Y while X is not defended, and so on. As we will discuss more thoroughly in the next chapters, all these risks are direct consequences of X's decision to trust Y.

- As for the terms **Competence** and **Willingness**, they identify the two basic prototypical features of 'active'[3] *trust in Y*, i.e. the two necessary components of the positive evaluation of Y that qualify trust:

  - The belief of X (evaluation and expectation) that Y is *competent* (able, informed, expert, skilled) for effectively doing α and produce O;
  - The belief of X (evaluation and expectation) that Y is *willing* to do α, intends and is committed to do α – and notice that this is precisely what makes an agent Y predictable and reliable for X. Obviously this feature holds only when Y is a cognitive, intentional agent. It is in fact just a specification of a more abstract component that is Y's *predictability*: the belief that 'Y will actually do α and/or produce O', contrasted with merely having the potentiality for doing so.

In sum, a good definition of trust, and the related analytical model that supports it, must be able to explicitly account for two kinds of relationships between the different components of this multi-layered notion: *conceptual/logical links*, and *process/causal links*. A mere list of relevant features is not enough, not even when complemented with frequency patterns.

More specifically, a satisfactory definition should be able to answer the following questions:

1. What are the relevant connections between the overall phenomenon of trust and its specific ingredients? Why are the latter within the former, and how does the former emerge from the latter?
2. What are the pair-wise relations between different features of trust? For instance, how do *belief* and *expectation*, or *outcome* and *reliance*, interact with each other?
3. What is the conceptual link and the process relationship between trust as attitude (*belief*, *evaluation*, *expectation*) and trust as decision and action (relying on, accepting, making oneself vulnerable, depending, etc.)?

1.2 Missed Components and Obscure Links

The content analysis of 72 definitions presented in the previous section reveals some relevant gaps in such definitions, as well as several notions that remain largely or completely implicit.

An aspect absolutely necessary but frequently ignored (or at least left unstated) is *the goal, the need*, relative to which and for the achievement of which the trustor counts upon the trustee.

---

[3] As we will discuss later on (Chapter 2, Section 2.4), we distinguish between *active trust* and *passive trust*. The former is related to the delegation of a positive action to Y, and to the expectation of obtaining the desired outcome from this action. The latter, instead, is just reduced to the expectation of receiving no harm from Y, no aggression: it is the belief that Y will not do anything dangerous for me, hence I do not need to be alerted, to monitor Y's behavior, to avoid something, to protect myself. This passive trust has a third, more primitive component: the idea or feeling that "there is nothing to worry about", "I am/feel safe with Y".

This is implicit when a given definition of trust mentions some 'positive' result/outcome, or the 'welfare' or 'interest' of the trustor, and also whenever the notions of 'dependence', 'reliance', or 'vulnerability' are invoked. In fact, something can be 'positive' for an agent only when this agent has some *concern, need, desire, task,* or *intention* (more generally, a goal), because 'positive' means exactly that the event or state or action is favorable to or realizes such a goal – whereas 'negative' means the opposite, i.e. a threat or frustration of some goal.

Analogously, whenever it is observed that the trustor makes her/himself vulnerable to the trustee (see for instance (Mayer *et al.,* 1995)), the unavoidable question is – vulnerable for what? Alongside other costs that are intrinsic in any act of reliance, the trustor becomes especially vulnerable to the trustee in terms of potential failure of the expected action and result: the trustee may not perform the action α or the action may not have the *desired* result O.

Moreover, it is precisely with reference to the desired action/result that the trustor is 'dependent on' and relies upon the trustee. Also in the famous definitions provided by (Deutsch, 1985) where trust is relative to an entity 'on which my *welfare* depends', the goal of the trustor is clearly presupposed, since the notion of 'welfare' refers to the satisfaction of her needs and desires.

Building on these observations, in the following we will extensively argue for the necessity of the trustor to be actively *concerned,* i.e. to have goals at play in the decision to trust someone or something, as well as full *expectations*[4] rather than mere beliefs on the future (forecasts).

Another aspect frequently missed is that trust is an **evaluation**, and more exactly a *positive evaluation about Y*. In the cognitive architecture developed by (Miceli and Castelfranchi, 2000; Castelfranchi, 2000), an evaluation (which is also an attitude) is a *belief about some power* (capacity, ability, aptitude, quality; or lack of capacity . . .) *of Y relative to some goal*. Thus the beliefs about *Y* being able and willing to do his share for achieving *O* are in fact evaluations of *Y*: positive when there is enough trust, negative when there is mistrust and diffidence.

Here it is important to appreciate the intimate relation between 'beliefs', 'expectations' and 'evaluations'. In the case of trust, the *beliefs* on the competence and willingness of *Y* are both, and at the same time, *parts of expectations* (since they are about the future) and *evaluations* (since they are about *Y*'s powers and inclinations); moreover, they are *positive* both as expectations and as evaluations, insofar as agent *X* is expecting from and attributing to *Y* an attitude and a subsequent performance that is in *X*'s best interest. In addition, *X* might ground the decision to trust also on other positive evaluations of *Y*, for example, intelligence, honesty, persistency (on this point, see also Chapter 2).

It is worth noticing that the characterization of *trust as a structure of mental dispositions* is not in contrast with the analysis of *trust as a decisional process culminating into an action* – quite the contrary. Indeed, it is rather pointless to dispute whether trust is a *belief* or an *act,* opposing the view of trust as an *evaluation* or *expectation* to the idea that trust is a *decision* or a *behavior* (for example, of making oneself vulnerable, of risking, of betting on *Y*). The point is rather that *trust has both these meanings,* which stand in a specific structural relation with each

---

[4] As detailed in Chapter 2 (see also (Castelfranchi, 2005)), by expectation we mean the functional integration of a belief on the future (forecast) with a goal, i.e. a motivational state. In short, an expectation (either positive or negative) is defined as the prediction that a state of the world which constitutes one of the agent's goals will either be realized or not in the future. Thus, both the goal that *p* and the belief that, at some time in the future, *p* will (or will not) be the case are necessary components for expecting that *p* (or *not-p*).

other: more precisely, trust as decision/action presupposes trust as evaluation/expectation, as we shall argue in Chapter 2.

Yet another issue that remains often underestimated or confused is the *behavioral aspect of trust*, i.e. all the different types of actions that distinct actors (roles) have to perform, in order for trust to be a felicitous choice.

As for ***the act of the trustee***, what is frequently missed is that:

(a) The act can be *non-intentional*. First, the trustee might not be aware of my reliance on him/it,[5] or he/it may not know or intend the specific result of its action that I intend to exploit (see for instance the famous anecdote of Kant and his neighbor, where the latter was relying on the former and trusting him for his punctuality and precision, without Kant being aware of such a reliance). Second, the act that I exploit can be a non-intentional act by definition: e.g. a reactive behavior, or a routine. Third, if we endorse a very general notion where one can also trust natural processes or artifacts (see (c)), then of course the exploited process and the expected result that we delegate to a natural event or to an artifact are not intentional.

(b) Also *omissions* may be relevant in this context: e.g. 'doing nothing', 'not doing α', 'abstaining from α'. Obviously, omissions can be acts, even intentional ones – in which case, they are the result of a decision. In addition, omissions can also be the outcome of some more procedural choice mechanism, or of a merely reactive process, as well as just the static and passive maintenance of a previous state (i.e. they are not even proper 'acts' in the latter sense). Regardless the specific nature of the omission, the trustor might precisely expect, desire, and rely on the fact that Y will not do the specific action α, or more generally that Y will not do anything at all (See note 3).

(c) The trustee is *not necessarily a cognitive system*, or an animated or autonomous agent. Trust can be about a lot of things we rely upon in our daily activity: rules, procedures, conventions, infrastructures, technology and artifacts in general, tools, authorities and institutions, environmental regularities, and so on. Reducing trust to 'trust *in* somebody' is not only an arbitrary self-limitation, but may also bias the proper interpretation of the phenomenon, insofar as it hides the fact that, even when we trust somebody's action, we are necessarily trusting also some external and/or environmental conditions and processes (on this point, see Chapter 2). However, it remains obviously very important to precisely characterize *social trust*, i.e. trust in another agent as an agent, and the so called 'genuine' or typical trust in another human (see Chapter 2).

As for **the act of the trustor**, the more frequent shortcomings and confusions in the literature are the following:

(a) It is often missing a clear (and basic) distinction between the act of the trustor and the act of the trustee: for instance, Castaldo does not clearly disentangle the occurrences of the two different acts within the definitions covered by his survey. Obviously, the act of the trustor consists in the very act of 'trusting', of counting upon and deciding to rely on Y.

(b) Much more importantly, it is not always emphasized enough that the 'act' of trusting is a necessary ingredient of one notion of trust (i.e. as a decision and a subsequent action), but

---

[5] "It" since it can even be a tool, an inanimate active entity (agent).

not of the other notion of trust as a preliminary mental attitude and evaluation. Since this crucial distinction is considered as a semantic ambiguity, rather than a valuable insight into the internal articulation of this complex phenomenon, it is still lacking the theory of the logical and causal relationships between these two aspects of trust: the mental attitude and the decision to act upon it (see Chapter 2).

1.3 Intentional Action and Lack of Controllability: Relying on What Is Beyond Our Power

In any intentional action α exerted upon the external world, there is one part of the causal process triggered by the action and necessary for producing its intended and defining result (the goal of the action) which *is beyond the direct executive control of the agent* of α. Whenever an agent is performing α in the real world, there is always some external condition or process P that must hold for the action to be successful, and the agent does not have direct executive control over such an external feature – although he might have foreseen, exploited, or even indirectly produced it. Therefore, the agent while performing α is objectively making reliance on these processes in order to successfully realize the whole action and thus achieve his goal.

This objective reliance holds in both of the following cases:

- when the agent is aware of this fact, models this act of reliance in his mind, and even expects it;
- when the agent does not understand the process, he is not aware of it, or at least he does not explicitly represent it in his plan (although in principle he might be able to do so).

When the subject is aware of the reliance that he is making for his action α on some external process, and counts upon such a process P, which does not depend completely and directly on him, we can say that the reliance has become *delegation*. Delegation (Castelfranchi and Falcone, 1998; Falcone and Castelfranchi, 2001) is a type of reliance, which is subjective (i.e. aware) and decided by the agent; it consists of the act of 'counting upon' something or someone, which is both a mental disposition and a practical conduct.[6]

In contrast, *reliance* in general can be merely objective or also subjective, e.g. like in delegation. When reliance is subjective, it can (like delegation) be correct or wrong and illusory: e.g., the beliefs on which it is based may be false, it may not be true that that expected process will be there or that it is responsible for the desired effect (like happens, for instance, with a placebo).

It is worth noticing that, although the presence of P (due to Y) is a necessary process/condition for the achievement of X's own goal, it is not sufficient. X has also *to do (or abstain from doing) something of his own*, and thus he has *to decide something*: regardless that X is counting on Y for P or not, he still has to take his own decision on whether to pursue

---

[6] Here we use 'delegation' in its broader and non-organizational meaning: not only and not necessarily as delegation of powers, or delegation of institutional/organizational tasks from one role to another (involving obligations, permissions, and other deontic notions). Our use of delegation is more basic, although strictly related to the other: to delegate here means *to allocate, in my mind and with reference to my own plan, a given action that is part of the plan to another agent*, and therefore *relying on the performance of such an action by the other agent* for the successful realization of my own plan.

the goal in the first place, i.e. on whether to engage in action α (that would make X dependent upon Y for P) or not.

The link between delegation and trust is deep and should not be underestimated: delegation necessarily requires some trust, and trust as decision and action is essentially about delegation. This also means that *trust implies that X has not complete power and control on the agent/process they are relying and counting upon*. Trust is a case of limited power, of 'dependence' on someone or something else. Although the notion of 'reliance' and 'reliable' is present in several of the definitions analyzed by Castaldo, the theory of this strange relation, and its *active* aspect of deciding to depend, deciding to count on, to invest, to delegate, is not yet well developed in the literature on trust. For instance, several authors consider a crucial and necessary aspect of trust the fact that while relying on the behavior of another person we take a risk because of the lack or limit of '*controllability*' and because the other's behavior *cannot be under coercion*, so that our expectation on the other's actions cannot be fully certain. This intuition is correct, and it just follows from our previous analysis. In any act of *trusting in Y* there is some *quid* delegated to another agent Y, and, especially when this agent is viewed as purposive, goal-oriented (be it Nature, a tool, an animal, or a person), the delegated process that consists of Y's performance is *beyond our direct control*. Y has some autonomy, some *internal degree of freedom*, and also for this – not only for external interferences – it is not fully predictable and reliable.

When Y is an autonomous cognitive agent this *perceived* degree of freedom and autonomy mainly consists in Y's choice: Y can *decide* to do or not to do the expected action. With this kind of agent (i.e. within the domain of social trust), we in fact trust Y for deciding and be willing to do what Y 'has to' do (for us) – even against possible conflicting goals that may arise for Y at the very moment of the expected performance. In other words, we trust (in) Y's motivation, decision, and intention.

This characteristic feature of social trust strictly derives from the more basic notion of trust as involving reliance upon some non-directly controlled process and agent, on the perception of this lack of controllability, and its associated risk; on the distinction between trust 'in' Y, and global trust; and, in the end, on the very idea of 'delegation', as the decision to count upon such a process/agent for the pursuit of my own goal. If I have not decided to depend on this, I would have no reason to care for any non-controllability of such a process or agent.

Finally, as we briefly mentioned before, aside from *Competence* and *Willingness*, there is a third dimension in evaluating the trustworthiness of Y: Y must be perceived as non threatening (passive trust), i.e. *harmless*. There is no danger to be expected from Y's side; it is 'safe' to rely on Y and to restrain from fully monitoring Y's conduct (see Section 2.4 for more details).

In a sense 'feeling safe' could be taken as the basic nucleus of trust in and by itself, seemingly without any additional component (*passive trust* – see note 3). However, looking more carefully we can identify other core components. Clearly positive *evaluations* and *expectations* (beliefs) are there and play a role in my feeling of safety. If I do not worry and do not suspect any harm from you, this means that I evaluate you positively (in the sense that you are 'good for me', not to be avoided, at least harmless), since not being harmed is one of my goals. I can be relaxed (as for you) and this is also pleasant to me. Moreover, this feeling/belief is an expectation about you: I do not expect damage from you, and this constitutes a passive, weak form of positive expectation. Perhaps I do not expect that you will actively help me realize a goal of mine that I am pursuing; but at least I expect that you will not compromise a goal that I have already achieved, i.e. you will not damage me, but rather let me continue to have what I have. (This is why we call 'passive' this kind of expectation).

One of the clusters of definitions found by Castaldo is the definition of trust as a 'Belief about future actions' (27, 8%) that 'makes no reference to the concept of Expectations'. This is in our interpretation either due to the fact that the authors adopt a very weak notion of 'expectation' and thus simply analyze it in terms of 'a belief about the future'; or, instead, they have a richer and stronger notion of 'expectation' but expunge this notion from the definition of trust, ultimately reducing trust to some *grounded prediction*. In both cases, this position is unsatisfactory; trust, even when considered as a pure mental attitude before and without any decision and act, cannot be reduced to a simple forecast (although it contains this element of prediction within its kernel). Computers, if adequately instructed, can make excellent weather forecasts, but they do not have 'expectations' about the weather, nor do they put their 'trust' in it – indeed, they remain incapable of trusting anything or anyone, as long as they are mere forecasting machines. In what follows, we will take Andrew Jones' analysis of trust (Jones, 2002) as a good example of this tendency to reduce trust to an epistemic attitude, to grounded prediction, and discuss it in order to show why such an analysis necessarily misses some very important trait of the notion of trust (see Section 1.4 and Chapter 2).

Some of the most frequent terms highlighted by Castaldo's content analysis, like ***Cooperate, Mutually, Exchange, Honesty, Commitment, Shared Values***, are clearly valid for describing trust only in specific domains and situations, e.g. commerce and organization. *Mutuality*, for instance, is not necessary at all: in most contexts, trust can be just unilateral – and in fact later on we will criticize this same bias in philosophical, economic and game theoretic theories of trust. Meanwhile, it is a real distortion of the game theoretic perspective to use 'trusting' as a synonym of 'cooperating' (see below and Chapter 8). Analogously, the terms ***Customer, Company, Salesperson, Firm*** (gathered by Castaldo in the category of 'Subjects') are clearly domain-specific. The same would hold for the term ***Security*** in the growing domain of 'Trust and Security' in Information and Communication Technologies (see Chapter 12).

1.4   Two Intertwined Notions of Trust: Trust as Attitude vs. Trust as Act

Although we are aiming for a unified, covering, general and possibly shared definition of trust, this result will not be achieved by looking for just one unique monolithic definition. We also do not want to gather just a list of different meanings and uses that share with each other only some features; we can accept this 'family resemblance' as a possible result of the conceptual analysis, but not as its starting assumption and its ultimate objective.

Ideally, what we will try to identify is a *kernel concept*: *few common and truly fundamental features*. In doing so, what is needed – as is often the case (see for instance Castelfranchi's definition of 'agent' in (Castelfranchi, 1998; Castelfranchi, 2000a; Castelfranchi and Falcone, 2003)) – is a *layered definition* or a hierarchy of definitions, with an explicit account of the conceptual relationships between the different but connected meanings of the same notion.

The common sense term *trust* (at least in many of the languages used by the scientific community, like English, German, French, Spanish, Italian) covers various phenomena structurally connected with each other. As we said, it is crucial to distinguish at least between two kinds

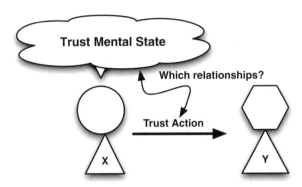

Figure 1.1 The double nature of Trust, as psychological attitude and as act. The relationship between these two elements needs much exploration

and meanings of trust (see Figure 1.1):

(a) Trust as *psychological attitude* of X towards Y relative to some possible desirable behavior or feature.
(b) Trust as the *decision and the act of relying on*, counting on, depending on Y.

In our theory there is a twofold connection between these two types of trust:

- The *conceptual link* in that the *intension* of *(b)* contains *(a)*, i.e. trust as an attitude is part of the *concept* of trust as a decision/action. Obviously, the *extension* of *(a)*, i.e. the set of cases where there is such a mental attitude, includes the extension of *(b)*, i.e. the set of acts of trusting someone or something.
- The *process/causal link* in that *(a)* is the temporal presupposition (an antecedent) and a con-cause of *(b)*. The process for arriving at the act of trusting entails the formation of such a positive expectation on and evaluation of Y.[7]

This provides us with the dyadic general notion of trust, capable of connecting together the two basic meanings of this phenomenon.[8] Revolving around this common core, there are then several domain-related specifications of trust: subcategories and/or situational forms of trust, due to the sub-specification of its arguments (the kind of trustor or of trustee; the kind of action; the kind of results), or to different contexts that make relevant some specific reason of trust (like interest, commitment, or contract) or some specific quality of the trustee (like sympathy, honesty, friendship, common values).

---

[7] Without the psychological attitude (a) there might be delegation but not trust: An obliged, constrained, needed, (may be) desperate delegation/reliance but not trust.

[8] One might more subtly distinguish (see (Falcone and Castelfranchi, 2001b)) between the mere decision to trust and the actual act of trusting. In this case the definition of trust becomes three-layered, involving a three-steps process (Chapter 2). One might also argue that there is trust as a social relation, in consequence of the act of trusting or of a trustful or distrustful attitude towards someone else. This is certainly true, so one might enrich a layered definition of trust along the lines suggested by these considerations.

1.5   A Critique of Some Significant Definitions of Trust

At the beginning of this chapter we introduced our aim to review the predominant definitions of trust in the literature, motivated by the purpose of showing that, in cognitive and social sciences, there is not yet a shared or prevailing, and clear and convincing notion of *trust*. After a content analysis of a larger number of definitions, let us now consider some specific definitions in order to show how they frequently are incomplete, vague or obscure, domain specific, divergent. To show this, we will be a bit simplistic and not completely fair with the authors, just discussing and criticizing their definitions of trust. This is useful for comparing them, and for stressing confusion, inaccuracy, and ad hoc features; but it is quite ungenerous, since sometimes the analysis or theory of the author is more rich and correct than their 'definition'. However, several authors are more extensively discussed here or in other chapters of this book.

1.5.1   Gambetta: Is Trust Only About Predictability?

Let us first consider the definition of trust provided in the classic book of Gambetta and accepted by the great majority of other authors (Gambetta, 1988): 'Trust is the *subjective probability* by which an individual, *A*, *expects* that another individual, *B*, performs a given action on which *its welfare depends*' (translation from Italian).

In our view, this definition is correct, insofar as it stresses that trust is basically an estimation, an opinion, an expectation, i.e. a belief. We also find commendable that there is no reference to exchange, cooperation, mutuality, and *B*'s awareness of being trusted by *A*, since none of these features is, in our view, part of the core notion of trust.

However, Gambetta's definition is also too restricted in a variety of respects:

- It just refers to one dimension of trust (*predictability*), while ignoring the *competence* dimension.
- It does not account for the meaning of '*A* trusts *B*' where there is also the *decision* to rely on *B*.
- It does not explain what such an evaluation is made of and based on, since the measure of *subjective probability* collapses together too many important parameters and beliefs, which each has its own role and relevance in social reasoning.
- It fails to make explicit the 'evaluative' character of trust.
- Finally, reducing trust to the notion of 'subjective probability' is quite risky, since it may result in making superfluous the very notion of 'trust' (on this point, see Williamson's criticism (Williamson, 1993) as well as Chapter 8).

1.5.2   Mayer, Davis, & Schoorman: Is Trust Only Willingness, for Any Kind of Vulnerability?

Mayer, Davis, and Schoorman provide an interesting and insightful (albeit somehow limited) definition of trust, as follows: 'The *willingness* of a party *to be vulnerable* to the actions of another party based on the *expectation* that the other party will perform a particular action *important* to the trustor, irrespective of the ability to monitor or control that other party' (Mayer *et al.*, 1995).

This definition of trust as *the decision to make oneself vulnerable* deserves some consideration. On the one hand, it is strange that it focuses only on trust as a decision, without considering the background of such a decision, and thus missing basic uses of the term like in 'I trust John but not enough'.

On the other hand, in our view it identifies and expresses very well an important property of trust, but without providing a really good definition of the general phenomenon. To begin with, insofar as the definition is mainly applicable to trust as decision and action, it seems to allude to vulnerability only in relation to a transition of state, whereas one might also say that a disposition of trust or a relation of trust is enough to make the trustor vulnerable, although in a static sense, not as state-transition.

More importantly, the idea of equating trust with self-decided vulnerability is, as a definition, both too broad and too vague, since there are a lot of states (including psychological states) and acts that share the same property of making oneself vulnerable to others; for example, *lack of attention and concentration, excess of focus and single-mindedness, tiredness, wrong beliefs about dangers* (e.g. concerning exposition to an enemy, being hated, inferiority, etc.), and so on. Moreover, some of these states and acts can be due to a decision of the subject: for example, the decision to elicit envy, or to provoke someone. In all these cases, the subject is deciding to make themselves vulnerable to someone or something else, and yet no trust is involved at all.

Therefore, the problem is not solved by defining trust as the decision to make oneself vulnerable, although one should characterize trust in such a way that can *explain* and *predict* these important effects and consequences of trust. For instance, it is worth emphasizing all the dangers implicit in the decision to trust, in terms of:

- Considering sufficient and reliable enough the current information about the trustee and about the relevant situation. This implies that the trustor does not perceive too much uncertainty and ignorance on these matters, although their estimate is, of course, subjective and fallible (and ignorance or wrong certainty can be dangerous indeed).
- Having enough good evaluations and predictions about the trustee; but these might be wrong, or the negative evaluations and foreseen dangers can be poorly estimated, and false predictions based upon misleading evaluations may be extremely noxious.
- Relying upon Y, counting on Y to help realize a given goal; i.e. for the realization of goal G, agent X depends (accepts, decides to depend) on Y. This – analytically in our model– gives Y the power of frustrating X's goal, thus X makes oneself vulnerable to Y for G; moreover the actual decision to trust further increases X's dependence.

So, in order to explain and predict trust-related vulnerability and the fact that the trustor's welfare comes to depend upon the trustee's action (as mentioned in Gambetta's definition), a model of trust – as we said – must at least integrate:

- beliefs about the trustee's internal attitudes and future conduct (more or less complete; more or less grounded on evidence and rationally justified; more or less correct);
- the subjective propensity of the trustor to accept a given degree of uncertainty and of ignorance, and a given perceived amount of risk;

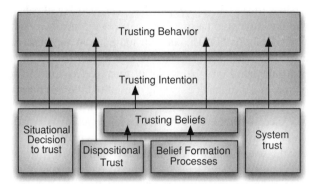

Figure 1.2 Relationships among Trust Constructs (arrows represent relationships and mediated relationships). (Reproduced with kind permission of Springer Science+Business Media)

- the trustor's decision to rely upon the action of another entity for the realization of a goal, and the expectations upon which such a decision is based;
- the relationships of dependence and power between the trustor and the trustee with respect to the intended goal of the former.

1.5.3 McKnight: The Black Boxes of Trust

A very apt and frequently cited approach to the nature and internal dynamics of trust is McKnight's model (McKnight and Chervany, 2001). In Figure 1.2 one can see the general schema of this model.

This model is rather comprehensive, since it takes into account several important aspects and some of their mutual interactions. For example, the authors are able to distinguish between the belief-component and the decisional and behavioral aspects of trust, and to explain that the latter depends on the former; they also recognize the role of situational and system components in determining trust behavior. However, this is just a black-boxes model, without much insight on what is supposed to be going on *within* each box. Moreover, the semantics of the arrows is undefined and definitely non-uniform. The specific nature, organization, structure and process of the content of the various boxes are not well specified. There is no deeper characterization of trust beliefs in terms of 'expectations' and 'evaluations', nor is there an explicit model of the critical factors involved in the decision process.

On the whole, this remains a 'factors' model (typical of psychology), where the authors just capture correlations and mutual influences, but the precise nature and 'mechanics' of the process are not defined. In sum, it is indeed important that they identify different kinds and levels of trust (as beliefs, as intention, as behavior), and that they connect each one with the others. However, a much more *analytic*, *process-oriented*, and *relational* (not only mental) model is needed.

1.5.4 Marsh: Is a Mere Expectation Enough for Modeling Trust?

As for the growing trust literature in Information Technology and Artificial Intelligence (see Chapter 12 in this book and 'Trust: Theory and Technology group' site (http://www.istc. cnr.it/T3/), let us cite here only Marsh's thesis, which has been in fact the first attempt.

According to Marsh, X trusts Y if and only if 'X *expects* that Y will behave according to X's best interest, and will not attempt to harm X' (Marsh, 1994a; Marsh, 1994b).

This is a rather good definition of trust, but with some significant limits. First and foremost, it only refers to an attitude, the expectation of X that Y will behave according to X's own interest. Therefore, in this definition, the notion of trust as decision and act, as reliance and counting on Y, is missed; and the idea of exposing oneself to failure or harm, of making oneself vulnerable by Y has also been overlooked.

Moreover, it is not clear whether Y's behavior is (expected to be) intentional or not necessarily intentional. The first part of the definition is ambiguous about this: 'Y will behave according . . .' – does this mean intentionally or accidentally? Will Y intentionally help X, or will his action just be factually exploited by X? The second sentence instead seems to refer explicitly to an intentional act, since it uses the term 'attempt'; thus the possible harm that X does not expect from Y would be intended, not accidental. However, Marsh's definition leaves open the possibility that two independent conditions for X to trust Y are there: on the one hand, X must expect Y to behave in a certain way, either intentionally or unintentionally; in addition, X must expect Y not to attempt (hence intentionally, by definition) to bring X any harm. In this interpretation, the second sentence of Marsh's definition would refer to something akin to the feeling of safety that is frequently linked with the notion of trust. However, it remains manifest that the exceeding ambiguity of this definition of trust in turn impinges on its applicability.

Finally, it is worth noting that the notion of 'interest' is not defined by Marsh,[9] so we can only assume it to be somehow close to other notions like welfare and goals.

1.5.5   *Yamagishi: Mixing up the Act of Trusting and the Act of Cooperating*

Following Yamagishi's interpretation of his comparative results between American and Japanese culture (Yamagishi and Yamagishi, 1994) (Yamagishi, 2003), what characterizes the latter is *assurance* rather than proper trust.

This means that the Japanese are more trusting (let us mark this use of the notion of trust as ω), i.e. more disposed to rely on the others (X's side), than the Americans, when and if they feel protected by *institutional mechanisms* (authorities and sanctions).

Moreover, according to Yamagishi, the Japanese would tend to trust (in a different sense, that we shall indicate with λ) only when it is better for them to do so, because of the institutional or social costs associated with being 'untrusting' (in Yamagishi's own words), i.e. only for avoiding sanctions (Y's side).

First of all, notice that here there is a misleading use of the term 'trust': in the first case (ω), it means that X *trusts in* Y *to do* α, i.e. X believes that Y is trustworthy and relies on Y; in the second use (λ), to trust is an act of cooperation, the proper contribution the agent should make to a well-codified social interaction.

These two uses must be distinguished. Obviously they are related, since in Japan X contributes/cooperates since he worries about institutional sanctions, and he *trusts* the others because he ascribes to them the same cultural sensibility and worries. Nonetheless, the two perspectives are very different: expectations about the others' behavior, on the one hand, and

---

[9] For a definition about the notion of *interest* as different from *goal* (and on trust relative to *interest protection*) see Chapter 2.

my own behavior of contributing, on the other, must be distinguished. We cannot simply label them both 'trust'.

Second, the confusion between 'tendency to trust' and 'tendency to cooperate/contribute', and between 'not trusting' and 'not cooperating/contributing' is misleading per se. If X co-operates just in order to avoid possible sanctions from the authority or group, trust is not involved. X does not contribute because he either trusts or does not trust the others, but rather for fear of sanctions – in fact, the only thing that X is trusting are the social authorities and their capacity for monitoring and sanctioning his conduct ((Castelfranchi and Falcone, 1998) (Falcone and Castelfranchi, 2001)). Calling this cognitive attitude 'tendency to trust' may be quite confusing. This is not simply a problem of terminology and conceptual confusion; it is a problem of *behavioral* notions that are proposed as *psychological* ones.[10]

Finally, here the concept of 'trusting' ends up losing its meaning completely. By missing the fundamental elements of having a positive evaluation of others and good expectations about their behavior, and because of these reasons relying on them and becoming vulnerable to them, this notion of trust comes to mean just to cooperate (in a game theoretical sense), to contribute to the collective welfare and to risk for whatever reason. The resulting equation 'Trust = to contribute/cooperate'; 'untrust = do not contribute/cooperate' is wrong in both senses: *there are cooperative behaviors without any trust in the others, as well as there being trust in the others in non-cooperative situations.*

We have to say that simply cooperating for whatever reason is not to trust. The idea that this *behavior* necessarily denotes trust by the agent and is based on this, so that the behavior can be used as a synonym of the attitude, is wrong. For example, as we have already said, worrying about institutional sanctions from the authority has nothing to do with putting one's trust *in* the other. Confusion between these two attitudes is fostered, among other things, by the fact that, usually, it is not specified in whom and about what a given subject trusts another, and based on what expectations and evaluations one has on the other. One should be clear in distinguishing between X *trusting the others* (possibly because he believes that they worry about the social authority and its possible sanctions), and X *doing something pro-collectivity* just because he worries about sanctions, not because he trusts the others.

Furthermore, it is wrong to assume that trust coincides with cooperation in these kinds of social dilemmas, and it is misleading to propose that trust consists in betting on some reciprocation or symmetric behavior. Here Yamagishi is clearly influenced by economists and their mental framework (see Chapter 8). As we will explain, trust also operates in completely different social situations.

Trust is not the feeling and disposition of the 'helper', but rather of the 'receiver' of some expected contribution to one's own actions and goals. Trust is the feeling of the helper only if the help (goal-adoption) is *instrumental* to triggering some action by the other (for example, some reciprocation). In this case, X is cooperating towards Y and trusting Y, but only because he is expecting something from Y. More precisely, the claim of interest for the economists is that X *is 'cooperating' because he is trusting* (in view of some reciprocation); he wouldn't cooperate without such a trust in Y.[11]

---

[10] Calling this behavior "trust behavior" is rather problematic for other reasons: it can be a behavior just relying in fact on the others' concurrent behavior, but unconsciously, without any awareness of 'cooperation'; as is the case – for a large majority of people – with paying taxes.

[11] In other words here we have a *double* and *symmetric* structure (at least in X mind) of goal-adoption and reliance (see later).

However, this is just a very peculiar case, and it is mistaken to take it as prototypical for founding the notion and the theory of 'trust' and of 'cooperation'. In general, a symmetric and reciprocal view of trust is unprincipled. In contrast, we can (and should) distinguish between *two basic constituents and moves of pro-social relations*:

- On the one side, *goal-adoption*, i.e. the disposition (and eventually the decision) of doing something *for* the other, in order to favor him.
- On the other side, *delegation*, i.e. the disposition (and eventually the decision) to count on the other, to delegate to him the realization of our goals and welfare (Castelfranchi and Falcone, 1998).

It is important to realize that this basic pro-social structure, which constitutes the nucleus of cooperation, exchange, and many other social relations, is *bilateral but not symmetrical*. In other words, pro-social bilateral relations do not start with reciprocation (which would entail some symmetry), nor with any form of exchange. The basic structure is instead composed by a social disposition and act of counting on the other and being dependent on him, thus expecting adoption from him (i.e. *trust*); this pro-social attitude will hopefully be matched by a disposition and an act of doing something for the other, of goal-adoption (see on this point Spinoza's notion of *benevolence*).[12]

Notice that the anti-social corresponding bilateral structure is composed of *hostility*, i.e. the disposition not to help or to harm, paired with *distrust* and *diffidence* from the other actor.

As this analysis should have made clear, benevolence and trust are not at all the same move or disposition (although both are pro-social and often combine together); they belong to and characterize two different although complementary actors and roles. Benevolence and trust are *complementary* and closely related, but they are also in part independent: they can occur without each other and can just be 'unilateral'. X can rely on Y, and trust him, without Y being benevolent towards X. Not only in the sense that X's expectation is wrong and he will be disappointed by Y; but in the sense that X can successfully rely on Y and exploit Y's 'help' without any awareness or adoption by Y. On the other hand, Y can unilaterally adopt X's goals without any expectation from X, and even any awareness of such a help.

Moreover, both *trust and benevolence do not necessarily involve symmetric relations* between agents. It is possible to have cases of asymmetric trust[13] where only X trusts Y, while Y does not trust X (although he knows that X trusts him and X knows that Y does not trust her). And this holds both for trust about a specific kind of service or performance, as well as for generalized trust.

In addition, *trust does not presuppose any equality among the agents*, since there can be asymmetric power relationships between the trustor and the trustee: X can have much more power over Y, than Y over X (like it happens between parents and children). Analogously, goal-adoption can be fully asymmetrical, whenever X does something for Y, but not vice versa.

When there is a bilateral, symmetrical, and possibly reciprocal goal-adoption (i.e. the contribution of X to Y's interests is also due to the help of Y towards the realization of X's goals,

---

[12] On the contrary, 'justice' either is the rule of providing adoption or (if interpreted as 'fairness') it is also the rule of exchange, and thus it presupposes some reciprocity.

[13] This is for example in contrast with May Tuomela's account of Trust (Tuomela, 2003).

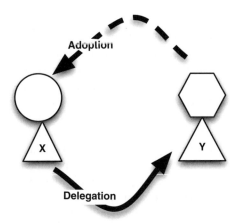

Figure 1.3   *X*'s Delegation meets *Y*'s Adoption

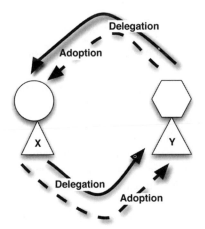

Figure 1.4   *X*'s Delegation meets *Y*'s Adoption, and vice versa: reciprocal trust

and vice versa), the structure presented in Figure 1.3 is doubled, as indicated in Figure 1.4. In this case, there is in fact trust/reliance from both sides and adoption from both sides.

Finally, it is worth noticing that even in *asynchronous* exchanges, when *X* acts *before Y* and *Y* acts only after *X*'s 'help', *Y* is trusting *X*. Not necessarily at the very moment of doing his own share, but before, when *Y* decides to accept *X*'s help and to rely on it.[14] Of course, in asynchronous 'exchanges' where *X* moves first, *X*'s trust in *Y* is broader and more risky: *X* has additionally to believe (before obtaining concrete evidence) that *Y* will do the expected action, whereas *Y* already has some evidence of *X*'s contribution (although this might be deceptive).

---

[14] For instance, *Y* has to believe that *X*'s help is good, is as needed, is convenient, is stable enough (it will not be taken back by *X*), and so on.

1.5.6   Trust as Based on Reciprocity

There is another important and recent family of definitions of trust, in which *trust denotes a behavior based on an expectation* (a view that we share), but both the behavior and the expectation are defined in a very restricted way. Let us take as the prototype of this approach the definition of trust provided by Elinor Omstrom and James Walker in their book *Trust and Reciprocity* (Omstrom and Walker, 2003): they define trust as 'the willingness to take some risk in relation to other individuals on the expectation that the others will reciprocate' (p. 382). This view in fact restricts the act of trusting to the act of 'cooperating', contributing, sustaining some cost in view of a future advantage that depends, in a strategic framework, also on the other's conduct; and it restricts the expectation to the expectation of reciprocation. By doing so, they exclude from the very notion and from the relations of trust all those cases that are not based at all on some exchange or cooperation; where X just counts on Y's adoption of her goals, even in an asymmetric relationship, like in a son-mother relation, or in a request for help. We will extensively discuss these views in Chapter 8 about the notion of trust in economics and game theory.

1.5.7   Hardin: Trust as Encapsulated Interest

In Russell Hardin's view (Hardin, 2002), in part based on Baier's theory (Baier, 1986)):

'I trust you because I think it is in your interest to take my interests in the relevant matter seriously'.

This is the view of trust as *encapsulated interest*. I believe that you will take care of my interests, but I also believe that you will be rational, i.e. you will just follow your own interests; on such a basis I predict your favorable behavior: 'Any expectations I have are grounded in an understanding (perhaps mistaken) of your interests specifically with respect to me'. Expectations are not enough: 'The expectation must be grounded in the trustee's *concern* with the trustor's interests'[15].

With this very interesting notion of 'embedded interests', Hardin arrives close to capturing the crucial phenomenon of *social goal-adoption*, which really is foundational for any form of pro-sociality: from exchange to cooperation, from altruism to taking care of, and so on (Castelfranchi, 1989), (Conte and Castelfranchi, 1995), (Castelfranchi, 1998). In our vocabulary the fact 'that you encapsulate my interests in your own interests' means that you adopt my goals (for some reason of yours). Social goal-adoption is precisely the idea that another agent takes into account in his mind my goals (needs, desires, interests, projects, etc.), in order to satisfy them; he 'adopts' them as goals of himself, since he is an autonomous agent, i.e. self-driven and self-motivated (which does *not* necessarily mean being 'selfish'!), and he is not a hetero-directed agent, so that he can only act in view of and be driven by some internal purposive representation. Therefore, if such an internally represented (i.e. adopted) goal will be preferred to others, then he will happen to be self-regulated by my goal; for some motive of his own, he will act in order to realize my goal.

---

[15] See note 14.

As we shall discuss in a while, we agree with Hardin that there is a restricted notion of social trust which is based on *the expectation of adoption*, not just on the prediction of a favorable behavior of *Y*. When *X* trusts *Y* in a strict social sense and counts on him, she expects that *Y* will *adopt* her goal and that this goal will prevail, in case of conflict with other active goals of his. That is, *X* not only expects an *adoptive goal* by *Y*, but also an *adoptive decision and intention*. A simple regularity-based prediction or an expectation grounded on some social role or normative behavior of *Y* are not enough (and we agree with Hardin on this) for characterizing what he calls 'trust in a strong sense', which is the central nature of trust (Hardin, 2002), what we call *genuine social trust*.

However, to our mind, Hardin still fails to conceive a broad theory of goal-adoption, so that his notion of 'encapsulated interests' provides us with only a restricted and reductive view of it. Indeed, the theory of adoption is crucial, although not yet well understood, in sociology, economics and game theory (Tummolini, 2006), and in cooperation theory (Castelfranchi, 1997), (Tuomela, 1988, 1993). The fundamental point is to realize that *there are different kinds of goal-adoption depending on Y's motives (higher goals) for doing something 'for X'*, for spending resources in order to realize another agent's goal. Let us list these different cases:

1. Adoption can be just *instrumental* to *Y*'s personal and non-social goals; completely *selfish*. Like when we satisfy the need of chickens for food, only in order to later kill and eat them to our best satisfaction; or like when we enter a *do ut des* relation, like an economic exchange in Adam Smith's view (Smith, 1776).
2. Adoption can be *cooperative* in a strict sense. *X* and *Y* are *reciprocally dependent on each other* but just for one and the same goal, which constitutes their common goal. This is very different from other social situations, e.g. exchange, where there are two different and private/personal goals. Here instead the agents care for the same result in the world, and they need each other for achieving it. For this reason (being part of a necessarily common plan), each of them favors the actions and goals of the other within that plan, since he needs them and relies on them. In a sense, this adoption may be considered a sub-case of *instrumental* adoption, but it is definitely better to clearly distinguish (2) from (1). In fact, in (1) a rational agent should try to cheat, to avoid his contribution to the other: especially after having received *Y*'s service or commodity, and assuming no iteration of the interaction, *X* has no reason for doing her share, for giving *Y* what *Y* expects. On the contrary, in strict cooperative situations, based on real reciprocal dependence for the same objective, to cheat is self-defeating; without doing her share, *X* will not achieve her own (and common) goal (Conte and Castelfranchi, 1995), (Castelfranchi, Cesta, Conte, Miceli, 1993).
3. Finally, there is also non-instrumental, *terminal*, or altruistic adoption. The good of *X*, the realization of *X*'s needs, desires, and interests is an end *per se*, i.e. it does not need to be motivated by higher personal goals.

It is then just an empirical matter whether a behavior and cognitive structure as postulated in (3) does really exist in humans, or if humans are always motivated by selfish motives (although reduced to internal hidden rewards, like self-approval and esteem, or avoiding feelings of guilt or pity), so that only 'pseudo-altruism' actually exists (Batson, 1991), (Lorini, 2005).[16] What really matters here is the well-defined *theoretical* possibility of having such an altruistic mind.

On this point, unfortunately Hardin seems to remain within the cultural framework of economics and game theory. It seems that his notion of 'self-interest' and 'rationality' necessarily collapse in 'selfishness'. In fact, missing a clear and explicit theory of purposive (internally goal-driven) behavior and thus of autonomous purposive agents, and an explicit theory of the goals of the agents (instead of just considering their quantitative dimension, i.e. utility), these approaches are not able to disentangle the fact that by definition such agents act for *their own* (internally represented and regulating) goals, choosing whatever options happen to maximize the achievement of those goals (that is, 'rationally'), from the wrong conclusion that *thus* they can only be 'selfish'. This is completely mistaken, since selfishness does <u>not</u> refer to autonomy, to non-exogenous but endogenous regulation, to being regulated by 'my own' goals; it refers instead to *a specific sub-family of the agent's goals*, which are non-adoptive, i.e. they do not aim to realize the goal of another agent.[17]

For example, one might build a robot, fully autonomous and rational in choosing among its goals, self-regulated by *its own* internal goals that, however, might just consist in the goal of helping its user, realizing all the desires or requests of the other, independently of any cost (including 'life') or 'personal' risks (which *subjectively* would not in fact exist at all, not having 'personal' goals, like safety or saving resources).

An important corollary of this view is that rationality has nothing to do with the nature or content of my goals; it cannot prescribe me my motives. A suicide or an altruist can take a perfectly subjectively rational decision, given their *unobjectionable* preferences. *Rationality is only instrumental: it is all about how we order and choose among our (conflicting) goals.*

In Hardin's perspective it seems that, whenever Y adopts X's goals and does something *for X*, so that, being autonomous, he is adopting X's goal for some reason, i.e. some of his *own* goals (Conte and Castelfranchi, 1995; Castelfranchi, 1998) this necessarily means that X's goals are subordinated to Y's personal and selfish convenience – an untenable claim, as discussed above.

---

[16] However, we do not believe it to be so. In our view, psychology still fails to model a very fundamental and obvious distinction: the fact that I have a positive/likeable expectation while deciding to do a given action does *not* entail that I am doing that action *for* that outcome, in order to achieve it, being motivated by it. As it happens I am not motivated by bad *expected* results, it happens also that I am not necessarily motivated by good *expected* results. It is true that my motivating goal (the goal in view of which I act, and which is both 'necessary' and 'sufficient' for intending to act; (Lorini, Marzo, Castelfranchi, 2005) must be among the positive (goal-satisfying) expected results, but it is not true that all the positive expected results motivate me, being necessary or sufficient for deciding to perform that action. Seneca had clarified this point already 2000 years ago. It would be nice if psychology would at last acknowledge and model such a distinction. More importantly, Seneca's claim effectively defeats the 'pseudo-altruism' thesis and opens the way – at least in principle – to true psychological altruism. "*Sed tu quoque' inquit 'uirtutem non ob aliud colis quam quia aliquam ex illa speras uoluptatem. Primum non, si uoluptatem praestatura uirtus est, ideo propter hanc petitur; non enim hanc praestat, sed et hanc, nec huic laborat, sed labor eius, quamuis aliud petat, hoc quoque adsequetur*" (*De vita beata*, IX). In a few words: Do we cultivate our virtues just because we wish to obtain some pleasure? The fact that virtue gives some pleasure does not mean that we follow it *because of* this. The pleasure is just an additional result, not our aim or motive. We will get it while pursuing another aim, that is, virtue.

[17] Otherwise, *any* act would be 'selfish', egoistic. But this is *not* our usual notion of selfishness at all, and it does not correspond to our intuitions about our own and each other motives. Rather, this is the expression of a restricted philosophical view, embedded in a cynical understanding of human nature.

In Hardin there is also a strong but quite arbitrary restriction (although typical of the game-theoretic framework[18]) on Y's *motives* for caring about X's interests and doing something for X, to Y's desire to maintain that relationship: 'I trust you because I think it is your interest to attend to my interests in the relevant matter. This is not merely to say that you and I have the *same* interests. Rather, it is to say that you have an interest in attending to *my* interest because, *typically, you want our relationship to continue*' (Hardin, 2005, Ch.1).

Here we agree with Hardin on the claim that trust is a three-part relation: X *trusts Y to do* α (this relates also to the analysis of Baier (Baier, 1986), and Luhmann (Luhmann, 1979)). In our own terminology, this would translate as: X *trusts Y as for something*, e.g. doing α (we call α the *delegated task* (Castelfranchi and Falcone, 1998)). Many authors do not realize the importance of the third element (indeed, even the context deserves to be modeled explicitly: X *trusts Y in context C to do* α) and this creates several problems; for example, they cannot develop a theory of disomogeneous trust evaluation (X trusts Y for one task but not for another one), or a theory of how we transfer trust in Y for one performance to trust in Y for another performance – an issue of paramount importance in marketing (see Chapter 8). Thus we agree with Hardin that there is no two-part trust or one-part trust; they are just elliptical instances of a more complex relation.

However, we disagree about the rigid use that Hardin makes of this triadic view. He seems to reject any theory of trust generalization, transfer, and abstraction, and focus instead only on a specific personal relation (i.e. trust in a *specific* person, not for example in a 'category'), and only for a specific action α (not for other similar actions, or for a class of actions). This is probably due to his quest for an understanding of Y's specific incentives and interests in attending X's interest. So, although Hardin admits the importance and the advantages of a 'general atmosphere of trustworthiness', in his model this would necessarily and only be a rich network of specific trust relationships among particular individuals: lots of people, each trusting other 'particular' people (Hardin, 2005 (p. 179)). But this, of course, is not what it is usually meant by 'generalized trust'.

In contrast, our approach will be shown adequate to account for trust transmission, generalization, and vagueness (for any value of Y; for any value of X; for any value of α; see Chapter 6 for details). We deny that the theory of generalized or diffuse trust (the so called 'social capital') is another and different model from the theory of interpersonal trust. In our perspective, the former must be built upon and derived from the latter. On the one side, trust capital is a macro, emerging phenomenon; but it must be understood also in terms of its micro foundations. On the other side, conceptually speaking, the notion of trust is just one and the same notion, at different levels of abstraction and generalization.

Another significant disagreement that we have with Hardin is that he decided to consider as 'trust' only the epistemic/doxastic mental attitude of the trustor, i.e. his beliefs. Hardin's book begins with this statement: 'To say that I trust you (. . .) simply means that I *think* you will be trustworthy toward me'; 'To say that I trust you (. . .) is to say nothing more than that I *know or believe certain things about you*' (Hardin, 2005).

Moreover, he claims that the declarations *I believe you are trustworthy* and *I trust you* are equivalent. This is clearly false. The first sentence is the paraphrase of only one of at least two

---

[18] Hardin cites Thomas Schelling (1960, 1980, 134-5): "Trust is often achieved simply by the continuity of the relation between parties and the recognition by each that what he might gain by cheating in a given instance is outweighed by the value of the tradition of trust that makes possible a long sequence of future agreement".

different readings of the second sentence. *I trust you* can mean *I believe you are trustworthy*, but it can also mean much more than this: in particular, it can mean *I have decided to rely on you, to depend and count on you, since I believe you to be trustworthy*.[19] Trust in our model is not only a doxastic mental attitude, but also a *decision* and *intention*, and the *subsequent action* based upon these mental dispositions. There is no contradiction between the so-called 'cognitive' and 'behavioral' notion (acting on trust). In contrast, for Hardin, trust 'if it is cognitive is not behavioral' (Hardin, 2005).[20] In our model the two notions are embedded one into the other; and the explicit theory of the relations between these layers and of the transition from one to the other is important.

Based on this view is the idea that in principle one cannot really *decide* to trust somebody else; trust 'is not a matter of choice', insofar as we refer to trust as a doxastic attitude. This is because, as a matter of fact, one cannot decide to believe something[21]; but this fact is not true for the second layer of trust, concerning the decision to delegate and the act based upon this decision. While trusting you I can really 'decide' to take some risk (which I mentally consider) and to make myself vulnerable to you. So much so that, indeed, I can later come to regret my decision, and blame myself for that choice.

It follows that trust can be (and usually is) rational on both these two levels: as a belief, and as a decision and action. *Epistemic rationality* consists in forming a justified belief on the basis of good evidence and reliable sources. *Decision rationality* consists in taking a decision on the basis of a correct calculation of values (outcomes) and of their (limited) optimization. In our model, trust can be rational at both levels: grounded as a belief (evaluation and prediction), and optimal as a choice.

The last important difference between Hardin's analysis and our model is that Hardin completely puts aside the issue of 'competence', overlooking all those cases in which we put our trust and reliance on the trustee's ability, expertise, quality of service, and so on. In his own words: 'I will usually assume through this book that competence is not at issue in the trust relationships under discussion' (Hardin, 2005, Introduction). Unfortunately, this assumption is untenable, since the competence aspect cannot be either marginalized or rigidly separated from trust in Y's reliability (see Section 2.2.5 for more details).

1.5.8   Rousseau: What Kind of Intention is 'Trust'?

A very good (indeed, our favorite) definition of trust, based on a large interdisciplinary literature and on the identification of fundamental and convergent elements, is the following: '[Trust is] a psychological state of a trustor comprising the intention to accept vulnerability in a situation involving risk, based on positive expectations of the intentions or behavior of the trustee' (Rousseau *et al.*, 1998).

---

[19] This is the meaning of 'trust' in expressions like: "While trusting Y you expose yourself, you risk very much!"; "My poor friend, how could you trust him?!"; "Trust me! You can trust me!" "OK, I trust you!".

[20] This claim can be correct if 'behavioral' means a behavioristic definition (like in Yamagishi's approach); but it is wrong if it is aimed at excluding trust as decision and action. Actually 'Action' is a cognitive notion.

[21] Nonetheless, there may be trust attitudes based on 'acceptances' and not on 'beliefs'; and 'acceptances' can be intentionally assumed. In this perspective, I can also 'decide' to trust you, in the sense that I decide to presume that you are reliable and competent, and to act on the basis of such an assumption, i.e. "as if" I believed you to be reliable and competent (Cohen, 1992) (Engel, 1998).

What we like here is the idea of a composite psychological state (although it is not fully characterized), which does not only include 'positive expectations', but where these expectations are the *base* for the intention to make oneself vulnerable. This crucial link is made explicit in this definition, so it is less important that other beliefs are ignored (like the fact that trust is also an appraisal), that the 'competence' and 'ability' of *Y* remain implicit, or that there seems to be no trust before and without intention. Notice that – given this definition – since trust necessarily includes the 'intention', the idea of trust might not be enough to entrust *Y*, would be contradictory. One could never say: 'I trust *Y* but not enough', or 'I trust *Y* but I do not intend to rely on him'.

This brief survey of various definitions of trust from different disciplines was just for highlighting the main limits of the current (mis)understanding of this notion, but also a lot of very important intuitions that – although partial and even contradictory – deserve to be preserved, well defined, and coherently organized.

On the basis of these results, it is now time to start introducing in a more explicit and systematic way our layered model of trust, which is what we shall do in the next chapter.[22] We will encounter and discuss other definitions and models throughout the book.

References

Baier, A. Trust and antitrust, *Ethics*, 96 (2): 231–260, 1986.

Batson, C.D. (1991) *The Altruism Question: Towards a social social-psychological answer*, Hillsdale, NJ: Erlbaum.

Busacca, B. and Castaldo, S. Trust in market relationships: an interpretative model, *Sinergie*, 20 (58): 191–227, 2002..

Castaldo, S. (2002) *Fiducia e relazioni di mercato*. Bologna: Il Mulino.

Castelfranchi, C. Paradisi artificiali: prolegomeni alla simulazione della interazione sociale. *Sistemi Intelligenti*, I (1): 123–165, 1989.

Castelfranchi, C. (1997) Principles of (individual) social action. In Tuomela, R. & Hintikka, G. (eds) *Contemporary Action Theory*. Kluwer.

Castelfranchi, C. Towards an agent ontology: autonomy, delegation, adaptivity. *AI*IA Notizie*. 11 (3): 45–50, 1998; Special issue on 'Autonomous Intelligent Agents', Italian Association for Artificial Intelligence, Roma.

Castelfranchi, C. Again on agents' autonomy: a homage to Alan Turing. *ATAL 2000*: 339–342.

Castelfranchi, C. (2000) Affective appraisal vs cognitive evaluation in social emotions and interactions. In A. Paiva (ed.) *Affective Interactions. Towards a New Generation of Computer Interfaces*. Heidelberg, Springer, LNAI 1814, 76–106.

Castelfranchi, C. Modelling social action for AI agents. *Artificial Intelligence*, 103, 157–182, 1998.

Castelfranchi, C. Mind as an anticipatory device: for a theory of expectations. In *'Brain, vision, and artificial intelligence' Proceedings of the First International Symposium, BVAI 2005, Naples, Italy, October 19-21*, 2005. Lecture notes in computer science ISSN 0302-9743 vol. 3704, pp. 258–276.

Castelfranchi, C., Cesta, A., Conte, R., Miceli, M. Foundations for interaction: the dependency theory, *Advances in Artificial Intelligence*. Lecture Notes in Computer Science; Vol. 728: 59–64, 1993. ISBN:3-540-57292-9.

Castelfranchi, C. and Falcone, R. (1998) Towards a theory of delegation for agent-based systems, *Robotics and Autonomous Systems*, Special issue on Multi-Agent Rationality, Elsevier Editor, 24 (3-4): 141–157.

Castelfranchi, C. and Falcone, R. Founding autonomy: the dialectics between (social) environment and agent's architecture and powers. *Agents and Computational Autonomy*, 40–54, 2003.

Castelfranchi, C., Giardini, F., Lorini, E., Tummolini, L. (2003) The prescriptive destiny of predictive attitudes: from expectations to norms via conventions, in R. Alterman, D. Kirsh (eds) *Proceedings of the 25th Annual Meeting of the Cognitive Science Society*, Boston, MA.

---

[22] Later, (see Chapter 8), we come back to the scientific literature, in order to dedicate special attention to economic and game-theoretic accounts of trust, considering both recent developments in this domain, and the old Deutsch's definition of this notion.

Castelfranchi, C. and Lorini, E. (2003) Cognitive anatomy and functions of expectations. In *Proceedings of IJCAI'03 Workshop on Cognitive Modeling of Agents and Multi-Agent Interactions*, Acapulco, Mexico, August 9–11, 2003.

Cohen, J. (1992) *An Essay on Belief and Acceptance*, Oxford University Press, Oxford.

Conte, R. and Castelfranchi, C. (1995) Minds is not enough. Pre-cognitive bases of social interaction. In: Gilbert, N., Doran, J., *Simulating Societies*, UCL, London.

Cummings, L. L. and Bromiley, P. (1996) The organizational trust inventory (OTI): development and validation. In R. Kramer, and T. Tyler (eds.), *Trust in Organizations* (pp. 302–330). Thousand Oaks, CA: Sage.

Dennett, D.C. (1989) *The Intentional Stance*. Cambridge, MA: MIT Press.

Deutsch, M. (1985) *The Resolution of Conflict: Constructive and destructive processes*. New Haven, CT: Yale University Press.

Elster, J. (1979) *Ulysses and the Sirens*. Cambridge: Cambridge University Press.

Engel, P. Believing, holding true, and accepting, *Philosophical Explorations* 1: 140–151, 1998.

Falcone, R., Singh, M., and Tan, Y. (2001) Bringing together humans and artificial agents in cyber-societies: A new field of trust research; in *Trust in Cyber-societies: Integrating the Human and Artificial Perspectives* R. Falcone, M. Singh, and Y. Tan (eds.), LNAI 2246 Springer. pp. 1–7.

Falcone, R. and Castelfranchi, C. (2001) The human in the loop of a delegated agent: The theory of adjustable social autonomy, *IEEE Transactions on Systems, Man, and Cybernetics, Part A: Systems and Humans*, Special Issue on 'Socially Intelligent Agents - the Human in the Loop, 31 (5): 406–418, September 2001.

Falcone, R. and Castelfranchi, C. (2001) Social trust: a cognitive approach, in *Trust and Deception in Virtual Societies* by Castelfranchi, C. and Yao-Hua, Tan (eds), Kluwer Academic Publishers, pp. 55-90.

Falcone, R. and Castelfranchi, C. (2001) 'The socio-cognitive dynamics of trust: does trust create trust?' In R. Falcone, M. Singh, Y.H. Tan, eds., *Trust in Cyber-societies. Integrating the Human and Artificial Perspectives*, Heidelberg, Springer, LNAI 2246, 2001, pp. 55–72.

Falcone, R. and Castelfranchi, C. (2003) A belief-based model of trust, in Maija-Leena Huotari and Mirja Iivonen (eds) *Trust in Knowledge Management and Systems in Organizations*, Idea Group Publishing, pp. 306–343.

Gambetta, D. (1988) 'Can we trust trust?' In *Trust: Making and Breaking Cooperative Relations*, edited by D. Gambetta. Oxford, Blackwell.

Hardin, R. (2002) *Trust and Trustworthiness*, New York: Russell Sage Foundation.

Hertzberg, L. (1988) On the Attitude of Trust. *Inquiry* 31 (3): 307–322.

Johannisson, B. (2001) Trust between organizations: state of the art and challenges for future research, paper presented at the EIASM Workshop on Trust Within and Between Organizations, Vrije Universiteit Amsterdam.

Jones, A. On the concept of trust, *Decision Support Systems*, 33 (3): 225–232, 2002. Special issue: Formal modeling and electronic commerce.

Jones, K. Trust as an affective attitude, *Ethics* 107: 4–25, 1996.

Lorini, E., Marzo, F., Castelfranchi, C. (2005) A cognitive model of the altruistic mind. Boicho Kokinov (eds.), *Advances in Cognitive Economics*, NBU Press, Sofia, Bulgaria, pp. 282–294.

Luhmann, N. (1979) *Trust and Power*, John Wiley & Sons Ltd, New York.

Mayer, R. C., Davis, J. H., and Schoorman, F. D. (1995) An integrative model of organizational trust. *Academy of Management Review*, 20 (3): 709–734.

Marsh, S. P. (1994) Formalising trust as a computational concept. PhD thesis, University of Stirling. Available at: http://www.nr.no/abie/papers/TR133.pdf.

McKnight, D. H. and Chervany, N. L. Trust and distrust definitions: one bite at a time. In *Trust n Cyber-societies*, Volume 2246 of Lecture Notes in Computer Science, pages 27–54, Springer, 2001.

Miceli, M. and Castelfranchi, C. (2000) The role of evaluation in cognition and social interaction. In K. Dautenhahn (ed.) *Human Cognition and Social Agent Technology*. John Benjamins, Amsterdam.

Miceli, M. and Castelfranchi, C. (2002) The mind and the future: The (negative) power of expectations. *Theory & Psychology*. 12 (3): 335–366.

Moorman, C., Zaltman, G., and Deshpandé, R. (1992) Relationships between providers and users of market research: the dynamics of trust within and between organizations. *Journal of Marketing Research* 29 (3) : 314–328, August, 1992.

Mutti, A. (1987) La fiducia. Un concetto fragile, una solida realtà, in Rassegna italiana di sociologia, pp. 223–247.

Ostrom E. and Walker J. (eds.) (2003) *Trust and Reciprocity: Interdisciplinary Lessons from Experimental Research*, Russell Sage Foundation, New York, NY, 409 pp., ISBN 0-87154-647-7.

Rousseau, D. M., Burt, R. S., and Camerer, C. Not so different after all: a cross-discipline view of trust. *Journal of Academy Management Review*, 23 (3): 393–404, 1998.

Seneca, *De vita beata*, IX.

Schelling, T. (1960, 1980) *The Strategy of Conflict*, Harvard University Press.

Smith, A. (1776) *An Inquiry into the Nature and Causes of the Wealth of Nations*, London: Methuen & Co., Ltd.

Tummolini, L. and Castelfranchi, C. (2006) The cognitive and behavioral mediation of institutions: Towards an account of institutional actions. *Cognitive Systems Research*, 7 (2–3).

Tuomela, R. and Miller, K. We-Intentions, *Philosophical Studies*, 53: 115–137, 1988.

Tuomela, R. What is cooperation. *Erkenntnis*, 38: 87–101, 1993.

Tuomela, M. A collective's rational trust in a collective's action. In Understanding the Social II: Philosophy of Sociality, *Protosociology*. 18–19: 87–126, 2003.

Williamson, O. E. (1993) Calculativeness, trust, and economic organization, *Journal of Law and Economics*, XXXVI: 453–486, April 1993.

Yamagishi, T. and Yamagishi, M. (1994) Trust and commitment in the United States and Japan. *Motivation and Emotion*. 18, 129–166.

Yamagishi, T. (2003) Cross-societal experimentation on trust: A comparison of the United States and Japan. In E. Omstrom and J. Walker, eds., *Trust, Reciprocity: Interdisciplinary Lessons from Experimental Research* NY, Sage, pp. 352–370.

2

Socio-Cognitive Model of Trust: Basic Ingredients

Trust means different things, but they are systematically related one with the other. In particular we analyze *three crucial concepts* that have been recognized and distinguished in the scientific literature. Trust is:

- A mere *mental attitude* (prediction and evaluation) towards an other agent; a simple *(pre)disposition*.[1]
  This mental attitude is in fact an opinion, a judgment, a preventive evaluation about specific and relevant 'virtues' needed for relying on the potential trustee, but that might remain separated from the actual exercise of trust.
- A *decision* to rely upon the other, i.e. an *intention* to delegate and trust, which makes the trustor 'vulnerable' (Mayer *et al.*, 1995).
  This is again a mental attitude, but it is the result of a complex comparison and match among the preventive evaluations of the different potential trustees, about the risks and the costs, and about the applicability of these evaluations to the actual environments and context.
- A *behavior*, i.e. the intentional *act* of (en)trusting, and the consequent overt and practical *relation* between the trustor and the trustee.
  This is the consequent act, behavior, of the trustor, generally coordinated and coherent with the previous decision; and the public 'announcement' and social relation.

Trust is in general all these things together.

In order to understand this concept and its real meaning we have to analyze it, and show the complex relations that exist between different terms and constituents.

---

[1] Or even just a feeling, an affective disposition where those 'beliefs' are just implicit (see Chapter 5).

Trust Theory   Cristiano Castelfranchi and Rino Falcone
© 2010 John Wiley & Sons, Ltd

2.1   A Five-Part Relation and a Layered Model

In our view trust is a *relational* construct between:

- An agent X (*the trustor*) (we will name Xania, a woman). This agent X is necessarily an 'intentional entity' (see intentional stance (Dennett, 1989)); let's say a 'cognitive agent', that is, an agent with mental ingredients (beliefs, goals, intentions, and so on). Trust must be (also) be extended in cognitive terms, as X's specific mental attitudes towards other agents and about given tasks.
- An addressee Y (*the trustee*) that is an 'agent' in the broader sense of this term (Castelfranchi, 1998; 2000a): an entity able to cause some effect (outcome) in the world; the outcome X is waiting for. When Y is an intentional agent: we will name it Yody, and he is a man.
- Such a 'causal' process (*the act, or performance*) and its result; that is, an act α of Y possibly producing the outcome p; which is positive or desirable because it includes (or corresponds to) the content of a goal of X ($Goal_X(g)=g_X$), the specific goal for which X is trusting Y. We call this act: Y's *task: τ*. τ is the couple (α, p), with g included in p or in some cases $p \equiv g$.
- $g_X = Goal_X(\boldsymbol{g})$ is in fact a crucial element of the trust concept and relation, frequently omitted.
- A context ($\boldsymbol{C}$) or situation or environment where X takes into account Y (thus affecting X's evaluation and choice of Y) and/or where Y is supposed to act and to realize the task (thus affecting the possibility of success).

In other words, in our model trust is a five-part relation (at least):

$$TRUST\,(X\ Y\ C\ \tau\ g_X) \tag{2.1}$$

that can be read as X trusts (<u>*in*</u>) Y in context C *for* performing action α (executing task τ) and realizing the result p (that includes or corresponds to her goal $Goal_X(g)=g_X$).

A deep analysis of each component is needed, and a theory of their relationships and variations.

2.1.1   A Layered Notion

As we said, we consider and analyze trust as a composed and *'layered' notion*, where the various more or less complex meanings are not just in a 'family resemblance' relation, but are embedded one into the other, and it is important the explicit theory of the relations between those layers and of the *transition* from one to the other; since the most simple form can be there without the richer one, but not vice versa.

As we said, in our theory there is a double link between these forms of trust (from the dispositional one to the active one): a conceptual link and a process/causal link (see Section 1.4).

Trust as a *layered notion* (see Figure 2.1) means:

- in its basic sense just a mental (cognitive and affective[2]) *attitude* and *disposition* towards Y (*beliefs*: *evaluations* and *expectations*); this already is a social *relation*;

---

[2] We will put aside here the affective nature of trust; trust as an intuitive disposition, as a feeling. Here we consider trust as an explicit (and usually grounded) judgment. We dedicate the entirety of Chapter 5 to affective trust dispositions

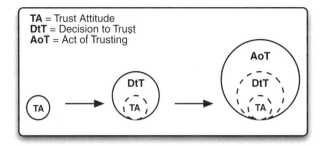

Figure 2.1 Trust stages and layers

- in its richer use, a *decision* and *intention* based on that disposition;
- and then the *act* of *relying* upon *Y*'s expected behavior;
- and the consequent overt social *interaction* and *relation* with *Y*.

The trust attitude/disposition is a determinant and precursor of the decision, which is a determinant and a precondition of the action; however, the disposition and the intention remain as necessary *mental constituents* during the next steps.

Let us introduce an example to better clarify the different concepts: Xania is the new director of a firm; Mary and Yody are employees, possible candidates for the role of Xania's secretary. Now, consider these three different aspects (and components) of Xania's trust towards them:

1) Xania evaluates Mary on the basis of her personal experience with Mary, if any, of Mary's CV and of what the others say about Mary; on such a basis she forms her own opinion and *evaluation* about Mary: how much she considers Mary trustworthy as a secretary? Will this trust be enough for choosing Mary and deciding to bet on her?
2) Now Xania has also considered Yody's CV, reputation, etc.; she knows that there are no other candidates and *decides* to count on Yody as her secretary; i.e. she has the *intention* of delegating to Yody this job/task when it comes to reorganizing her staff.
3) One week later, in fact, Xania nominates Yody as her secretary and uses him, thus actually she trusts him for this job: the trust relationship is established.

In situation (1) Xania has to formulate some kind of trust that is in fact an *evaluation* about Mary as a secretary. This trust might be sufficient or not in that situation. Since she knows that there is also another candidate she cannot decide to choose Mary (she has to wait for this) but she can just express a mere evaluation on her, and perhaps this evaluation is not good enough for being chosen (even if it has been made before next candidate's evaluation).

In situation (2) Xania arrives at a decision that is based on trust and is the decision to trust. It is also possible that – in Xania's opinion – neither Mary nor Yody are suitable persons with respect to the secretary's role and she might take the decision not to trust them, and to consider different hypotheses: i) to send a new call for a secretary or ii) to choose the least

---

and impulses. Consider however, that those two 'forms' of trust can coexist: there may be an explicit evaluation joined with (eliciting or based on) a feeling (Castelfranchi, 2000b).

worst between them for a brief period, looking for new candidates at the same time; or iii) to manage for a period without any secretary; and so on.

In situation (3) Xania expresses her trust through an (official) *delegation*, in fact an act of communication (or more in general, through observable external behavior). In general, this kind of relationship does not necessarily have to be official, it is simply known to the two agents (as we will see in the following, in special cases, it could also be unknown to the trustee).

In the case of autonomous agents who have no external constraints conditioning their freedom of making a reliance,[3] we can say that a sufficient value[4] of core trust is a necessary but not sufficient condition for a positive decision to trust (see Chapter 3); vice versa, a freely chosen delegation (Castelfranchi and Falcone, 1998) implies the decision to trust, and the decision to trust implies a sufficient value of core trust. We will specify which beliefs and which goals characterize the trust of the *Trustor* (X) in another agent (the *Trustee, Y*) about Y's behavior/action relevant for a given result p corresponding to (or included in) the goal of X, g_X. Given the overlap between trust and (mental) reliance/delegation, we need also to clarify their relationships.

Before starting the analysis of the basic kernel, let us analyze the relationship between p and g_X.

2.1.2   Goal State and Side Effects

As we said: $g_X \subseteq p$ (the set of the results of Y's action α contains not only X's goal but also a set of side effects). In other words, sometimes the achievement of the goal g_X is conditioned (given the specific trustee and his own features, the contextual situation, and so on) by the realization of other results in the world that could compromise other interests or goals of X.

I can trust my dentist to be able to solve my problem with my tooth, but – while going there – I know (expect) that although he will also be careful and honest I will feel some pain and will spend a lot of money. Or, I know that if I trust John – which is actually reasonable and enough for what I need –he will become too familiar, he will take liberties with me, and maybe I won't like this.

So the analysis and the knowledge of these side effects are really relevant when deciding upon trusting agents, and often a merely qualitative analysis is not sufficient (see Chapter 3).

2.2   Trust as Mental Attitude: a Belief-Based and Goal-Based Model

Our main claim is that: *only a cognitive agent can trust another agent; only an agent endowed with goals and beliefs.*

First, one trusts another only relative to a goal, i.e. for something s/he wants to achieve, that s/he desires or needs. If I don't potentially have goals, I cannot really decide, nor care

---

[3] Absence of external constraints is an ideal condition: in fact, also in case of autonomous agents some constraints are always present. In the case of the previous example, a constraint is given from the impossibility to evaluate all the potential candidates available in the world.

[4] In fact, trust is also a quantitative notion: it is constituted by different ingredients to which it is possible/necessary to attribute a value.

about something ('welfare'): *I cannot subjectively trust somebody.* These goals could also be maintenance goals (of a situation, a state), not necessarily achievement goals.

Second, trust itself basically consists of (implicit or explicit) beliefs.

The root of trust is *a mental state*, a complex *mental attitude* of an agent X towards another agent Y, in context C, about the behavior/action α relevant for the result (goal) g_X.

Since Y's action is useful to X, and X is relying on it, this means that X is *delegating* to Y some action/goal in her own mental plan. This is the strict relation between trust and reliance and delegation: *Trust is the mental counter-part of reliance and delegation.*[5]

This mental attitude is based on and consists of *beliefs* (or in general of doxastic representations[6]), about the trustee and his behavior. And in fact they may be wrong. X can have false (although well grounded and subjectively justified; rational) beliefs about Y's qualities, skills, and behavior. In short, the main beliefs are:

i) X believes that Y is able and well disposed (willing) to do the needed action;
ii) X believes that in fact Y will appropriately do the action, as she wishes;
iii) X believes that Y is not dangerous; therefore she will be safe in the relation with Y, and can make herself less defended and more vulnerable.

The first (and the third) family of beliefs is *'evaluations'* about Y: to trust Y means to have a good evaluation of him. Trust implies some appraisal.

The second (and the third) family of beliefs is *'expectations'*, that is (quite firm) predictions about Y's behavior, relevant for X's goal: X both wishes and forecasts a given action α of Y, and excludes bad actions; she feels safe.

The basic nucleus of trust – as a mental disposition towards Y – is a positive expectation based on a positive evaluation; plus the idea that X might need Y's action.

Let us carefully consider these various roles of the beliefs in a trust mental state, and these two facets of trust: as valuation, as expectation.

2.2.1   Trust as Positive Evaluation

An explicit positive evaluation is a judgment, a belief about the goodness of Y for the achievement of a certain goal. 'Y is good' actually means 'Y is good for...' (Miceli & Castelfranchi, 2000). Sometimes we do not specify *for* what Y is good, just because it is included in the very concept of Y ('This is a good knife/pen/car/... mechanic/doctor/father/...'), or because it is clear in the context of the evaluation ('On which mountain we can do some climbing?' 'That mountain is good!'). These are direct explicit evaluations: 'Y is good, is OK, is apt, able, useful', and so on.

The abstract content of these predicates is that: *given the goal g_X, Y is able to realize it or to help X in realizing it* (a tool). In other words, Y has the *power of* realizing g_X.[7]

---

[5] Given this strict relation and the foundational role of reliance and delegation (see Section 2.6) we need to define delegation and its levels; and to clarify also differences between reliance, delegation, and trust.

[6] See Section 2.2.8 on Trust and Acceptance and Chapter 1, note 21.

[7] The evaluation of Y is about the *internal powers* for g_X (internal resources, capabilities, competences, intelligence, willingness, etc.), but for relying on Y for g_X external conditions and the control on the external conditions might also be necessary: Y may also have (or not) the *external powers*, the external conditions and resources for realizing g_X; Y

However, trust is not only a belief that Y 'is good for τ', it is also a set of beliefs about the needed qualities or 'virtues' of Y. Trust, as evaluation, is in fact a model of Y's *qualities* and *defects* (which define his *trustworthiness dimensions*).

As for any kind of explicit evaluation, with trust we are not satisfied by the mere belief that Y is OK, is 'good for', has the power to achieve goal g_X, to execute the action/task α/τ (delegated to him). We try to *understand* why Y is good for this (while Z is not); to have a *theory* of what makes Y able.[8] In other terms, we try to know which kind of characteristics are useful or required for effectively performing τ or achieving g_X. And many of them are just hidden, internal (and mental): *kripta* (Section 2.2.7).

Qualities and Standards

This applies in general to the Evaluation theory (Miceli and Castelfranchi, 2000).

Given that Y is 'good for' τ (for example, this knife Y is good for cutting the bread), to which features of Y should this be ascribed? In the case of the knife: To its plastic or wooden handle? To its sharpening? To its color? To its being serrate and long? And so on. In several cases this implies a true *causal model* (although naive and commonsensical) of τ, of Y, and of what effects it produces. In the example of the knife, it is clear that the plastic handle or the color are irrelevant, while a good knife for bread needs to be long, serrate, and sharpened.

Those features (F) to whom the 'goodness of Y for...' is ascribed are Y's *qualities* (Q). *Defects* (D) are those features of Y's to which is attributed the lack of power, the inadequacy or dangerousness of Y for α/τ.

Notice that a feature of Y that is a 'quality' relative to task τ, can be a 'defect' relative to another task τ', and vice versa (see Figure 2.2).

Let us also take note of how this theory of 'qualities' and 'defects' is cognitive and pragmatically very crucial and effective. In fact, while buying a new knife in a store we are not allowed to have with us a piece of bread and experimentally evaluate which knife is 'good for' it. Thus, how can we choose a 'good' knife without trying it? Just because we have a theory of the needed qualities, of what makes a knife a good knife for bread. We just look for these characteristics; we compare that knife with the 'standards' for a good bread-knife.

Standards are in fact just *qualities generalized* to the class of the object; the ideal properties that such a kind of object O *must* possess in order to be good as an O.

Qualities (and standards) are necessary not only for *recognizing* and choosing a good instance of object O; that is they are *signs* of its being good, reliable for τ (trustworthy) (see Section 2.2.7 on Signs); but they:

- are fundamental also for generalizing trust from one task to another (are for τ needed more or less the same qualities than for τ'?), or from one agent to another: 'has Z the relevant qualities remarked in Y?' (see Chapter 6); and thus they;
- are also fundamental for predictions based on general models and classes.

---

is both 'able' and 'in condition' for realizing g_X. For example, in J. J. Meyer's logic (Meyer, 1992) Y 'CanDo' when both Y is 'Able' and 'In Condition'.

[8] While we put the other dimension (the evaluation of the conditions and external resources) in the 'environmental trust' (see Section 2.10).

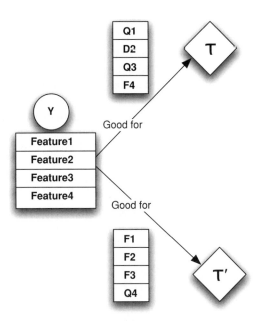

Figure 2.2 From Features (F) to Qualities (Q) through Evaluation: in particular (Feature 1 and Feature 3) are evaluated as good for the task τ while (Feature 4) is good for the task τ'

X trusts Y – that is, she has a positive evaluation of him for τ (and she generates positive expectations of him), also because she ascribes to him certain *qualities*.

Then, we can say that if (see Figure 2.3):

SetQ is the set of qualities (powers) needed for α/τ, then for p (g_X); in case Y possesses all the elements in *SetQ* then X could trust Y for achieving g_X.

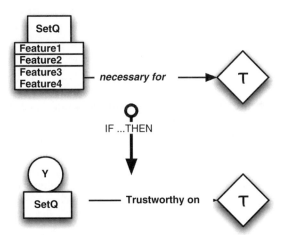

Figure 2.3 Given a Set of Abstract Qualities for a task, how to select the right Agent for that task

For any Y, the more the elements in $SetQ$ are owned by Y the more trustworthy is Y for α/τ.

So we can say that trust is constituted of what we call 'implicit and indirect evaluations', that is of specific features (like 'sharpening; long; serrate', or for a person like 'tall, bold, agile, intelligent, ...') which only apparently are just *descriptive* predicates and usually are *evaluative* predicates. These are the specific and analytic beliefs on which the global positive or negative judgment is grounded.

Reason-based trust is in fact a *theory* about Y, and about his qualities and defects that make him more or less apt and reliable for α/τ; on this basis we make predictions and build expectations about Y's performance and its outcome; and we explain the reasons of success or failure; like for any *theory*.

This also means that the feedback on trust due to the observed outcome of Y's performance (see Chapter 6), not only can be attributed to Y, that is to internal factors (and not to external accidents or conditions), but in some cases can be specifically ascribed to a sub-feature of this causal model of Y's *power on* τ. And one can revise this specific piece of evaluation and trust: 'Y is not so sharp as assumed'; 'Y is not so agile as supposed', and so on.

Trust and Powers

An evaluation is a judgment about the possible powers of Y. In our abstract vocabulary Y *can* (has the *power of*) perform a given action or realize a given goal. It is relative to this that Y is *good for*. Correspondently, *qualities* are just *powers of* (to be *strong*, to be *smart*) or conditions for *powers of* (to be *entitled*, to be *prepared*).

This is about not only personal powers (be strong, intelligent, skilled, etc.) but also social powers: be influent, prestigious, being a friend of Z (and thus able to obtain from Z something), being sexually appealing, etc.; and also *institutional powers*, the new capacities that a given institutional role gives to its players: like nominating somebody, or proclaiming, or officially signing, or to marry, etc.[9]

So there is a complex relation between powers and trust. On the one side powers – and in particular perceived and appreciated powers – make us trustworthy for the others (they can need us and wish to rely on us), and this is why our *social image* and the *signals* that we give of our qualities are so important. As we will see, trust (being trusted in a given community, and in particular being better evaluated than others) is a *capital* (see Chapter 10).

On the other side, this is a capital precisely because it increases our powers; provides us new powers. Since we (being perceived as trustworthy) are appreciated and demanded, we can enter in exchange relations, we can receive cooperation, we can influence other people, etc. This means that we greatly enlarge our *practical powers* (the possibility to achieve – thanks to the other's cooperation – our goals) and *social powers*: powers on the others. Also for being invested by institutional roles (and receive new special powers) we need some competence and reliability, or even better we need some trust from the others on this. In sum, *powers give trustworthiness and trust from the others; and trust gives powers.*

---

[9] See (Searle, 1995); as for the relationships between personal and institutional powers, see (Castelfranchi, 2003).

The Recursive Nature of Trust as Mental Representation

One might claim that in a strict sense trust just is or implies the positive *expectation*, which in its turn is based on positive evaluations; but that (i) *the evaluations are not trust or* (ii) *they are not implied by trust* and *they are not strictly necessary*: One might trust Y without specific and explicit beliefs on his skills or reliability, just on the base of experience and learning.[10]

However, this is not fully true.

As for (ii), Y's trustworthiness or the trust *in* Y always imply some (perhaps vague) attribution of internal qualities (such that the task could be achieved/realized); some 'kripta' which makes Y apt and reliable (Section 2.2.5).

As for (i), the evaluations of Y, when used as the bases and reasons for a trust expectation and decision, are *subsumed* and *rewritten* as aspects and forms of 'trust'. Given that on the basis of those features (*qualities*) X trusts Y, then X trusts those features of Y. 'I trust (in) his competence', 'I trust (in) his benevolence', 'I trust (in) his morality', 'I trust (in) his punctuality', and so on.

Let us ground this on the general theory of goals and actions. Given her goal g_X (X's aim or end) X searches for some possible action α (usually of X) able to achieve g_X (where g_X is included in the outcomes p, the post-conditions of α). Given an action α useful for g_X, this action in order to be successfully performed requires some condition C to be true. If C holds the subject can perform α; if C does not hold it becomes the new goal g^1_X of X, subordinated and instrumental to g_X: a sub-goal. This obviously is the essential abstract principle of planning.

Now, given that X has the goal of realizing g_X, and that she is dependent on Y and needs an action α_Y of Y, she has the sub-goal that Y successfully performs α_Y. However, certain conditions are needed for both (i) Y successfully performing α_Y; (ii) X can decide to count on this and counts on this.

Y's valuation is precisely the epistemic activity aimed at those conditions; and the same hold for X's predictions about Y doing or not α_Y. As we have underlined, evaluations are about goals (something is or isn't 'good *for*'), and predictions in trust are in fact not simple 'predictions' (beliefs about the future) but more rich 'expectations' (involving goals). Actually, since X wishes to achieve g_X through Y's action, and has the goal that Y be able, in condition, and predictable in performing α_Y, all these necessary conditions (also those 'internal' to Y) are new (sub)goals for X: X *wishes* that Y is skilled enough, competent enough, not hostile or dangerous, willing and reliable, and so on. Relative to those sub-goals she evaluates Y and has (or not) trust in Y. She trusts Y *for* being competent, *for* being persistent, etc.

So trust in Y as for action α_Y for goal g_X, (at least implicitly – but frequently explicitly) entails sub-trust supporting the broad trust about the action; in a recursive way. Since any new goal about Y might have its sub-conditions and needed sub-qualities on which X relies; thus potential sub-trusts are generated. For example, X can trust Y for being really willing to cooperate with her, because she knows and *trusts in* Y's friendship, or because she knows and *trusts in* Y's generosity and empathy, or because Y is morally obliged and she knows and *trusts in* Y's morality.

This is not only the real use of the word and the commonsense on trust; it is a logical and principled use, as we have just explained.

---

[10] We thank Fabio Paglieri for this observation. This is also one of Andrew Jones' criticisms to our model (Jones, 2002); see Section 2.2.3 for our reply.

2.2.2   The 'Motivational' Side of Trust

In our opinion, part of the core of trust is the 'prediction' component: a belief about a future event or state. However, this core element although *necessary* is not *sufficient* in our view (as for example claimed by A. Jones (A. Jones, 2006)).

Many definitions in the literature (Chapter 1) (even those mentioned by Jones) either explicitly or implicitly also contain the idea that trust is relative to some reliance upon, to some welfare and interest. In our analysis this means that the trustor has some 'goal', not only beliefs. Some author contexts precisely this point, that for us is fundamental: the other core (motivational).

Is Trust Reducible to a (Grounded) Belief, a Regularity-Based Prediction?

Certainly, one can establish a conventional, technical meaning of 'trust' far from its natural language meaning and psycho-sociological use; but in our view this is not particularly useful. It would be more heuristic to abstract from common-sense meaning and to enlighten and identify (and formalize) the basic, necessary and sufficient conceptual constituent (useful for a rich and explanatory theory of the psycho-social phenomena). In this case we think that one cannot miss the fact that when X trusts someone or something for something X is concerned, is involved; X cannot be neutral and indifferent about what will happen. In other words, X has not simply a prediction, but a full 'expectation'. In our analysis an 'expectation' is more than a simple forecast or belief about the future (based on some experience, law, regularity, or whatever). Trust in our model is composed of 'expectations' about Y, his behavior, and a desired result. A mere 'regularity' does not produce 'trust' (it can even produce 'fear'). Even to produce Luhman's 'confidence' (just based on regularities without choice) something more is necessary. In fact, confidence (which in fact is just a simpler form of trust) is a positive, pleasant feeling; it implies some implicit or explicit goal and its realization (for example to avoid stress, surprise, problems, anxiety). When apparently a mere prediction or perceived regularity determines a feeling or attitude of trust or confidence it is because X not only believes but desires that the world goes in such a regular, predictable way: see the need for 'predictability' ((Bandura, 1986) our theory of expectations, etc.). The issue is how predictions become prescriptions (Miceli and Castelfranchi, 2002) and (Castelfranchi *et al.*, 2003).

When we set up or activate protection or surveillance systems, precisely because we know that there is a high probability of rapine (ex. banks) or of aggressions, we do not have 'trust' they will rapinate or aggress us! And when we institute the firemen organization we 'expect' but we do not 'trust' that there will be fires! We 'trust' that firemen will be able to put out the fire (as desired!).

There are computer making/producing weather forecasts, but they do not have 'expectations' about the weather (although they could check whether the prediction were correct and fulfilled, in order to learn and adjust); even less likely do they have 'trust' about a sunny day (or rainy!).

Thus a mere belief (prediction) (even regularity based) is not enough. Nor it is 'necessary'. In fact there can be trust and bet/reliance on non-regular, exceptional events (perhaps an 'irrational' trust, but trust; *'This time it will be different! I'm sure, I feel so: it cannot happen again!'*).

Are Expectations and Trust Always Based on Regularities?

There can be expectation about conformity to rules, but also expectations about 'exceptions' or violations (they are not an oxymoron).

I'm worrying about being aggressed and robbed; actually I know (they said) that this has never happened here, but since I feel weak, alone, and without defense I really worry and expect some aggression.

Moreover, there can be new, unexplored circumstances and problems and usually – that's true –we are more careful, diffident; but we can also be fully convinced and trustful in our creative solutions and on some intuitive or reasoned prediction.

Previsions, expectations, and trust are not (for us) always and necessarily based on rules, norms, regularities (except to postulate that any 'inference', or association and learning, is by definition based on an explicit rule we trust.[11]

In conclusion, (regularity-based) beliefs/predictions are:

- Neither *necessary* (predictions are not necessarily regularity/rule based; moreover even when there is a regularity or rule I can expect and trust an exceptional behavior, event, like winning my first election or the lotto).
- Nor *sufficient*: the goal component (some wish, concern, practical reliance) is implied for us in any 'expectation' and a fortiori in any form of 'trust' which contains positive expectation.[12]

2.2.3   The Crucial Notion of 'Goal'

For our non-reductive theory of trust (not simply epistemic but motivational and pragmatic) it is crucial to make clear the central notion of the 'motivational' dimension: 'Goal'.[13]

Trust attitude is not just a (grounded) belief, a prediction; it is not just a subjective probability of an event, because this belief structure is motivated, is charged of value, is anchored to a goal of X.

In trust X is interested, concerned; the event is a 'favorable' one; X's 'welfare' is involved. An 'expectation' is not a 'prediction' or 'forecast'; when X trusts somebody this implies a positive evaluation of him. Also affects and emotions can be involved, as the real basis of a trust disposition or complementary to the judgment and decision (for example, X will not just be surprised but she will be disappointed or even feel betrayed); but, in fact, *no affective reactions or emotions are possible without involved goals.*

---

[11] See also Section 9.6 on Norms.

[12] Thus the relationship between our model and A. Jones' model is not a relation of abstraction or inclusion (where Jones' core would be more abstract, pure, and contained in our definition: vice-versa the extension of our trust would be included in Jones' trust extension); but, it is a relation of partial overlapping: the common constituent being the prediction belief, the diverging constituents being the 'regularity belief' (not necessary for us) and the wish/goal component (non necessary for Jones). As for another Jones' critic to our model (that is: that we explain what it means to "trust", but not "why" we trust something/somebody), we reject this critic by modeling Y's trustworthiness (ascribed 'virtues' and 'powers), the perceived success and risk, the personal or social reasons Y's should have for behaving as expected, etc., as the very basis of the decision to trust.

[13] An interesting and remarkable reference to this component is given by Good (Good, 2000).

Under all these crucial aspects a motivational component is presupposed in X, and makes clear 'expectations', 'evaluations', 'concern' and 'interest', 'affects', etc.

Actually, when we examine all these phenomena, we explain them in terms of important *motives, needs, projects, desires, preferences, objectives,* and so on, for the realization of which the agent is evaluating other agents and the possibility to rely on (exploit) them. The abstract category we use *for all these motivational terms* and categories is 'goal'. However, it must be very clear what a 'Goal' is in cognitive psychology, on the basis of the cybernetic definition and of the following psychological models.

'Goal': What is This?

'Goal' is a perfect term but not in the most superficial and typical English use, where it usually refers to a *pursued external objective to be actively reached*; some step in my active plan, driving my action.[14]

The right general notion of 'goal' and of 'goal-directed or driven behavior' has been provided by cybernetics and control-theory many years ago (Rosenbleuth Wiener in 'Purposive Systems'), and has been imported in psychology and Cognitive Science in the 1960s with the TOTE model by Miller, Galanter and Pribram (Miller, Galanter and Pribram, 1960).

A 'Goal' is the mental representation that 'evaluates' the world (we evaluate the world against it); if the world does not match with, we can be activated for changing the world and realize that state; but, before this, we examine it to see if it is the case or not: is the goal self-realizing or impossible? Or: is it up to us to realize it? Do we have the appropriate actions for this? Are there the necessary conditions for acting successfully? Are there more important goals to be preferred? After all these tests the Goal may become our pursued objective. *But, it already is a 'Goal' in all the other conditions and previous steps.*

It is a 'Goal' even before we believe or know that it is realized or not; before we decide that it depends on us to (attempt to) realize it; that we can and should act. Then it can become an 'intention' (if chosen, if I have a plan for it, if I'm ready to; if I have decided to act for realizing it) and be 'pursued' (Castelfranchi and Paglieri, 2007).

In sum, a Goal is a Goal even before or without being pursued: happiness is due to goal-realization and sufferance is due to goal frustration, but not necessarily to our *active* successes or failures: we are crying because our mother has died, or happy because, without asking, doing or expecting anything, she gave us a kiss.

Given this – not vague, common sense, or reductive – notion, we can make clear that:

---

[14] Consider for example that one of Andrew Jones' central objections to our model of trust is precisely due to such a peculiar use of the notion of "goal" (Jones, 2002): *"While it is true to say that a goal-component of this sort is often present, this is by no means always so. For example, x might trust y to pay his (y's) taxes . . . , even though it is not a goal of x that y pays. Also, x might trust y when y asserts that p, even though x does not have it as a goal to find out whether p is the case."* (p. 229). On the contrary, the general notion of "Goal" precisely covers also those cases. Of course X has the goal that the other guys pay their taxes! Only in this sense he *"trusts"* them as for paying their taxes". Obviously, this does not mean that X is personally doing something in order to produce this result in the world; she is not actively pursuing that goal. But not all our goals are or must be personally pursued. X is wishing, desiring, prescribing, expecting, ... that the others will pay taxes. That's why, just in case, she will be not just surprised but frustrated and upset, and will blame them, and so on. Analogously, *"x does not have it as a goal to find out whether p is the case"*, sure! However, X has the goal (wish, prescription, and so on) that Y says the truth; precisely in this sense *"X .. trusts Y when Y asserts that p"*.

- In Trust as 'disposition', 'attitude', evaluation of Y (before deciding to delegate or not, before delegating) *Goals are not yet or necessarily pursued*: I evaluate 'potential' partners or goods or services relative to a *possible* goal of mine, or relative to a goal of mine that I have not yet decided if I want to achieve or pursue.
  Saying that X trusts Y relatively to a given (possible) goal of her, does not necessarily mean that she is actively *pursuing* a goal: on the one hand, she can be just evaluating a potential delegation; on the other hand, she can be completely passive, just waiting and expecting.
- In Trust, as decision and action, clearly that goal has become not only 'active' but 'pursued': I want to realize it. However, it is pursued in a strange way: thanks to the action of another agent; parasitically. It is *indirectly* 'pursued'; perhaps I'm doing nothing, just expecting and waiting for it to be realized thanks to Y's behavior.
- When I do actively and personally pursue a Goal, this is a Goal in the reductive sense, and I trust myself (self-trust, self-confidence; feeling able and competent, etc.).
  This is the right, broad, scientific notion of 'goal' needed for the theory of Trust.

2.2.4   Trust Versus Trustworthiness

It is also fundamental to make clear the relationships between *trustworthiness* and *trust*; they are frequently mixed up; in many cases we tend to use trust (T) instead of trustworthiness (TW). TW is a property of Y (but in *relation* to a *potential* evaluator/partner X); while T is a property of X (but in *relation* to Y).

On the one hand, there is an *objective* TW of Y (what he is actually able and willing to do in standard conditions; his actual reliability on a more or less specific task, and so on). It is just relative to this reliability that X's T can be *misplaced* and X's evaluation (belief) can be *wrong*. On the other hand, there is a *perceived*, or evaluated, or *subjective* TW of Y for X: $_XTW_Y$.

Now, the relation between $_XTW_Y$ and X's T in Y: $Trust(X\ Y\ \tau)$[15] is not simple. They cannot be identified. $_XTW_Y$ is one of the bases of T, but the latter cannot be reduced to the former. T is a direct function of $_XTW_Y$, but not only of it. T is also a function of a factor of X's personality and general trustworthy disposition (see (McKnight and Chervany, 2001)). Moreover, T depends on the favorable or unfavorable ascription of the *plausibility* gap (lack of evidences); so the perceived $_XTW_Y$ is only one component of the positive evaluation (T attitude).

Of course, $_XTW_Y$ is even more insufficient for defining and determining T's potential decision and T's actual decision and act (see below and Chapter 3).

TW is *multidimensional*; thus $_XTW_Y$ is also a multidimensional evaluation and profile of Y, and cannot be collapsed in just one number or measure. The same holds for trust. We will in fact present a multi-dimensional model of T (and TW) (Section 2.2.6).

2.2.5   Two Main Components: Competence Versus Predictability

The (*positive*) *evaluation* of Y has different aspects and is about different *qualities* of Y. The most important dimensions for trust, that is, of Y's trustworthiness, are the following ones.

---

[15] We consider $p=g_X$ and do not consider here the context C.

Competence

Competence is the set of qualities that makes Y *able* for τ; Y's internal powers: skills, know how, expertise, knowledge, self-esteem and self-confidence,[16] and so on. When X trusts Y for τ, she ascribes to Y some competence.

Competence – as we claimed – cannot be put aside (like in many models and definitions; see Chapter 1) and cannot be separated from trust in Y's reliability.

First of all, it is an important issue in rejecting the fully normative foundation of trust – pursued by many authors (like (Elster, 1979); (Hertzberg, 1988); (Jones, 2002)), which cannot be extended in a simple way Y's skills and competences.[17]

Moreover, competente and reliability are not fully independent dimensions. Even Y's cooperative (*adoptive*) attitude towards X may require some skill and competence. For example, Y may be more or less competent and able in *understanding* X's needs or in comprehending X's interests (even beyond X's own understanding); it is not just a matter of 'concern' or of good will. For example, Y's competence, ability, expertise, can be the basis for his self-confidence and sense of mastering, and this can be crucial in Y's willingness and intention to adopt X's goal, in Y's persistence in this intention, which are crucial aspects of Y's 'reliability'. And so on.

Predictability and Willingness

The second fundamental dimension is not about Y's potential and abstract capability of doing τ, but about his actual behavior; the fact that Y is reliable, predictable, one can count on him; he not only *is able to do*, but *will actually do* the needed action.

Applied to a cognitive Y, this means that Y is *willing* (really has the intention to do α for g_X) and *persistent* (Castelfranchi and Falcone, 1998a), (Falcone and Castelfranchi, 2001b). Also, in this case we have to consider on the one hand, the abstract predictability and willingness of Y not related to other elements and, on the other hand, its relevant correlations with the specific tasks and the different trustors.

These (*Competence and Willingness*) are the prototypical components of trust as an attitude towards Y. They will be enriched and supported by other beliefs depending on different kinds of delegation and different kinds of agents; however, they are the real cognitive kernel of trust. As we will see later, even the goal can be varied (in negative expectation and in aversive forms of 'trust'), but not these beliefs.

Those (evaluative) beliefs are not enough; other important beliefs are necessary, especially for moving towards the decision to trust, and the intention and action of trusting.

Using Meyer, van Linder, van der Hoek *et al.*'s logics ((Meyer, 1992), (van Linder, 1996)), and introducing some 'ad hoc' predicate (like WillDo)[18] we can summarize and simplify the mental ingredients of trust as follows:

---

[16] In rational beings (which decide to act on the basis of what they believe about the possibility of achieving the goal) there is a strange paradox of power: It is not enough 'to be able to'; in order to really be able, having the power of, the agent must also believe (be aware) of having the 'power of', otherwise they will renounce, they will nor exploit their skills or resources. (Castelfranchi, 2003)

[17] Although this can be made simpler precisely by our theory of 'standards' as the needed, expected, and thus 'prescribed' qualities that a given instance of class O must possess in order to be a good or regular O. However, in any case, one has to distinguish 'normative' from 'moral'.

[18] This is a simplification. Before being a belief that "Y will do" this is a belief about a potential delegation: "Y (in case) would do" "Y would be able and willing to...", "X might rely on Y". (See Section 2.3.2).

Potential Goal	G_0 : $Goal_X(g) = g_X$ with $g_X \subseteq n$	
Potential Expectation	B_1 : $Bel_X (Can_Y (\alpha,p))$ G_1 : $Will_X (Can_Y (\alpha,p))$	**(Competence)**
Potential Expectation	B_2 : $Bel_X (<WillDo_Y (\alpha)>p)$ G_2 : $Will_X (<WillDo_Y (\alpha)>p)$	**(Disposition)** **Core Trust**

Figure 2.4 Mental Ingredients of the 'core trust'

In Figure 2.4 we simplify and summarize the kernel of trust as potential evaluation and expectation for a potential delegation.

Notice that Competence, Willingness (Predictability), and also Safety are three necessary components and *Dimensions* of trust and trustworthiness. This doesn't mean that in order to trust *Y* (and possibly and eventually to decide to trust him) *X* should necessarily have a good evaluation of *Y*'s competence and of *Y*'s willingness. As will be clear later, after introducing the degree of believing as the basis for the degree of trust, trust is not a yes/no object; only a trust decision eventually is a yes/no choice, and clearly needs some threshold. So *X*'s trust in *Y* (evaluation of trustworthiness) must be just *sufficient* (and frequently just in a comparative way) for taking a risk on him. Perhaps competence is perceived as rather low, but altogether the positive evaluation and expectation is enough. Of course there might also be a specific threshold for a given dimension: 'no less than this'; and in this case *X* must focus on this and have an explicit perception or evaluation of it.

For example, we assume that we have a threshold of *risk acceptance*: although convenient, we may refuse a choice which involves more than *a certain* risk. What we claim is just that (explicitly or implicitly) these *dimensions* about *Y*'s ability and know how, about his predictability and reliability, about their safety, are there, in the very disposition to trust and entrust *Y*.

2.2.6 Trustworthiness (and trust) as Multidimensional Evaluative Profiles

As we have seen, both while explaining the theory of *qualities*, and when analyzing the basic constituents or dimensions of trust evaluation (*competence* and *willingness*), which can be further decomposed and supported: Trustworthiness is not just a simple mono-dimensional quality. It is the result of several dimensions. We can, for example, consider two rather independent dimensions under the notion of *competence*: the *skills* or *abilities* of *Y* versus his *know how* (knowledge of recipes, techniques, plans for)[19]; and other rather independent dimensions around the *willingness* in social trust: *Y*'s *concern* and *certainty of adoption* versus his *persistence in intending* (Figure 2.5).

---

[19] One can have a perfect knowledge about "how to do" something, but not be able to, since one lacks the necessary skills and abilities; vice versa, one might in principle be able to successfully perform a given activity, but lack the necessary "know how": the instructions or recipes about how to do it.

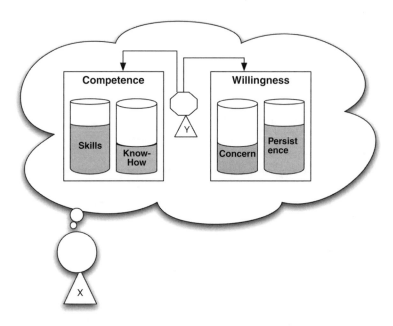

Figure 2.5 Dimensions of Trust Evaluation and their sub-components

2.2.7 The Inherently Attributional Nature of Trust

The very attitude, act, and relation of trust of X in Y (for a given performance) implies X's *causal internal attribution* of the possibility of success. Trust is not simply a prediction (although it implies a prediction). In particular, it is not a prediction of a given behavior based on some observed frequency, on some estimated probability, or on some regularity and norm. It requires the grounding of such a prediction (and hope) on *an internal attribution to Y*.

This is why one trusts *in* Y, and trusts Y. Also the 'trust *that* (event)' (that something will happily happen) cognitively and procedurally implies a 'trust in' something or somebody on which one relies, that is, the assumption that this entity will produce the desired event.

Trusting (in) Y presupposes a possibly primitive and vague 'causal mental model' (Johnson-Laird, 1983) one that produces the expected result. Even when trust is non-social, when for example we decide to trust a *weak chair* by sitting on it, we assume that its material, or its mechanics, or its structure will resist under our weight. Trust presupposes at least a *design* stance (to use Dennet's terminology[20] (Dennet, 1989)).

This is the deepest meaning of the *competence* ascribed to Y, of the internal *power of* appropriately executing the task; this is the real meaning of the predicate *Able* used for its decomposition. It is different from the other component. In order to perform the task Y must be both *Able* and *in condition* but while *Able* is an internal attribution, *in condition* can be external and contextual.

---

[20] Although we believe that our theory of 'functioning' versus function (Miceli and Castelfranchi, 1983) is somewhat clearer.

Moreover, this is also the deep meaning of the other component: *willingness, intention, disposition, motivation* of Y that makes that prediction grounded.

In other words, in social trust the necessarily internal causal attribution of the trust requires an *intentional stance* towards Y (Dennett, 1989). *Trust is a disposition (and an affect) and a social decision only possible in agents able to assume an 'intentional stance', that is to ascribe a mind and mental stuff to the other agent.* The prediction of the other's behavior is based on some theory of mind or on some projection/ascription of *internal* factors (including moods and emotions) playing a causal role in the activation and control of the behavior.

The fact is that in humans those *qualities* are *internal* in a psychic sense; thus – by definition – they are *unobservable*. They can just be aduced by external signs. They are 'kripta' ((Bacharach and Gambetta, 2000), (Bacharach and Gambetta, 2001)); but it is precisely in 'kripta' that we trust and on 'kripta' that we count on when we 'trust' somebody; not the 'manifesta' (visible signs). Trust is a hermeneutic and semiotic act. We trust the signs only for being reliable, for informing us about the 'kripta', but *at the end it is of /on/for the 'kripta' that we trust.*

'Manifesta', signs, are not only the direct observable behaviors or markers of Y, but also all the other *sources* of trust (of trust beliefs) – like *experience, reputation, certificates, category, role*, and so on. All of them are direct or indirect *manifestations* of how Y is or 'works', of what makes him trustworthy.

Our favorite example of *weak delegation* (Section 2.6.1) makes our claim clear: I'm running to take the bus, the driver and the people cannot see me, but I see that there are people at the bus stop apparently 'waiting for the bus' (notice that this is just an mental ascription), on such a basis – attributing them with the intention to take the bus and thus the intention to stop the bus – I rely on them, I trust them for this.

While a *simple prediction* that somebody standing at the bus stop might raise their arm would be based on simple observed frequency and probability, we deny that trust in Y to stop the bus is just this. It is based on the interpretation of Y as 'waiting for the bus' and *on the ascription of the intention that* will cause the action on which one relies. This is the – never clarified – gap between mere subjective probability and trust.

In Chapters 3 and 8 we will criticize the reduction of trust to subjective probability, arguing about the importance of explicitly distinguishing internal and external components of the prediction; and also explaining how crucial and inherent for real trust are the bases (beliefs) supporting the prediction. We do agree with Williamson (see Chapter 8): if trust is just a euphemism for 'subjective probability' we do not need this term, we already have a strong theory of it, and a new vague and merely evocative term is just a confusing layer. On the contrary, we believe that trust is a specific, well-defined, *mental and social construct.*

On the basis of this 'internal attribution' ('design' or 'intentional' stance) *foundation* of trust we are able to account for several things.

For example in Section 2.7.2 we argue that there is internal versus external trust and explain why it is important to differentiate them. Consider a user working (collaborating, negotiating, etc.) on the web with other users, she has to distinguish her trust in these other potential partners from her trust in the actual context: the infrastructure with its internal constraints and rules. We also show that trustee and context have different dynamics; etc.

However, it is important to make clear that a given trust is internal or external only 'relative' to the assumed entity. While changing the target, again the distinction applies and again a form of *internality* is needed. If we consider for example our trust in the technical infrastructure (that relative to the partner Y, was an 'external' condition), we are now necessarily doing some 'causal internal attribution' relative to its (good) working. In general: *if I trust an entity E (of any kind) I'm always ascribing to E some internal properties, some virtues, on which the possible success depends*, and I depend on these 'virtues'.

This 'internal attribution' foundation of trust explains why it is trivial and false that the failure of Y necessarily produces a decrement of X's trust in Y, and a success of Y should necessarily increase or cannot reduce X's trust in Y. The effect of the failure or of the success again depends on its 'attribution': How much and for which aspect is it ascribed to Y? For which aspect is it ascribed to an external (from Y) circumstance? Only internal attribution to Y affects trust *in* Y, since trust holds upon this; while an external attribution to C (say the environment, the infrastructure, etc.) obviously affects the trust in C (of course, it is really important also to understand the relation and correlation between Y and C) (see Chapter 6 for more details).

Mistrust and diffidence (Chapter 4) are *negative* forms of trust. They too entail an internal causal attribution of the inadequate or bad behavior of Y. It is for some internal 'virtue' that Y is poor or harmful; there is not simply something lacking such that I do not trust Y (enough); but, positively, I attribute to Y some 'defect'; I think something bad of him.

Trust as an *External Goal* on Y

When X trusts Y, an *external goal* is put on Y (Castelfranchi, 2000c). Moreover, Y is assumed to respond to this impinging goal:

(i) either, with an *internalization* of it; that is by an internal goal, copied by the external one; by 'goal adoption' (or goal-adhesion) (Section 2.8); of course, this is possible only if Y is an intentional agent;

(ii) or with some internal mediator of the external function; some 'mechanism', some 'functioning' satisfying/performing that function.

2.2.8   Trust, Positive Evaluation and Positive Expectation

Trust is not Reducible to a Positive Evaluation

That *trust* is a *positive evaluation* is also confirmed by the fact that expressing trust towards Y is a way of *appreciating Y*. To express trust in Y is an indirect/implicit positive evaluation act (even a *compliment*) towards him; and this can be highly appreciated, and is one of the reasons for *reciprocation* (Chapter 8).

However, (as we will see) trust cannot be reduced to a positive evaluation. This is so for two main reasons. First of all, because there is much more than evaluations in trust mental attitude: there are also other kinds of beliefs, for example, expectations. Second, a positive valuation about Y is not *per se* trust in Y, or a trust attitude towards Y. It is only a possibility and a *potential* for trust. Only relatively to a possible dependence of X on Y, and to a *possible* delegation/counting of X on Y, that evaluative beliefs become a trust attitude.

Given X's goal that Y bring it about that $g_X \subseteq p$, as a means for X's achieving g_X, the beliefs about Y's *qualities* for this acquire the color of trust.

Good evaluations, positive beliefs, good reputation, are just *potential* trust. Trust also implies a (potential or actual) decision of 'counting on' Y, of risking something while betting on it. It cannot just be a positive evaluation where X is not concerned and involved. X can evaluate that Y has such a good quality (X likes it), or has done or even will do such a good action, but this doesn't mean that X trusts Y (for that quality or action): *X has to consider the possibility to rely on this action and quality for achieving some goal of hers* in the future. Trust is about the future and about *potentially* exploiting something (that's why X 'likes' Y). (See Section 2.6.2).

To be more precise, *a positive evaluation of Y is not 'trust' per se*, even as a simple trust attitude or disposition, because it must be considered within a possible *frame*. It is a matter of the 'gestalt' nature of complex mental states. The side of a square is a linear segment; but: is a segment the side of a square? Not per se; only if considered, imagined, within that figure, as a component of a larger configuration that changes its meaning/role. Analogously: trust is based on and implies a positive evaluation, and when there is not yet a decision/intention it just consists in this, but *only if viewed in a perspective of the potential larger mental state*. Given X's positive beliefs (evaluation) about Y (as for something), if it is the case X might decide to rely on Y. In this sense those evaluations are a pre-disposition to trust, a trust attitude towards Y.

The same holds for the 'prediction' about Y's behavior. It is not yet trust. It is trust only as a possible 'positive expectation', that is in relation to a goal of X and in the perspective of a possible reliance on it (see below).

Decomposing a 'gestalt' is not reducing it to its components.

So the correct representation of the trust 'core' would be the insertion of the basic square (Figure 2.4) within the broad potential phenomenon.

Notice that this can just be the view of an 'observer': I see in X a trust attitude, predisposition, potential, towards Y. Actually in X's mind there is just a good evaluation of Y.

We can arrive at a true/full trust *disposition* of X towards Y if this thought has been formulated in X's mind. X not only has a positive evaluation of Y, but she has explicitly considered this as a potential, a base for a possible (non excluded) delegation to Y: *'If I want I could count on Y as for. . .'*, *'If it will be the case I might rely on Y, since . . . .'*. (For a clear distinction – on such a basis – between mere potential attitude and a real 'disposition' see later Section 2.3.2).

This is the psychological relationship between a mere positive evaluation of Y and a positive evaluation as a trust component or basic trust attitude.

Trust as Positive Expectation

On the basis of her positive beliefs about Y's powers and willingness (or actualization) X formulates a prediction about Y's behavior and outcomes. This is why a lot of scholars define trust in terms of an *expectation*. However, an expectation is not simply a prediction (or a strong prediction).

So trust is a positive expectation. Where a *positive expectation* is the combination of a *goal* and of a *belief* about the future (prediction). X in fact both believes that Y will do the action α and desires/wishes/plans so. And she both believes and wants that the goal g_X will be realized (thanks to Y).[21] Moreover, X is 'expecting', that is, waiting and checking

---

[21] The fact that when X trusts Y, X has a positive expectation, explains why there is an important relationship between trust and hope, since *hope* implies some positive expectation (although weaker and passive: it does not necessarily depend on X, X cannot do anything else to induce the desired behavior); and why trust can be 'disappointed'.

for something. She is concerned with her prediction; and has the *goal to know* whether the expected event will really happen (Castelfranchi, 2005) (Miceli and Castelfranchi, 2002); (Lorini and Castelfranchi, 2003).

We have to introduce, briefly, our theory of expectations as peculiar mental representation, because it predicts and explains a lot of the features and behavior of 'trust' (like its complement or counterpart, like its exposure to disappointment, like its structural link and ambivalence towards fear (for example towards the authorities), and so on).

2.3 Expectations: Their Nature and Cognitive Anatomy

'Expectations' are not just 'predictions'; they are not fully synonyms. And we do not want to use 'expectations' (like in the literature) just to mean 'predictions', that is, epistemic representations about the future. We consider, in particular, a 'forecast' as a mere belief about a future state of the world and we distinguish it from a simple 'hypothesis'. The difference is in terms of *degree of certainty*: a hypothesis may involve the belief that future *p* is possible while in a forecast the belief that future *p* is probable. A forecast implies that the chance threshold has been exceeded (domain of probability).

Putting aside the degree of confidence (we need a general term covering weak and strong predictions), for us 'expectations' has a more restricted meaning (and this is why a computer can produce weather 'predictions' or 'forecasts' but does not have 'expectations'). In 'expectations':

 (i) the prediction is *relevant* for the predictor; he is *concerned*, *interested*, and that is why
 (ii) he is 'expecting', that is the prediction is aimed at being verified; he is *waiting* in order to
     know whether the prediction is true or not.[22]

Expectation is a suspended state *after* the formulation of a prediction.[23] If there is an expectation then there is a prediction, but not the other way round.

2.3.1 Epistemic Goals and Activity

First of all, *X* has the goal of knowing whether the predicted event or state really happens (epistemic goal). She is 'waiting for' this; at least out of curiosity. This concept of 'waiting for' and of 'looking for' is necessarily related to the notion of expecting and expectation, but not to the notion of prediction.

---

[22] Notice that the first two meanings of 'to expect' in an English dictionary are the following ones:

 − *to believe with confidence, or think it likely, that an event will happen in the future*
 − *to wait for, or look forward to, something that you believe is going to happen or arrive*

   While the definition of 'to forecast' is as follows:

 − *to predict or work out something that is likely to happen, for example, the weather conditions for the days
   ahead*

(Encarta® World English Dictionary © 1999 Microsoft Corporation). Notice, the second component of 'expecting' meaning (absent in 'forecasting'): *wait for, or look forward to*. But also the idea that there is some 'confidence' in expectation: the agent *counts on* that.

[23] 'Prediction' is the result of the action of predicting; but 'expectation' is not the result of the action of expecting; it is that action or the outcome of a prediction relevant to goals, the basis of such an action.

Either X is actively monitoring what is happening and comparing the incoming information (for example perception) to the internal mental representation; or X is doing this cyclically and regularly; or X will in any case at the moment of the future event or state compare what happens with her prediction (epistemic actions). Because in any case she has the goal to know whether the world actually is as anticipated, and if the prediction was correct. Schematically:[24]

$$(\textbf{\textit{Expectation X p}}) => (Bel_X{}^{t'} (Will\text{-}be\text{-}True^{t''} p)) \wedge (Goal_X{}^{period(t',t''')})$$
$$(KnowWhether_X (p\ OR\ Not\ p)^{t''})) \text{ where } t''' \geq t'' > t'.$$

X has the expectation p if X believes (at the time t') that p will be true (at the time t'') and has the goal (for the period $t'\text{-}t'''$) to know if p is true. This really is 'expecting' and the true 'expectation'.

2.3.2 Content Goals

The epistemic/monitoring goal described above (to know if p will be true) is combined with *Goal that p*: the agent's need, desire, or 'intention that' the world should realize. This is really why and in which sense X is 'concerned' and not indifferent, and also why she is monitoring the world. She is an agent with interests, desires, needs, objectives on the world, not just a predictor. This is also why computers, that already make predictions, do not have expectations.

When the agent has a goal opposite to her prediction, she has a 'negative expectation'; when the agent has a goal equal to her prediction she has a 'positive expectation' (see Section 2.3.5).[25]

In sum, expectations (*Exp*) are axiological anticipatory mental representations, endowed with *Valence*: they are positive or negative or ambivalent or neutral; but in any case they are *evaluated against some concern, drive, motive, goal of the agent*. In *Exp* we have to distinguish two components:

- On the one hand, there is a mental anticipatory representation, the belief about a future state or event, the 'mental anticipation' of the fact, what we might also call the pre-vision (to for-see).

The format of this belief or pre-vision can be either propositional or imagery (or mental model of); this does not matter. Here, the function alone is pertinent.

- On the other hand, as we have just argued, there is a co-referent goal (wish, desire, intention, or any other motivational explicit representation).

---

[24] We will not use here a logical formalization; we will just use a self-explanatory and synthetic notation, useful for a schematic characterization of different combinations of beliefs and goals.

[25] To be true a goal equal to the prediction in expectation is always there, although frequently quite weak and secondary relative to the main concern. In fact, when X predicts p and monitors the world to know whether it is actually p, she also has the goal that p, just in order to not disconfirm her prediction, and to confirm she is a good predictor, to feel that the world is predictable and have a sense of 'control'. (See Section 7.1.2). We are referring to *predictability*, that is, the cognitive component of self-efficacy: the need to anticipate future events and the consequent need to find such an anticipation validated by facts. This need for prediction is functional in humans in order to avoid anxiety, disorientation and distress. (Cooper and Fazio, 1984:17) have experimentally proved that people act in order to find their forecasts (predictions) validated by facts and feel distressed by invalidation.

Given the resulting *amalgam* these representations of the future are charged of value, their intention or content has a 'valence': it is positive, or negative.[26] More precisely, expectations can be:

- **positive** *(goal conformable): [(Bel$_X^t$ p$^{t'}$) $\wedge$ (Goal$_X^t$ p$^{t'}$)] $\vee$ [(Bel$_X^t$ $\neg$ p$^{t'}$) $\wedge$ (Goal$_X^t$ $\neg$ p$^{t'}$)]*
- **negative** *(goal opposite):): [(Bel$_X^t$ p$^{t'}$) $\wedge$ (Goal$_X^t$ $\neg$ p$^{t'}$)] $\vee$ [(Bel$_X^t$ $\neg$ p$^{t'}$) $\wedge$ (Goal$_X^t$ p$^{t'}$)]*
- **neutral:** *[(Bel$_X^t$ p$^{t'}$) $\wedge$ $\neg$ (Goal$_X^t$ p$^{t'}$) $\wedge$ $\neg$ (Goal$_X^t$ $\neg$ p$^{t'}$)] $\vee$ [(Bel$_X^t$ $\neg$ p$^{t'}$) $\wedge \neg$ (Goal$_X^t$ p$^{t'}$) $\wedge \neg$ (Goal$_X^t$ $\neg$ p$^{t'}$)]*
- **ambivalent:** *[(Bel$_X^t$ p$^{t'}$) $\wedge$ (Goal$_X^t$ p$^{t'}$) $\wedge$ (Goal$_X^t$ $\neg$ p$^{t'}$)] $\vee$ [(Bel$_X^t$ $\neg$ p$^{t'}$) $\wedge$ (Goal$_X^t$ p$^{t'}$) $\wedge$ (Goal$_X^t$ $\neg$ p$^{t'}$)] where t$'$ > t.*

2.3.3   The Quantitative Aspects of Mental Attitudes

Decomposing in terms of beliefs and goals is not enough. We need 'quantitative' parameters. Frustration and pain have an *intensity*, can be more or less severe; the same holds for surprise, disappointment, relief, hope, joy, ... Since they are clearly related with what the agent believes, expects, likes, pursues, can we account for those dimensions on the basis of our (de)composition of those mental states, and of the basic epistemic and motivational representations? We claim so.

Given the two basic ingredients of any Exp (defined as different from simple forecast or prediction) Beliefs + Goals, we postulate that:

> **P1:** *Beliefs* and *Goals* have specific quantitative dimensions; which are basically independent from each other.

Beliefs have strength, a degree of subjective certainty; the subject is more or less sure and committed about their content. *Goals have a value, a subjective importance for the agent.*

To simplify, we may have very important goals combined with uncertain predictions; pretty sure forecasts for not very relevant objectives; etc. Thus, we should explicitly represent these

---

[26] ● Either, the expectation entails a cognitive evaluation. In fact, since the realization of *p* coincides with a goal, it is "good"; while if the belief is the opposite of the goal, it implies a belief that the outcome of the world will be 'bad'.

- Or the expectation produces an implicit, intuitive appraisal, simply by activating associated affective responses or somatic markers; or both.

- Or the expected result will produce a *reward* for the agent, and – although not strictly driving its behavior, it is positive for it since it will satisfy a drive and reinforce the behavior.

We analyze here only the expectations in a strong sense, with an explicit goal; but we mentioned expectations in those forms of reactive, rule-based behaviors, first in order to stress how the notion of expectation always involves the idea of a *valence* and of the agent being concerned and monitoring the world; second, to give an idea of more elementary and forerunner forms of this construct. It is in fact the case of proto-expectations or expectations in 'Anticipatory-Classifiers' based behaviors, strictly conceived as reactive (not really goal-driven) behaviors, but based on anticipatory representation of the outcomes (Butz and Hoffman, 2002), (Castelfranchi, Tummolini and Pezzulo, 2005), (Butz, 2002), (Drescher, 1991), (Pezzulo et al., 2008).

dimensions of goals and beliefs:

$$^{\%}Bel_X \; p^t; \quad ^{\%}Goal_X \; p^t$$

Where % in goals represents their subjective importance or value; while in beliefs % represents their subjective credibility, their certainty.

An *Exp* (putting aside the epistemic goal) will be like this:

$$^{\%}Bel_X \; p^t \wedge ^{\%}Goal_X \; [\neg] \; p^t$$

The subjective *quality* of those 'configurations' or macro-attitudes will be very different precisely depending on those parameters. Also, the effects of the invalidation of an *Exp* are very different depending on:

(i) the positive or negative character of the *Exp*;
(ii) the strengths of the components.

We also postulate that:

P2: The dynamics and the degree of the emergent configuration, of the macro-attitude are strictly a function of the dynamics and strength of its micro-components.

For example, anxiety will probably be greater when the goal is very important and the uncertainty high, than when the goal is not so crucial or the certainty is high. Let us characterize a bit some of these emergent macro-attitudes.

Hope and fear. 'Hope' is in our account (Miceli and Castelfranchi, 2010), (Miceli and Castelfranchi, 2002) a peculiar kind of 'positive Exp' where the goal is rather relevant for the subject while the prediction is not sure at all but rather weak and uncertain.[27]

$$^{low}Bel_X \; p^t \wedge ^{high} Goal_X \; p^t$$

Correspondingly one might characterize being afraid, 'fear', as an *Exp* of something bad, i.e. against our wishes:

$$^{\%}Bel_X \; p^t \wedge ^{\%} Goal_X \neg p^t$$

but it seems that there can be 'fear' at any degree of certainty and of importance.[28]

Of course, these representations are seriously incomplete. We are ignoring their 'affective' and 'felt' component, which is definitely crucial. We are just providing their cognitive skeleton.

---

[27] To be more precise, 'hope' contains just the belief that the event is 'possible', not that it is 'probable'.

[28] To characterize *fear* another component would be very relevant: the goal of avoiding the foreseen danger; that is, the goal of *doing* something such that Not p. This is a goal activated while feeling fear; fear 'conative' and 'impulsive' aspect. But it is also a component of a complete fear mental state, not just a follower or a consequence of fear. This goal can be quite a specified action (motor reaction) (a cry; the impulse to escape; etc.); or a generic goal 'doing something' ("my God!! What can I do?!") (Miceli and Castelfranchi, 2005). The more intense the felt fear, the more important the activate goal of avoidance.

2.3.4   The Implicit Counterpart of Expectations

Since we introduce a quantification of the degree of subjective certainty and reliability of belief about the future (the forecast) we get a hidden, strange but nice consequence. There are other implicit opposite beliefs and thus implicit Exp. For 'implicit' belief we mean here a belief that is not 'written', is not contained in any 'data base' (short term, working, or long term memory) but is only potentially known by the subject since it can be simply derived from actual beliefs (see Section 8.2.1 for more details). See also Figure 8.3 and the following discussion for a more specific analysis about implicit expectations.

2.3.5   Emotional Response to Expectation is Specific: the Strength of Disappointment

As we said, the effects of the *invalidation* of an expectation are also very different depending on: a) the positive or negative character of the expectation; b) the strengths of the components. Given the fact that X has previous expectations, how does this change her evaluation of and reaction to a given event?

Invalidated Expectations

We call invalidated expectation an expectation that happens to be wrong: i.e. while expecting that p at time t', X now beliefs that *NOT p* at time t'.

$$(Bel_X^t \; p^{t'}) \wedge (Bel_X^{t''} \neg p^{t'}) \; where \, (t'' > t)$$

This crucial belief is the *'invalidating'* belief.

- Relative to the goal component it represents 'frustration', 'goal-failure' (is the *frustrating* belief): I desire, wish, want that p but I know that *not p*.
  FRUSTRATION: $(Goal_X \; p^{t'}) \wedge (Bel_X \neg p^{t'})$
- Relative to the prediction belief, it represents 'falsification', 'prediction-failure':
  INVALIDATION: $(Bel_X^t \; p^{t'}) \wedge (Bel_X^{t''} \neg p^{t'})$; where $(t'>t)$ and $(t''>t)$
  $(Bel_X^t \; p^{t'})$ represents the former illusion or delusion (X illusorily believed at time t that at t' p would be true).

This configuration provides also the cognitive basis and the components of '**surprise**': *the more certain the prediction the more intense the surprise.* Given positive and negative expectations and the answer of the world, that is the *frustrating* or *gratifying* belief, we have: the configuration shown in Table 2.1.

Disappointment

Relative to the whole mental state of 'positively expecting' that p, the *invalidating & frustrating* belief produces 'disappointment' that is based on this basic configuration (plus the affective and cognitive reaction to it):

$$DISAPPOINTMENT : ({}^{\%}Goal_X^{period(t,t')} \; p^{t'}) \wedge ({}^{\%}Bel_X^t \, p^{t'}) \wedge ({}^{\%}Bel_X^{t'} \neg p^{t'})$$

Table 2.1 Relationships between Expectation and Surprise

	p	$\neg p$
$(Bel_X{}^t\ p^{t'})^{t<t'} \wedge (Goal_X\ p^{t'})$	No surprise + achievement	*surprise + frustration* **disappointment**
$(Bel_X{}^t \neg p^{t'})^{t<t'} \wedge (Goal_X\ p^{t'})$	*surprise + non-frustration* **relief**	no surprise + frustration

At t X believes that at t' (later) p will be true; but now – at t' – she knows that *Not p*, while she continues to want that p. Disappointment contains goal-frustration and forecast failure, surprise. It entails a greater *sufferance* than simple frustration for several reasons: (i) for the additional failure; (ii) for the fact that this impacts also on the self-esteem as epistemic agent (Badura's 'predictability' and related 'controllability') and is disorienting; (iii) for the fact that losses of a pre-existing fortune are worse than missed gains (see below), and a long expected and surely desired situation are so familiar and 'sure' that we feel a sense of loss.

The stronger and well-grounded the belief, the more disorienting and restructuring is the *surprise* (and the stronger the consequences on our sense of predictability) (Lorini *et al.*, 2007). The more important the goal is, the more *frustrated* the subject.

In disappointment, these effects are combined: *the more sure the subject is about the outcome and the more important the outcome is for her, the more disappointed the subject will be.*

- The degree of disappointment seems to be a function of both dimensions and components[29]. It seems to be felt as a unitary effect:

  'How much are you disappointed?' 'I'm very disappointed: I was <u>sure</u> *to succeed'*
  'How much are you disappointed?' 'I'm very disappointed: it was very <u>important</u> *for me'*
  'How much are you disappointed?' 'Not at all: it was not <u>important</u> *for me'*
  'How much are you disappointed?' 'Not at all: I have just <u>tried; I was</u> <u>expecting</u> *a failure'.*

Obviously, worst disappointments are those which place great value on the goal and a high degree of certainty. However, the *surprise* component and the *frustration* component remain perceivable and a function of their specific variables.

Relief

Relief is based on a 'negative' expectation that results in being wrong. The prediction is invalidated but the goal is realized. There is no frustration but surprise. In a sense relief is the opposite of disappointment: the subject was 'down' while expecting something bad, and now feels much better because this expectation is invalidated.

$$\text{RELIEF} : (Goal_X \neg p^{t'}) \wedge (Bel_X\ p^{t'}) \wedge (Bel_X{}^{t'} \neg p^{t'})$$

---

[29] As a first approximation of the degree of disappointment one might assume some sort of multiplication of the two factors: Goal-value * Belief-certainty. Similarly to 'Subjective Expected Utility': the greater the SEU the more intense the Disappointment. * = multiplication.

- *The harder the expected harm and the more sure the expectation (i.e. the more serious the subjective threat) the more intense the 'relief'.*

More precisely: the higher the worry, the threat, and the stronger the relief. The worry is already a function of the value of the harm and its certainty.

Analogously, **joy** seems to be more intense depending on the value of the goal, but also on how *unexpected* it is.

A more systematic analysis should distinguish between different kinds of surprise (based on different monitoring activities and on explicit versus implicit beliefs), and different kinds of disappointment and relief due to the distinction between 'maintenance' situations and 'change/achievement' situations.

More precisely (making the value of the goal constant) the case of loss is usually worse than simple non-achievement. This is coherent with the theory of psychic suffering (Miceli and Castelfranchi, 1997) that claims that pain is greater when there is not only frustration but disappointment (that is a previous *Exp*), and when there is 'loss', not just 'missed gains', that is when the frustrated goal is a maintenance goal not an achievement goal. However, the presence of *Exp* makes this even more complicated.

2.3.6 Trust is not Reducible to a Positive Expectation

Is trust reducible to a positive expectation? For example, to the estimated subjective probability of a favorable event? (as in many celebrated definitions). Trust as belief-structure is not just an 'expectation' (positive/favorable).

Let us put aside the fact that trust (at least implicitly) is trust *in* an agent; it is an expectation grounded on an 'internal attribution'. Even not considering 'trust that *Y* will ....' or 'trust in *Y*', but just 'trust that *p*' (for example, 'I trust that tomorrow it will be sunny') there is something more than a simple positive expectation. *X* is not only positively predicting, but is 'counting on', that *p* is *actively* concerned; *X has something to do or to achieve, such that p is a useful condition for that.* Moreover, such an expectation is rather sure: the perceived favorable chances are greater than the adverse ones, or the uncertainty (the plausible cases) is assumed as favorable. This is one of the differences between *trust* and *hope*; the difference between 'I trust that tomorrow will be sunny' and 'I hope that tomorrow will be sunny'. In the second one, I'm less certain, and just 'would like so'; in the first one, I am more sure about this, and that is why I (am ready to) *count on* this.

In fact, even non-social trust cannot be simply reduced to a favorable prediction. This is even clearer for the strict notion of 'social trust' (*'genuine' trust*) (Section 2.11): which is based on *the expectation of adoption*, not just on the prediction of a favorable behavior of *Y*.

2.4 'No Danger': Negative or Passive or Defensive Trust

As we said, in addition to *Competence* and *Willingness*, there is a third dimension in evaluating the trustworthiness of *Y*: *Y* should be perceived as not threatening, as *harmless*.

Either *Y* is *benevolent* towards *X* (for similarity, co-interest, sympathy, friendship, etc.), or there are strong internal (moral) or external (vigilance, sanctions) reasons for not harming *X*. This very important dimension appears to be missing in the definitions considered by

(Castaldo, 2002); however, it is present in others, for example those from social psychology or philosophy.[30]

Perhaps the most 'primitive' and original component (nucleus) of trust (especially of implicit and affective trust) is precisely the belief or feeling: 'no harm here/from . . .' and thus to *feel safe*, no *alarm*, no *hostility*, being *'open to'..*, *well disposed*. This is why trust usually implies no suspect, no arousal and alarm; being accessible and non-defended or diffident; and thus being relaxed. The idea of *'no danger'* is equivalent to *'the goals of mine will not be frustrated by Y'*; which – applied to animated entities (animals, humans, groups, and anthropomorphic entities) – is specified as the idea that *'Y has no the goal of harming me'*.

We call this elementary form of trust: α-*form* (Negative or Passive or Defensive Trust). In a sense 'feeling safe' can be the basic nucleus of trust and entire in itself; seemingly without any additional component. However, looking more carefully we can identify the other core components. Clearly positive *evaluations* and *expectations* (beliefs) are there. If I don't worry and do not suspect any harm from you, this means that I evaluate you positively (good for me; not to be avoided; at least harmless), since not being harmed is a goal of mine. Moreover, this feeling/belief is an expectation about you: I do not expect damage from you; which is a passive, weak form of positive expectation. Perhaps I do not expect that you might actively realize an achievement goal of mine; but I at least expect that you do not compromise a maintenance goal of mine: to continue to have what I have.

It is rather strange that this basic meaning of 'trusting' and this component is not captured in those definitions (except indirectly, for example, with the term 'confidence',[31] or marginally). This is for us the most 'primitive' and basic nucleus of trust, even before relying on Y for the active realization of a goal of X; just *passively* relying on Y to not be hostile or dangerous, non harming X.

Of course there is a stronger, richer, and more complete form of trust (β-*form*: that we call 'active', 'positive', 'achievement' trust) not only due to the idea/feeling (expectation) that the other will not harm me (lack of negative expectations); but including specific positive expectations: the idea/feeling that the other *will 'adopt' (some of) my achievement goals*, will be helpful; that Y's attitude is 'benevolent' in the sense that not only is it not hostile, noxious or indifferent, but that he can be disposed to adopt and realize my goal (or at least that Y can be useful for achieving my goals). I can count on Y, and make myself dependent on Y for realizing (some of my) goals.

In terms of the theory of 'interference' (the basic notion founding social relations and action (Castelfranchi, 1998), α-*form* is the assumption or feeling that 'there can/will not be *negative interferences*' from/by Y's side; while β-*form* is the assumption or the feeling that 'there can/will be *positive interferences* from/by Y' (Where 'by' means 'on purpose': goal-oriented or intentional).

To be 'full' and complete trust should contain both ideas (α-*form*) and (β-*form*); but this is not always true. Sometimes it is more elementary and seems just limited to (α-*form*); sometimes it is mainly (β-*form*).

---

[30] See, for example, (Hart, 1988): trust enables us to assume "benign or at least non-hostile intentions on the part of partners in social interactions".

[31] The English term *'confident'/'confidence'* seems mainly to capture this nucleus of trust.

In a sense – as we have said – α-form is always present, at least implicitly; also when there is the richer and more specific attitude of β-form. In fact, when applied to the same set of goals (β-form) implies (α-form):

> if Y will be useful or even helpful for the achievement of the goal g_X, he is not a problem, a threat for the goal g_X.

To favor goal g_X implies to not harm goal g_X, since the achievement of goal g_X implies the non frustration of goal g_X. However, this implication does not mean that when X trusts Y (as capable and responsible for certain goals) X is always and fully relaxed and has nothing to worry from Y. In fact: what about other possible goals of X?

Except when applied to the same sub-set of goals or when 'generalized' (i.e., applied to all X's possible goals) β-form in fact doesn't necessary imply α-form. If trust is relative to a sub-set of X's goals, it is perfectly possible that X trusts Y (in the sense of β-form) for certain goals, but X could worry about her other goals; or, vice versa, that X trusts Y (in the sense of α-form) which is unwarlike, not threatening, but X cannot trust (in the sense of β-form) him as able and helpful towards some goal of hers. Thus, β-form and α-form don't necessarily co-occur, except for the same subset of goals. To be true the α-form potentially entails the β-form since it is the presupposition for a <u>possible</u> reliance.

2.5   Weakening the Belief-Base: Implicit Beliefs, Acceptances, and Trust by-Default

To make things simpler, we assume in our model that trust is composed by and based on 'beliefs'. However, this is an *antonomasia*: trust is based on *doxastic attitudes*: beliefs, knowledge, but also just *acceptances* (in our vocabulary: *assumptions*). Beliefs are assumed to be true in the world; to match with the (current, future, previous) world, if/when tested; or at least they are produced with this function and perspective. But we also have different and weaker doxastic attitudes on mental representations; or better different functions and uses of them.

For example, a very important function for the theory of purposive behavior (and for the theory of trust) is the use of doxastic representations as *conditions* for actions and decisions. In order to decide to act and to act (and for choosing an action) some conditions must be true, or better: they must be *assumed* (but not necessarily verified or proved). These are *assumptions*. We can use beliefs as assumptions; but they can also be unreal beliefs.

We can base our actions or reasoning on simple or mere 'assumptions' (non-belief assumptions), which have not been tested or are not destined to be tested. They are just – implicitly and automatically or explicitly – 'given for granted' or used 'as if'. Only the success of the practical action based on them, will provide an unconscious and indirect feedback about their 'truth'; will 'confirm' and indirectly 'verify' them. *It is important to distinguish between mere-assumptions and beliefs because one cannot decide to believe something while one can decide to assume something* (Cohen, 1992) (Engel, 1998).

This is also very relevant for trust because sometimes we trust Y not on the basis of real – more or less certain – beliefs (based on experience or on inference), but just assuming something about Y, and acting 'as if'. It is even possible to explicitly 'decide' to trust Y. We do not have sufficient evidence; current evidence does not provide us with enough certainty ('degree of trust') for trusting Y, but we can overcome this situation not by waiting or searching for additional information and evidence, but just by making our decision: 'I have decided to trust you, (although . . .)'. This does not only mean that I have decided to rely on you (trust as *act*), but it can also mean that I am assuming something about you (expectations, evaluations), and that I am testing you out: precisely those assumptions will be confirmed or invalidated by your behavior (see Section 2.6). Sometimes we even delegate a task to *Y*, we take a risk, in order to get information about *Y*'s trustworthiness; we put him on test.

Trust beliefs obviously can be just *implicit*; not explicitly represented and considered in *X*'s mind. They can be just presupposed as logical conditions or logical entailment of some (explicit) belief. Suppose, for example, that *X* has such an evaluation about *Y*: '*Y* is a very good medical doctor' and suppose that this evaluation comes from *Z*'s recommendation (and *X* trusts very much *Z*), or from her direct practical experience. In this evaluation *X* implicitly assumes that *Y* is well prepared (competent on the subject), and also technically able to apply this doctrine, and also reliable in interaction, he takes care of you and of your problems. All these evaluations, or these possible pieces of *X*'s trust in *Y* (the fact that *X* trusts *Y* for being prepared, for taking care of, etc.; the fact that *X* trusts in *Y*'s expertise, attention, etc.), are just implicit beliefs; not necessarily explicitly derived and 'written' in *X*'s mind (just 'potential' (Levesque, 1984; Castelfranchi, 1996; 1997)), and/or not explicitly focused and taken into account.

It is also important to remark that there are many forms of trust which are not based on such an explicit and reason-based (argumentative) process we have presented in previous sections. They are rather automatic; not real 'decisions' or 'deliberations'. These are forms of *Trust* 'choices' and 'acts' based on some 'default rules' (positive evaluations are just implicit). The rule is: 'Except you have specific signals and specific reasons for do not trusting *Y* and rely on him, trust him'.

So the lack of distrust is the condition for trusting more than the explicit presence of trust evaluations (see Table 2.2; where 'Not (Believe q)' denotes the absence of such Belief in *X*'s mind).

These forms of automatic or by-default trust are very important, not only for characterizing generalized dispositions of people or affective trust attitudes, but also in other domains. For example, trust in our own natural information sources (our memory, our eyes, our reasoning) and, frequently even in social information sources, is suspect-less, is automatic, by-default, and doesn't need additional justification and meta-beliefs. (Castelfranchi, 1997).

Moreover, to 'believe' is in general not a real 'decision', but is certainly the result of some computation process based on sources, evidences, and data. But – in a sense – to 'come to believe' is an act of trust:

- trust in the belief (you rely on it and implicitly assume that it is valid, true), and
- Implicitly trust in the sources and bases of the belief; and
- (just procedurally and fully implicitly) even trust in the machinery or 'procedure' which outputs the belief.

Table 2.2 By-default Trust

IF	*Not (Believe X Not (Trustworthy Y τ C))*
THEN	*Trust (X Y C τ g_X) will be 'naturally' over the threshold for delegating*

2.6   From Disposition to Action

As we said, trust cannot be limited to a (positive) evaluation, an esteem of Y, and to a *potential* disposition to relying on him. *This potential can become an act.* On the basis of such a valuation and expectation, X can decide to entrust Y within a given 'task', that is to achieve a given goal thanks to Y's competent action. 'To trust' is also a *decision* and an *action*.

The decision to trust is *the decision* to depend on another person to achieve our own goals; the free *intention* to rely on the other, to *entrust* the other for our welfare. However, to pass from a mere potential evaluation to a reliance disposition, that is, to the beliefs supporting the decision and the act to rely upon Y, the kernel ingredients we just identified are not enough.

At least a third belief (a part from that on being safe) is necessary for this: a *Dependence Belief*.

In order to trust Y and delegate to him, X believes that either X needs him, X depends on him (*strong dependence*), or at least that it is better for X to rely than to not rely on Y (*weak dependence*).

In other words, when X trusts someone:

- X has an active goal (not just a potential one; see Section 2.6.2); and
- X is personally and not neutrally 'evaluating' Y; moreover;
- X is in a strategic situation (Deutsch, 1985): X believes that there is 'interference' (Castelfranchi, 1998) and that her rewards, the results of her projects, depend on the actions of another agent Y.

To express it more clearly, we could say that:

> *Strong dependence* (Sichman *et al.*, 1994), is when X is not able at all to achieve her goal; she lacks skills or (internal or external) resources, while Y is able and in condition to realize her goal. Y's action is a necessity for X.

> *Weak dependence* (Jennings, 1993), is when X would be able to realize her goal; however, she prefers to delegate to Y, to depend on Y. This actually means that X is strongly dependent on Y and needs Y for a broader outcome, which includes her original goal plus some additional advantage (like less effort, higher quality, etc.). This is why she prefers and decides to delegate to Y. In other words, X is reformulating her goal (which includes the original one: G_0), and, relative to this new goal, she is strongly dependent on Y. Then she formulates the instrumental sub-goal (plan) about 'Y doing the action α', and – of course –for this goal also she strictly depends on Y.

These dependence beliefs (plus the goal g_X) characterize X's 'trusting Y' or '*trust in Y*'[32] in delegation. However, another crucial belief arises in X's mental state – supported and implied by the previous ones – the *Fulfillment Belief*: X believes that g_X will be achieved and p will

---

[32] We are stressing now the internal attribution of trust and putting aside for the moment the external circumstances of the action (opportunities, obstacles, etc.). We will analyze this important distinction further in 2.10 about *social trust*, and in Chapter 3 on decision.

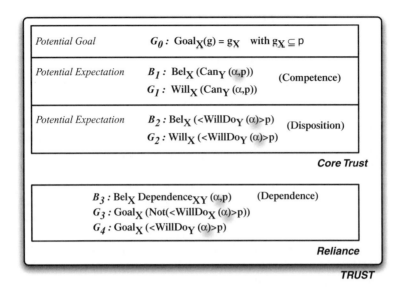

Figure 2.6 Mental State of the Decision to Trust

be true (thanks to Y in this case). This is the *'trust that' g_X (case in which $g_X = p$)*. That is, X's trust in Y about task τ, for goal that p, implies also some trust that p will be realized (thanks to Y).[33]

When X decides to trust, X has also the new goal that Y performs α, and X relies on Y's α in her plan (delegation) (for more on this decision see Chapter 3). In other words, on the basis of those beliefs about Y, X 'leans against', 'counts on', 'depends upon', 'relies on'; X *practically* 'trusts' Y. Where, notice, 'to trust' does not only mean those basic beliefs (the core: see Figure 2.4) but also the decision (the broad mental state) and the act of delegating.

To be more explicit: *on the basis of those beliefs about Y, X decides of not renouncing to g_X, not personally bringing it about, not searching for alternatives to Y, and to pursue g_X through Y. This decision is the second crucial component of the mental state of trust: let us call this part <u>reliance trust</u> (we called the first part <u>core trust</u>) and the whole picture mental state of trust and the <u>delegation</u> behavior.*

Also, once more using Meyer, van Linder, van der Hoek *et al.*'s logics ((Meyer, 1992), (van Linder, 1996)), we can summarize and simplify the mental ingredients of trust as in Figure 2.6.

Of course, there is a coherence relation between these two aspects of trust (core and reliance): the decision of betting and wagering on Y is grounded on and justified by these beliefs. More than this: the degree or strength (see Chapter 3) of trust must be sufficient to decide to rely and bet on Y ((Marsh, 1994), (Snijders, 1996)). The trustful beliefs about Y (core) are the presuppositions of the act of trusting Y.

---

[33] Is Section 2.6.1 we will be more precise and complete on the relationships between "Trust that" and "Trust in".

2.6.1   Trust That and Trust in

As we have seen, both analyzing the *fulfillment belief* and the *attributional nature* of trust, there are two notions of trust which are orthogonal to the other fundamental distinction between trust as mere attitude (beliefs) and trust as decision and action. We refer to the distinction between *Trust that* and *Trust in*.

Trust that is our trust in a given desired state of the world, event, outcome; while *Trust in* is the trust in a given agent for doing something: in sentences like 'I trust John (for that/doing that)', 'I have/feel/put trust in John (for that/doing that)', 'I entrust John within that', etc.). Necessary and systematic relationships have emerged between these two forms of trust: they imply each other.

On the one side, *Trust in Y* (both as an evaluation and potential expectation and decision, and as an actual decision and action) necessarily entails the *Trust that* the action will be performed, and –thus – the trust that p will be true, the goal g_X will be realized. In other words, the 'Trust in Y' is just the trust that Y will 'bring it about that p'.

Even more analytically: the *trust in Y*'s qualities: competence, willingness, etc. entails the *trust in Y* as for doing α, which entails that α will be correctly performed, which entails that the goal will be realized (p will be true).

But, on the other hand, as we said, any *trust that* a given event will happen, as trust, is more than a mere, quite firm and positive expectation, not only because X counts on this (she is betting on this for achieving something) but because it is based on the idea of some (although vague) active *process* realizing that result.

Any *Trust that* presupposes some *Trust in* some natural or social *agent*. Any '*Trust* that p will be true' (not just hope, not simply an expectation) presupposes a trust that some Y will bring it about that p.

An interesting example is Lula (the president of Brazil) interviewed before the vote with other candidates. While other candidates were saying 'I wish' 'I hope' or even 'I'm sure to win', Lula's response was: 'I trust to win'. The difference with the other sentences is not only about the degree of certainty ('hope' and 'wish' are weaker or less specified) or about the positive expectation ('sure' has a high degree but might also be negative); the difference is that while saying 'I trust that the result will be this' Lula is implicitly saying: 'I trust people, the voters' 'I trust my party', 'I trust myself as for being able to convince and attract'.

If I conceptualize the relation as 'trust', I implicitly assume that the result depends on some entity or process (some *agency*) and the trust is 'in' such an entity, which (I trust that) will bring about that p. So we can introduce a new operator *TRUST-That*, with just two explicit arguments (X and p) and show how it can be translated in the former *TRUST* operator:

$$\textbf{\textit{TRUST-that}} \; (X \; p) \; \textbf{\textit{implies}} \; \textbf{\textit{TRUST}} \; (X \; Y \; C \; \tau \; gx) \tag{2.2}$$

where X believes that there is any active Y (maybe he is unknown to X) that will bring it about that p through a behavior/action/task (maybe they are unknown to X too). In any case, X trusting the final achievement p, trusts also the agent performing the task.

We claim that this is true even with natural agents, in sentences like: 'I trust that it will rain'; there is something more than in 'I'm sure that...', 'I hope that....'; there is an implicit trust in some vague causal process that will produce the result 'it will rain'. On the other hand, as seen above:

$$TRUST\,(X\ Y\ C\ \tau\ gx)\ \textbf{\textit{implies}}\ \textbf{\textit{TRUST-that}}\,(X\ p) \tag{2.3}$$

The relation between 'Trust in' Y and in his action and trust that p is just one and the same relation (or better reflects the same basic action-theory relation) between the *Intention to do* a given action and the *Intention that* a given result holds: when I have the *Intention* that something holds, this necessarily implies that I believe that it depends on me, I have to act, and I *Intend to do* something; vice versa, if I have the *Intention to do* an action, this is certainly in order to achieve a given goal/result; thus I also *Intend that* this result be realized. When all there are not up to me, but delegated to another agent Y, we get all these forms of *Trust that* (outcome and performance) and the prerequisites of the trust *in* the agent, his virtues, and the virtues of his action. In sum, we can say that:

$$TRUST\,(X\ Y\ C\ \tau\ gx)\ <==>\ \textbf{\textit{TRUST-that}}\,(X\ p) \tag{2.4}$$

where $<==>$ means 'implies and is implied by'. There is a bidirectional relationship between *Trust-that* and *Trust-in*.

2.6.2 Trust Pre-disposition and Disposition: From Potential to Actual Trust

Frequently in Chapter 1 and in this chapter (and also later), for the sake of simplicity, we have been identifying and collapsing the notion of trust 'attitude' and 'disposition'. However, one might distinguish between the mere beliefs about Y (evaluations and expectations) and a true 'disposition', *which is something more than an evaluation*, but it is also something less than the actual decision and act. It is something in between; and preliminary to the actual decision.

Trust disposition[34] is the *potential* decision to *trust*, or better, the decision to possibly (en)trust Y. Not only X evaluates Y, but *she also perceives this evaluation as sufficient for* (if/when needed) trusting Y and relying on him. 'If it would be the case/when there will be the opportunity ... I will trust Y', 'One/I might trust Y'. X is *disposed* to trust Y (if/when it will be the case).

In *Trust disposition* in a strict sense the *expectations* are also conditional or *potential* (not actual); X has not the actual goal that Y does a given action and realizes a given result. X only has the prediction that 'if she would have the goal that p, Y will/would realize it; she might successfully rely on Y'. X is not actually concerned, actually waiting for something (expecting); she is only *potentially* expecting this, because she only potentially has the goal. Only this *potential* reliance actually makes the mere evaluation a possible expectation, a trust attitude (Section 2.2.1).

This specific mental attitude (*Trust disposition*) is in fact very important also for both selecting and ordering the preferences of the trustor; they contribute to a decision on which context and environment she has to situate herself in the (near or far) future. If Mary can decide to live in different environments where different goals of hers will be supported (she believes

---

[34] We will not consider the other notion of 'disposition' relevant for trust. The idea of a personality trait, or of mood, which make us generally open, well disposed, trustful towards the others; and increase the probability that we trust Y. This is the notion traditional in social psychology, and used – for example – in McKnight's model (McKnight and Chervany, 2001).

so) by different (artificial, human, institutional, and so on) trustees, she will decide to live in one in which her more relevant (with higher priority) potential goals will be better supported.

So we can say that *Trust disposition* is in fact a real, very important regulator of agents' behavior and goal selection.

Even the *decision* to trust can be conditional or hypothetical; I have already decided that, I have the future directed intention: '(if it will be the case) to address myself to Y, to trust, to rely on Y'. I have already decided, but, actually, I am not trusting him (as decision in action and act).

So, Figure 2.1, on trust stages and layers, should be even more articulated: in a potential trust attitude (*(pre)disposition*) versus a richer trust attitude (*disposition*) contained in a trust decision. We have to sophisticate a bit our analysis of the cognitive attitude of trust, by explaining that such a nucleus evolves in fact from its preliminary stage (before the decision) to its inclusion in the decision.

It is important to realize that the *disposition* kernel of the decision, intention, and action of (en)trusting includes or presupposes the first kind and nucleus of trust that we have just characterized (evaluation, prediction) but is broader, or better it is the actualization of it.

We pass from a *potential* evaluation/expectation to an *actual* one. There is a difference between the mere preliminary and potential judgment 'One can trust Y', 'Y is trustworthy', and the executive prediction that Y will actually (in that circumstance) do as expected and will realize the goal. X passes from the beliefs **2.5** and **2.6** (and the Belief that q) to the additional and derived belief (**2.7**):

$$Bel_X < Can_Y(\alpha) >_p \tag{2.5}$$

that means: X believes that Y is able and in condition to do α and the result of this action would be p true in the world (which is a *positive evaluation* of Y and of the context); and

$$Bel_X < (q \rightarrow Do_Y(\alpha)) >_p \tag{2.6}$$

that means: X believes that there is a condition q able to activate the performance α of Y; if it will be the case, Y will do α[35] (a *prediction* and just a '*potential*' expectation, since not necessarily X, while evaluating Y relative to the goal resulting from α, do currently have such a goal (Miceli and Castelfranchi, 2000).

$$Bel_X < Will\text{-}Do_Y(\alpha) >_p \tag{2.7}$$

that means: X believes that Y will do the action and will achieve p (which is in fact combined with the active goal that $Do_Y(\alpha)$ and thus is a real *expectation*), and also contains the expectation of α and of its desired outcome p (the goal X is relying on Y for).

We call the formulas (**2.5**) and (**2.6**) the *potential* evaluation and mental attitude towards Y: *trust pre-disposition*; and the mental attitude towards Y and (**2.7**) in the decision to rely on him: *trust disposition*.

---

[35] Where q is something like: "if X will need his help"; "if X will make the request", "if it will happen that . . . .", and so on.

2.6.3 The Decision and Act of Trust Implies the Decision to Rely on

Let us now come back to the relations between trust and reliance. Consider Holton's very nice example of the drama course (Holton, 1994) (pp. 63–64): *'If you have ever taken a drama course, you have probably played this game. You are blindfolded. You stand in the middle of a circle formed by the others. They turn you round till you lose your bearings. And then, with your arms by your sides and your legs straight, you let yourself fall. You let yourself fall because the others will catch you. Or at least that is what they told you they would do. You do not know that they will. You let yourself fall because you trust them to catch you'.*

We would like just to add to Holton's analysis a more subtle distinction. To decide to let yourself fall down is not the same as deciding to trust them. You can decide to let yourself fall down even if you do not trust them at all; you believe that they want to play a trick and to make fun of you, and you are ready to protect yourself at the last moment. If you decide to trust you *not only* decide to let yourself fall, but you *decide to count on them*, to act assuming that they will catch you. You decide to do your part of the plan *relying on* them doing their part of the plan. (Moreover – in 'genuine' social trust – you would count on them because you count on their motivations and their social-adoptive attitude towards you, and – following Holton – assuming a 'participant stance' towards them (p. 66) as persons treating you as a person).

Deciding to attempt, to try and see, is not deciding to rely/count on, and this is necessary in 'deciding to trust'; although also to decide to rely on is not enough for deciding to trust.

For many authors 'trust' is only social (and in a deep sense of 'social', Section 2.11); and they try to disentangle 'trust' from 'reliance' just on such a basis. See again (Holton, 1994) (p. 68): *'I have reason for simple reliance on an object if I need something done and reliance on it is my best bet for getting it done; likewise for simple reliance on a person. But in cases which involve not just simple reliance but trust, my reasons can be more complicated. Just because trust involves moving to a participant stance, I can have further reasons to trust, since that move can itself be something I value. Suppose we are rock climbing together. I have a choice between taking your hand, or taking the rope. I might think each equally reliable; but I can have a reason for taking your hand that I do not have for taking the rope. In taking your hand, I trust you; in so doing our relationship moves a little further forward.(..)'*[36]

In that statement Holton seems very close to (Baier, 1986), which claims that *trust must be distinguished from mere reliance, because it is a special kind of reliance: reliance on a person's goodwill towards me.* (p. 234)[37]

We agree that trust must be distinguished from mere reliance, but in our view, the real distinction is not directly based on 'sociality': intentional stance, or the richer 'participant stance', good will, or moral stuff. There is a preliminary distinction, before arriving at the special form of 'genuine' trust (Section 2.11).

---

[36] The text continues like this: "*. . .. This can itself be something I value. We need not imagine that you would be hurt if I chose the rope over your hand; you might be perfectly understanding of the needs of the neophyte climber. But our relationship would not progress.*" This is a nice issue: why does the act of trusting Y creates or improves a positive relationship with him? We examine this issue in Chapter 6.

[37] Baier's claim is based on the example of some safety and reliance assured by threats and intimidation on *Y*. If I count on *Y*'s fear of me or of my bodyguards, or on their protection, I do not really 'trust' *Y*. We disagree on this, because we claim that there are different kinds and layers of social trust; the one based on 'good will' or benevolence is only a sub-case of the one based on goal-adoption towards me for whatever reason (even avoiding sanctions or revenge) (See Section 2.8).

In natural language, I can 'trust' even the rope or the rock, but this is more than just 'relying on' it or deciding to grasp it.

Trust is (conscious and free, deliberated) reliance based on a judgment, on an evaluation of Y's virtues, and some explicit or entailed prediction/expectation: 'How could you *trust* that rock? It was clearly so friable!' 'No, I have tested it before; I evaluated it and I was convinced that it would have supported me!'

What is Reliance?

As showed in Section 1.3, in any (intentional, external) action α there is one part of the causal process triggered by the action and necessary for producing the result/goal of the action and defining it which *is beyond the direct executive control of the Agent (Ag) of α*. In performing α, Ag is making reliance on these processes and this is true in both cases:

- if Ag knows this, models this in his mind, and *expects* this;
- if Ag doesn't understand the process, is not aware of it, or at least doesn't explicitly represent it in his plan (although he might be able to do so).

As we said (Section 1.3), in the first case reliance becomes *delegation*. *'Delegation' would be the subjective and chosen reliance*. Counting upon: conceiving in $Ag's$ individual mind a multi-agent plan including (planning, expecting) the action of another autonomous agent. In the second case we have pure reliance.

In Delegation (at least) one part of the delegator's subjective plan for successfully accomplishing the intentional act α and achieving its goal, is 'allocated' to another agent either natural (like the sun when bronzing; or a coffee to feel awake) or social (like a waiter to bring food).

Let us clarify the concept: *Ag is making reliance upon Y/P* (where Y is another agent and P is a process) when: *there are actions (or inactions) in Ag's plan which are based on/rely upon Y/P, which depend on it for their efficacy* (in other words: that process P due to Agent Y creates some conditions for the performance or for the efficacy of those actions), *and Ag decides to perform those actions or directly performs them, Ag invests on Y/P (costs), Ag risks, Ag is relying on the fact that P will actually happen*.

P (due to Y) is a necessary process/condition for the achievement of $Ag's$ goal, but it is not sufficient: *Ag has to do (or abstain from doing) something, and thus Ag has to decide something: whether counting on Y/P or not, whether investing on it; Ag has to take her own decision of exploiting it or not*.

'Delegation' requires some trust, and trust as free decision and action is about delegation. This also means that *trust implies that X has not complete power and control over the agent/process Y, he is relying and counting upon*. Trust is a case of limited power, of 'dependence'.

When Y is an autonomous cognitive agent this *perceived* degree of freedom and autonomy consists in its 'choice': Y can *decide* to do or not to do the expected action. With this kind of agent (social trust) we in fact trust Y for deciding, being willing, to do – against possible conflicting goals at the very moment and in the circumstance of the performance – what Y 'has to' do (for us); we trust (in) Y's motivation, decision, and intention.

This feature of trust strictly derives from our founding trust on reliance on a non-directly controlled process and agent, on the perception of this 'non-complete control', and risk; on the distinction between trust 'in' Y, and global trust (internal attribution); on the idea of 'delegation': of deciding to count upon such a process/agent. If I do not decide to depend on this, I do not care about its non-controllability.

Reliance is a much broader phenomenon than trust. It even covers cases where the agent is unaware of the needed mediation. Let us consider the various cases and degrees before trust.

a) X does not understand or know whether to rely on a given agent or process. However, the positive result – due also to Y's action – reinforces and reproduces X's behavior and his reliance on Y.[38] We can call this *'confidence'*.

b) X is aware of the contribution of Y, but he *doesn't decide* to rely on Y; it is just so. It is, for example, when I just become aware of my confidence and reliance in a given support; I realize that it is only thanks to Y (that physical support? The obscure work of that guy?) that my activity was possible and effective.

c) X *decides* to rely on Y (not necessarily because he trusts Y; even without trusting Y; for example it is obliged to).

d) X *decides to count on* Y, but Y is *not an autonomous agent*; Y doesn't decide 'to do' what X needs (for example, I rely on the fact that – after this – she will be tired; or I decide to rely/bet on the fact that tomorrow it will be sunny).

e) X *decides to rely* on Y because she *trusts* Y (autonomous agent), but X does not rely on Y's adoption of his goal (not 'genuine' trust).

The 'act' of trust is not reducible to reliance; 'to trust' (as act) implies 'counting on', which implies 'to rely on', but is more than this.

'Counting on' is not just relying, it is first of all an (originally) conscious reliance; the agent knows to rely on a given process and entity. Moreover, this reliance is not simply the discovery of a state of fact, of a static given situation; it is a decision and the result of a decision, or at least a course of events, something that X expects will happen while 'doing' something. X is doing something (or deciding to do something) and she expects that this process will bring a good/desired result thanks to the action of another entity, that will create some necessary condition for the successful realization of the act or of the goal.

Counting on means to have in mind a multi-agent plan, where the *action* (the contribution) of Y is enclosed; and where X has to do her share, at least deciding to counting on, or deciding to do nothing (which is an action) and delegating and waiting for the result, or at least expecting for.[39]

Trust (as act) of course is not just counting on, it is counting on based on a good evaluation (of $Y's$ capacity and predictability) and on a good expectation. Moreover, to *count on* may be weaker – as a degree of certainty – than *to trust*, or less 'free'; trust in an autonomous decision based on an internal evaluation or feeling. One might 'count on' something even when pushed,

---

[38] While *walking* actually I'm implicitly and unconsciously relying on the floor; until I do not have some nasty surprise.

[39] In other words: *Counting on* it is not just to 'delegate', but is to do my share since and until I assume that Y will do his own share. Delegating is (deciding to) allocate/assign an action to Y in order – then – to count on this. They are two complementary moves and attitudes; two faces of the same complex relation.

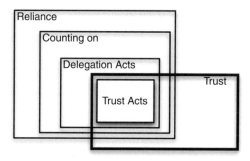

Figure 2.7   Relationships among Reliance, Counting on, Delegation, and Trust

obliged to do so, without really trusting it, and without a free decision.[40] If X's choice was free she wouldn't count on Y, precisely because she does not trust Y (enough).

Delegation is the free act of counting on; and precisely for this reason it normally presupposes some trust. So, trust includes 'counting on' (but is not just reducible to it) which includes 'relying on' (see Figure 2.7).

2.7   Can we *Decide* to Trust?

In the next chapter we will model the decision process and how trust enters into it. However, in relation to our layered notion of trust, and to a belief-based account of trust, a crucial preliminary question arises. Can one *decide* to trust? For example, Annette Baier in her work on 'Trust and Antitrust' (Baier, 1986) claimed that we can never decide to trust (p. 235). We disagree on this.

First of all, the question is not well addressed without immediately distinguishing between trust as belief-structure (attitude/disposition) and trust as decision and action.

(i) As for *Trust-act*, the answer is 'yes': I can decide to trust somebody, and even when the background *belief-trust* wouldn't be sufficient. I can in fact say 'I have decided to trust him (to entrust him), although I do not really trust him'. I may in fact have other reasons and other values such that (even with a risk perceived to be too high) I *accept* that ('as if') Y be reliable and I decide to delegate and *relying on Y* and to be vulnerable by Y. For example, I might do so in order to see whether I am right or wrong, whether Y is really untrustworthy; or in order to show my good-faith and good-will; and I am disposed to pay such a price for having this proof or for exhibiting my virtues.[41]

(ii) As for *Trust as beliefs* about Y (evaluations and expectations), like for any other *belief* I cannot *decide* about.

---

[40] Suppose that you don't trust at all a drunk guy as a driver, but you are forced by his gun to let him drive your car.

[41] Consider that in this case (*trust-act* with not sufficient *belief-trust*) the trustor's mental elements will be conditioned by this act: for example, X will have, after the act and caused by it, the goal that Y is able and willing to achieve the task.

However, as we have already seen in Section 2.5, those assumptions might not necessarily and always be true/full beliefs, but just *acceptances*. I act on the presumption that, since I do not really know/assume that it is true; I do not know, but it might be. This is another meaning of the expression: 'I have decided to trust him': 'I have decided to give for credible, to give *credit*'; and on such a basis I have decided to rely on him.

The two decisions are not fully independent. On the one side, obviously, if I can *decide* to believe (assume) that you are trustworthy, then – on such a basis – I can decide to trust you, as an action of reliance. But there is also a strange relation in the other sense. Following Festinger's model of 'cognitive dissonance' reduction (Festinger, 1957), after a decision automatically and unconsciously we adjust the strength and the value of our beliefs on which the decision is based, in order to feel consonant and coherent (probably the function is to make our intentions more stable). So we increase the value of the beliefs favorable to our preference and choice, and make weaker the contrasting beliefs (focused on costs, risks, and the value of alternative choices). In terms of trust, *after the decision to trust Y (as intention and action) we will adjust the attitude/disposition beliefs (evaluations and expectations about Y)*. So, the *decision* to *believe* obviously affects the *decision to trust* (counting on), but also the *decision to trust* may affect (feed back on) the *beliefs*.

Taking into account explicitly and consciously this effect in our decision-making would be irrational, since Y's trustworthiness is not enhanced at all; only our subjective perception of it is enhanced. We can be (after the choice) less anxious, but not safer.[42] This is quite different from the other prediction about Y's increased trustworthiness due to our action to trust him, as an actual effect on him (see Chapter 6).

2.8   Risk, Investment and Bet

Any act of trusting and counting on implies some bet and some risk. Trust is there precisely because the world is uncertain and risky (Luhmann, 1979). In fact, X is basing her behavior on uncertain expectations and predictions; and is making herself dependent on Y, and thus exposed to be vulnerable by Y. Also, because Y is not fully under X's control; especially when he is an autonomous agent, with his own mind and interest.[43] X might eventually be disappointed, deceived and betrayed by Y: her beliefs may be wrong. At the same time X bets something on Y.

First, X renounced to (search for) possible alternatives (for example, other partners) and X might have lost her opportunity: thus X is risking on Y the utility of her goal g_X (and of her whole plan).

Second, X had some cost in evaluating Y, in waiting for its actions, etc. and X wasted her own time and resources.

Third, perhaps X had some cost to induce Y to do what X wants or to have him at her disposal (for example, X has paid for Y or for his service); now this investment is a real bet (Deutsch, 1985) on Y.

---

[42] We can take into account this expected effect in our decision, not as an increased reliability of Y, but as a good outcome of the decision, to be evaluated.

[43] Trust is "a device for coping with the freedom of others" (Gambetta, 1988) (p.219), or better with their "autonomy".

Thus, to be precise we can say that:

> *With the decision to trust Y (in the sense of relying on Y), X makes herself*
> *both more dependent and more vulnerable, and is more exposed to risks and*
> *harms by Y.*

2.8.1 'Risk' Definition and Ontology

What is risk? We accept the traditional view that risks for X are possible dangers impinging on X in a given context or scenario. More precisely the 'risk' is the estimated gravity/entity of the harm *multiplied for* its likelihood: the greater the possible harm the greater the risk; the greater the likelihood the greater the risk.[44]

However, we believe that this characterization is not enough and that it is important for the theory of trust (and for the theory of intentional action) to have a good **ontology of risks**, introducing several additional distinctions.

Any intentional action, any decision exposes, makes us *vulnerable*, to some 'risk'. Any act of trust, of relying on actions of others, exposes us to risks just for this general principle. The additional feature in trust is that the incurred risks *come from the others*: depends on their possible misbehavior; just because one has decided to depend on them. The decision to trust is the decision to not fully protect ourselves from possible dangers from Y, of exposing ourselves to possible dangers from Y, at least as for the disappointment of our positive expectation, the failure of the 'delegated' action, but possibly also the other risks resulting from our non diffident attitude, good faith, non protection. But other distinctions are needed.

First of all, it is crucial to distinguish between *'objective'* risks (the risks that X incurs following the point of view of an external ideal observer) from the *'subjective'*, perceived risks. Second, it is important to build on Luhman's intuition about the useful distinction between the dangers to which one is exposed independently from his decisions and actions, and those to which one is exposed as a consequence of his own decision and action. He proposes to call 'risks' only the second ones: the risks we 'take'.[45]

Given a certain point of choice in time, with different possible 'futures', different paths, there are risks which are only in one path and not on the others; risks which are in all possible future worlds[46]. When we choose to go in one direction or in another (by doing or not doing something) we 'take' those specific risks. But in a given path (or on all paths) there might be risks which do not depend at all on us and our choices. For example, if our planet would collapse under the tremendous impact of an asteroid (and this 'risk' is in fact there) this will

---

[44] Actually the two 'components' are not fully independent – from a psychological point of view. One dimension can affect the other: the perceived 'value' of the threatened goal or of the pursued goal can be modified by its chance (also, for example, for additional activated goals, like avoiding or searching for excitement, hazard); while the same estimated probability can be perceived/valued in different ways depending on the importance of goal.

[45] Although, as we will show, it is better to restrict the notion of "taken" risks, and of "taking risks" to a subset of this.

[46] This is not exactly the same distinction. There might be risks present on every path which are nevertheless dependent on the fact that we have 'chosen' to go in that direction; for example, the risk "to be responsible" for our action/choice and for some possible bad consequence.

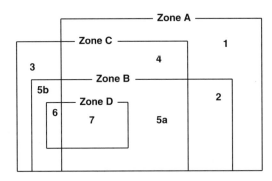

Figure 2.8   Risk Map

affect everybody living on earth no matter what their previous behavior was. It is important to combine these two dimensions: awareness and dependence.

The subject perceives, expects certain possible dangers: of course, he might be wrong. Thus some 'perceived', 'anticipated' risks are just 'imaginary': X believes that something is dangerous or that there is an incumbent risk, but he is simply wrong. Let's notice that frequently enough the lack of trust or distrust is due to an *over-estimation* of possible risks.

Some of those viewed, imagined risks are put aside, not taken into account in the decision, as unbelievable, too implausible, not to be considered. Others are, on the contrary, taken into account in and for the decision.

To 'take' a risk presupposes in our vocabulary that X assumes that such a risk is there and nevertheless decides to go in that direction. He knows (believes/assumes) and decides to expose himself to those dangers. Notice that X also 'takes' the imaginary risks; while he doesn't take the risks that don't realize at all, or that he has put aside. Trust is a subjective state and decision; thus what matters are not actual risks or safety, but the perceived safety and the believed risks. Not all the risks to whom X is exposed thanks to a given decision are 'taken'. We can resume the situation in the Figure 2.8.

This is where, about the specific intersections we have:

zone 1 represents *actual possible but unperceived dangers <u>not</u>* due to X's choice

zone 2 represents *actual possible but unperceived dangers* due to X's choice

zone 5 represents *Imagined (actual or unrealistic) dangers* not taken into account in the decision

zone 6 represents *Imaginary risks* (consequences of X's choice) *evaluated in the decision, and thus "taken"*

zone 7 represents *Actual risks* (consequences of X's choice) *evaluated in the decision, and thus "taken"*

zone 3 + **zone 5b** + **zone 6** represent *imaginary dangers*; perceived but not real

zone 4 + **zone 5a** + **zone 7** represent *perceived and realistic dangers*

Trust as decision has to do only with 'taken' risks: *perceived, imagined (true or false) risks to whom X believes to expose himself as a consequence of his decision/action.*

To 'feel confident' (we accept Luhman's proposal) has to do with any danger I (do not) feel exposed, independently from my own actions, in a given environment, context, situation (zones 3 and 4). However, trust as behavior and social phenomena (in economics, politics, etc.) also requires the theory of the 'objective' risks to which people or institutions are exposed thanks to their interdependence and reliance.

A particularly interesting case is risks X perceives (predicts) and in a sense 'chooses', but actually has no alternative: she has no real 'freedom of', she has no real responsibility in 'taking' that risk, since the alternative is even worst. Thus, even if X chooses a path which is not convenient at all, a bet which per se, in isolation, would be irrational, she is acting in a rational way: minimizing her risk and damage.

2.8.2 What Kinds of Taken Risks Characterize Trust Decisions?

When X trusts Y there are three risks:

a) the risk of failure, the frustration of g_X (*missed gains*) (possibly for ever, and possibly of the entire plan containing g_X);[47]
b) the risk of wasting efforts and investments (*losses*);
c) the risk of unexpected harms (frustration of others goals and/or interests of X).

As for the first risk (case *a*), the increment of X's dependence from Y is important.

Two typical cases are the dependence from time resources and trusted agents' resources. Maybe that after Y's failure there is no time for achieving g_X (it has a specific time expiration); maybe that initially X might have alternatives to Y (rely on Z or W) after her choice (and perhaps because of this choice) Z and W might be no more at her disposal (for example they might be busy); this means that X's alternative means (partners) for g_X are reduced and then X's dependence on Y has increased (Sichman *et al.*, 1994).

Given those (in part perceived) risks and thus the explicit or implicit additional goals of avoiding these harms, X becomes – relative to these additional goals – *more 'dependent' on Y*, since actually it is up to Y (after $X's$ decision to trust him and relying on him) do not cause those harms to X.

As for becoming more vulnerable (case *c*), since X expects some help from Y (as for goal g_X) X feels well disposed towards Y. The (implicit) idea that there is no danger from Y (as for g_X), reduces X's diffidence and vigilance; X feels confident towards Y, and this *generalizes* beyond g_X. This makes X – less suspicious and careful – more accessible and undefended. This is also due to a bit of *transitivity* in Positive Trust from one goal to other: if X trusts Y for g_X, X can be a bit prone to trust Y as for a goal g'_X (where g'_X is different from g_X) even if we have to consider all the limits of the transitivity applied to the trust concept (see Chapter 6).

---

[47] Moreover there might be not only the frustration of g_X, the missed gain, but there might be additional damages as effect of failure, negative side effects: the risks in case of failure are not the simple counterpart of gains in the case of success.

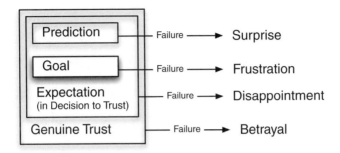

Figure 2.9   Potential Effects for the Failure of the Different Trust Elements

In sum, *Y* – after and because of *X*'s trust (attitude + act) in him – can harm *X* in several ways:

- By frustrating *X*'s goal (g_X) for which *X* relies on him. This also implies frustrating *X*'s expectations and hope: that is *disappointing X*. And this will impact on *X*'s self-esteem (as evaluator and/or decision maker).
- *Y* can also damage *X*'s general attitude and feeling towards the world; and so on (Miceli and Castelfranchi, 1997).
- Moreover, *X* may, not only, be *surprised, frustrated, disappointed*, but *X* can feel *resentment* (and even indignation) for moral violations, for being *betrayed* (see Figure 2.9).
- By frustrating other goals of *X* that she did not protect and defend from *Y*'s possible attack, being relaxed and non-diffident. This will imply analogous consequent frustrations.

It was necessary to immediately mention uncertainty and risk, in connection with the notions of reliance and expectations. However, we will deeply develop these issues (Chapter 3), after introducing the *degree* of certainty of beliefs, and the degree of trust in decision-making.

2.9   Trust and Delegation

What Delegation Is

As we said, *in Delegation the delegating agent (X) needs or likes an action of the delegated agent (Y) and includes it in her own plan: X relies, counts on Y. X plans to achieve g_X through Y. So, she is formulating in her mind not a single-agent but a multi-agent plan and Y has an allocated share in this plan: Y's delegated task is either a state-goal or an action-goal* (Castelfranchi, 1998) (see Figure 2.10).

To do this *X* has some trust both in *Y*'s ability and in *Y*'s predictability, and *X* should abstain from doing and from delegating to others the same task (Castelfranchi and Falcone, 1997).

We have classified delegation in three main categories: *weak, mild* and *strong delegation*.

(i) In *weak delegation* there is no influence from *X* to *Y*, no agreement: generally, *Y* is not aware of the fact that *X* is exploiting his action.

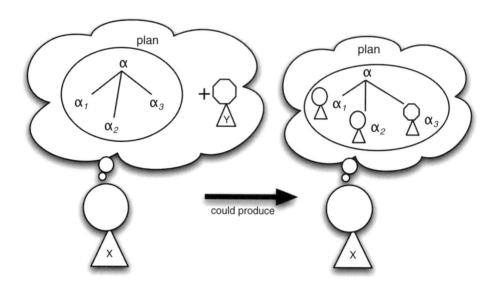

Figure 2.10   The potential for Delegation

　　As an example of weak and passive but already social delegation, which is the simplest form of social delegation, consider a hunter who is waiting and is ready to shoot an arrow at a bird flying towards its nest. In his plan the hunter includes an action of the bird: to fly in a specific direction; in fact, this is why he is not pointing at the bird but at where the bird will be in a second. He is delegating an action in his plan to the bird; and the bird is unconsciously and (of course) unintentionally collaborating with the hunter's plan.

(ii) In a slightly stronger form of delegation (*mild delegation*) X is herself eliciting, inducing the desired behavior of Y to exploit it. Depending on the reactive or deliberative character of Y, the induction is just based on some stimulus or is based on beliefs and complex types of influence.

(iii) *Strong delegation* is based on Y's awareness of X's intention to exploit his action; normally it is based on Y's adopting X's goal (for any reason: love, reciprocation, common interest, etc.), possibly after some negotiation (request, offer, etc.) concluded by some agreement and social commitment.

The Act of Delegation

Notice that weak delegation is just a mental operation or action, and a mental representation. X's external action is just waiting for or abstaining from doing the delegated action or doing her own part of the plan. On the contrary, in (ii) and (iii) delegation is an external action of X on Y, which affects Y and induces him to do the allocated task. Here to delegate means to bring it about that Y brings it about that p. If $E_X(p)$ represents the operator '*to bring it about that*', indicating with X the subject of the action and p the state resulting of X's action, we have the situation shown in Table 2.3.

Table 2.3 Role of X in the different Kind of Delegation

Weak Delegation	X *is exploiting* $E_Y(p)$
Mild Delegation	$E'_X(E_Y(p))$
Strong Delegation	$E_X(E_Y(p))$

2.9.1   Trust in Different Forms of Delegation

Although we claim that trust is the mental counter-part of delegation, i.e. that it is a structured set of mental attitudes characterizing the mind of a delegating agent/trustor, however, there are important differences, and some independence, between trust and delegation. Trust and delegation is not the same thing.

Delegation necessarily is an *action* (at least mental) and the result of a decision, while trust can be just a potential, a mental attitude. The external, observable behavior of delegating either consists of the action of provoking the desired behavior, of convincing and negotiating, of charging and empowering, or just consists of the action of doing nothing (omission) waiting for and exploiting the behavior of the other. Indeed, we will use trust and reliance only to denote the mental state preparing and underlying delegation (*trust* will be both: the small nucleus and the whole).[48]

There may be trust without delegation: either the level of trust is not sufficient to delegate; or the level of trust would be sufficient but there are other reasons preventing delegation (for example prohibitions); or trust is just potential, a predisposition: 'X will, would, might rely on Y, if/when . . .', but it is not (yet) the case. So, trust is normally *necessary* for delegation, but it is not *sufficient*: delegation requires a richer decision.

There may be delegation (or better just 'counting on') without trust: these are exceptional cases in which either the delegating agent is not free (coercive delegation[49]) or she has no information and no alternative to delegating, so that she must just make a trial (blind delegation). So, all trust decisions and acts imply an act of delegation, but not every act of delegation is an act of trust.

Moreover, the decision to delegate has no degrees: either X delegates or X does not delegate. Indeed trust has degrees: X trusts Y more or less relatively to α. And there is a threshold under which trust is not enough for delegating.

Trust in Weak Delegation

While considering the possible temporal gap between the decision to trust and the delegation we have to consider some other interesting mental elements. (The temporal gap ranges between 0 and ∞; 0 means that we have delegation at the same time as the decision to trust; ∞ means that delegation remains just a potential action). In particular we have in all the cases (weak, mild and strong delegation) X's *intention-that* Y will achieve the task (Grosz and Kraus, 1996). In every case this intention is composed through different intentions.

---

[48] In our previous works we used "reliance" as a synonym of "delegation", denoting the action of relying on; here we decide to use "reliance" for the (part of the) mental state, and only "delegation" for the *action* of relying and trusting.

[49] Consider the example of the drunk driver, in note 40.

Potential Goal	G_0 : $Goal_X(g) = g_X$ with $g_X \subseteq p$	
Potential Expectation	B_1 : $Bel_X (Can_Y (\alpha,p))$	(Competence)
	G_1 : $Will_X (Can_Y (\alpha,p))$	
Potential Expectation	B_2 : $Bel_X (<WillDo_Y (\alpha)>p)$	(Disposition)
	G_2 : $Will_X (<WillDo_Y (\alpha)>p)$	

B_3 : Bel_X Dependence$_{XY} (\alpha,p)$ (Dependence)

I : $Intend_X$ (Relay-upon$_{XY}$ τ)

I_1 : Intend-that$_X$ ($<Achieve_Y (\alpha)>p$)

I_2 : $Intend_X$ (Not ($<Achieve_{X \, or \, Z} (\alpha)>p$)) (where Z≠Y)

I_3 : $Intend_X$ (Not ($Do_X (\alpha')$))   (α' is an interfering action with α)

Weak Delegation

Figure 2.11   Mental Ingredients for Weak Delegation

In weak delegation, we have three additional intentions (I_1, I_2, and I_3 in Figure 2.11), respectively, the intention that Y achieves p by the action α; the intention to not to do (or do not delegate to others) that action; and the intention to not hinder Y's action with other interfering actions.

Trust in Mild Delegation

In mild delegation, in addition to I_1, I_2, and I_3 there is another intention (I_4), that is, X's intention to influence Y in order that Y will achieve τ (Figure 2.12).

Trust in Strong Delegation

In strong delegation X's intended action is an explicit request (followed by an acceptance) to Y about p (see Figure 2.13).

Consider that in mild and strong delegation the intentions are already present in the decisional phase and they are the result of an evaluation. For example, X has to evaluate if the delegation will be successful or not in the case of influence, request, etc.

2.9.2   Trust in Open Delegation Versus Trust in Closed Delegation

A very important distinction is also that between *open* and *closed* delegation. In *closed delegation* the task assigned to Y is fully specified. It is just a sequence of actions to be performed. The task is a merely executive task. The extreme case of this is the classical Tayloristic industrial organization of work, where the worker is explicitly forbidden to think

Potential Goal	G_0 : $Goal_X(g) = g_X$ with $g_X \subseteq p$
Potential Expectation	B_1 : $Bel_X (Can_Y (\alpha,p))$ (Competence)
	G_1 : $Will_X (Can_Y (\alpha,p))$
Potential Expectation	B_2 : $Bel_X (<WillDo_Y (\alpha)>p)$ (Disposition)
	G_2 : $Will_X (<WillDo_Y (\alpha)>p)$

B_3 : $Bel_X Dependence_{XY} (\alpha,p)$ (Dependence)

I : $Intend_X (Relay-upon_{XY} \tau)$

   I_1 : $Intend\text{-}that_X (<Achieve_Y (\alpha)>p)$

   I_2 : $Intend_X (Not (<Achieve_{X\ or\ Z} (\alpha)>p))$ (where $Z \neq Y$)

   I_3 : $Intend_X (Not (Do_X (\alpha')))$ (α' is an interfering action with α)

   I_4 : $Intend_X (Do_X (\alpha'')) --> (Do_Y \alpha))$ (in other terms: $(E_X (E_Y p))$)

(where α'' is not an explicit request to Y of doing α)

Mild Delegation

Figure 2.12   Mental Ingredients for Mild Delegation

about the delegated task (the conditions for realizing the task are constraining any potential free initiative in the job), and he has just to perform a repetitive and mechanic movement.

On the contrary, in (completely) *open delegation* the assigned task for Y is 'to bring it about that p', to achieve (in some way) a given result. Y has to find and choose 'how' to

Potential Goal	G_0 : $Goal_X(g) = g_X$ with $g_X \subseteq p$
Potential Expectation	B_1 : $Bel_X (Can_Y (\alpha,p))$ (Competence)
	G_1 : $Will_X (Can_Y (\alpha,p))$
Potential Expectation	B_2 : $Bel_X (<WillDo_Y (\alpha)>p)$ (Disposition)
	G_2 : $Will_X (<WillDo_Y (\alpha)>p)$

B_3 : $Bel_X Dependence_{XY} (\alpha,p)$ (Dependence)

I : $Intend_X (Relay-upon_{XY} \tau)$

   I_1 : $Intend\text{-}that_X (<Achieve_Y (\alpha)>p)$

   I_2 : $Intend_X (Not (<Achieve_{X\ or\ Z} (\alpha)>p))$ (where $Z \neq Y$)

   I_3 : $Intend_X (Not (Do_X (\alpha')))$ (α' is an interfering action with α)

   I_4 : $Intend_X (Do_X (\alpha'')) --> (Do_Y \alpha))$ (in other terms: $(E_X (E_Y p))$)

(where α'' is an explicit request to Y of doing α)

Strong Delegation

Figure 2.13   Mental Ingredients for Strong Delegation

realize his 'mission'. He has to use his local and timely information, his knowledge and experience, his reasoning, and so on. Of course, our evaluation and expectations about Y are very different in the two cases; and the trust that we have in Y refers to different things. In *closed delegation* we trust Y to be obedient, precise, and skilled; we trust him to 'execute' (with no personal contribution – if not minimal sensory-motor adaptation) the task, which is completely specified. In *open delegation* we trust Y not only in his practical skills, but also in his understanding of the real sense of the task, in his problem solving ability, in his competent (and even creative) solution.

We might even trust Y to violate our request and specific expectation in order to *over-help* us. Over-help (Castelfranchi and Falcone, 1998b) is when Y does more than requested (for example, he satisfies not only the required goal but also other goals of X in someway linked with it).

A special form of over-help is *critical help*: when Y is not able or in condition to do as requested, or understands that X's request is simply wrong (for her goal), or he has a better solution for her problem, and violates her specific request but in order to satisfy her higher goal. This is the best form of real *collaboration* (even if risky in some way); Y is really helpful towards X's goals; he is not a stupid executor.

Sometimes, we deeply trust Y for over-help or for critical-help. We confidently expect that – if needed – he will do more than requested or will even violate our request in order to realize the adopted our goal (or to guarantee our 'interests'). This is close to the most advanced trust: *tutorial trust* (Section 2.11.3).

2.10   The Other Parts of the Relation: the Delegated *Task* and the Context

2.10.1   Why Does X Trust Y?

It is not enough to stress that X trusts Y to do α (an action), or to do something. First of all, as we said, trust is not simply a (firm) prediction, but is a positive expectation. In other words X is interested in the performance of α, she expects some positive result; one of the outcomes of α is (considered as or is) a *goal* of X. As we have remarked, an aspect absolutely necessary, but frequently ignored, (or at least left implicit) is that of *the goal, the need,* relatively to whom and for the achievement of whom the trustor counts upon the trustee. This is implicit when some 'positive' result/outcome or the 'welfare' or 'interest' are mentioned in the definition, or some 'dependence' or 'reliance', or the 'vulnerability' are invoked. This is also why, when X trusts Y in something, he cannot only be surprised, but can be 'disappointed' (and even betrayed) by Y or by his performance. Moreover, as we have anticipated, the task allocated to Y – especially in *social trust* – is *delegated* to Y, since Y is an 'autonomous' agent, and in any case X is relying on a process which is beyond his power and control. Trust implies some (perceived) lack of controllability.

We call 'task' (τ) the delegated action α to be performed and the goal state p (corresponding or including g_X) to be realized by Y (in both cases in fact Y has to 'see it that . . .', to 'bring it about that . . .') (see Section 2.9.2 on *closed* versus *open* delegation), because it is something allocated to Y within a multi-agent plan; something *to be done*; something on which X counts on; and something that frequently Y has an obligation to do (due to a promise, a role, etc.). The theory of *tasks* is important (see *trust generalization* in Chapter 6).

2.10.2 The Role of the Context/Environment in Trust

Trust is a context-dependent phenomenon and notion. This means that X trusts Y for τ on the basis of a specific context; just changing the context (for the same τ and the same Y) X's attitude and decision might be different.

Consider X's trust attitude towards the same agent Y for the same task τ when:

- he (Y) is in two completely different contexts (maybe with different environmental and/or social conditions);
- she (X) is in two completely different contexts (maybe with different environmental and/or social conditions).

In fact, one should perhaps be more subtle, and clearly distinguish these two kinds of context:

- the context of X's evaluation and decision (affecting her mind) while feeling trust for Y and deciding to trust him or not (*evaluation context*); and
- the context of Y's performance of α (*execution context*).

They are not one and the same context. The execution context affects Y's objective trustworthiness; his possibility to really achieve the goal in a good way; and – as perceived by X ($_XTW_Y$) – affects X's expectation.

But the evaluation context is the social and mental environment of X's decision. This can affect:

- X's mood and basic social disposition;
- X's information and sources;
- the beliefs activated and taken into account by X;
- X's risk perception and acceptance;
- X's evaluation of the execution context; and so on.

Moreover, the evaluation and decision of X also depends on this complex *environmental trust*: X's trust in the *environment* where α will be executed, which can be more or less interfering or harmful; in the *supporting infrastructure* (execution tools, coordination and communication tools, etc.); in the *institutional context* (authorities, norms, and so on); in the *generalized atmosphere* and *social values*; and so on.

Environmental trust (*external attribution*) and trust 'in' Y (*internal attribution*) must be combined for a decision; and they are also non-independent one from the other (see also Section 8.3.3 for evaluating the importance of this decomposition with respect the subjective probability).

Table 2.4 Conditional Trust

IF	Event (e) = true
THEN	Trust (X Y C τ g_X) will be over the threshold for delegating

Not only the trust in Y as for τ is context dependent, but if the context (environment) in the mental model of X plays an active causal role, X has also to trust the context, as favorable or not too adverse or even hostile (to Y or to τ). But, of course, Y's capacity and reliability may vary with the more or less adverse nature of the context: it might decrease or even increase. On this we have developed a specific section (see Chapter 6). This is also very important in trust dynamics, since it is not true that a failure of Y necessarily will decrease Y's perceived trust-worthiness for X; it depends on the causal attribution of the failure. The same holds for success.

Another important way in which the context is relevant for trust, is that there can be different trusts about Y in different social contexts, related to the same task: for example, Y is a medical doctor, and he is very well reputed among the clients, but not at all among his colleagues. Or there can be different trusts in different social contexts because different *tasks* are relevant in those contexts. For example, Y is a well reputed guy within his university (as teacher and researcher), but has a very bad reputation in his apartment building (as an antisocial, not very polite or clean, noisy guy).

Trust can migrate from one task to another, from one trustor to another, from one trustee to another (see Chapter 6), and also from one social context to another. It depends on the connections between the two contexts: are they part of one another? Are they connected in a social network? Do they share people, values, tasks, etc.? So, trust is not only a context dependent and sensible phenomenon but is a context-dynamic phenomenon.

Moreover, not only is trust context-dependent but it can also be *conditional*: A special event (e) could be considered by X, in a given context and with respect to a specific trustee, as crucial for trusting Y (see Table 2.3).

Consider our example of a bus stop, in weak delegation. After Y raised his arm to stop the bus the driver is more sure that he will take the bus. In our view, this is not just simple 'conditional' probability (after *the first event*, or given condition C, the probability of *the second event* is greater or smaller). In real trust – given its attributional nature – the *first event* can be interpreted by X as a *signal*. For example, a given act or attitude or sentence of Y can be a sign for X of Y's capacity or of his internal disposition, which makes his doing τ more reliable.

2.11 *Genuine* Social Trust: Trust and *Adoption*

As we saw trust is not only a 'social' attitude. It can be directed towards an artifact or unanimated process. Someone would prefer another term, say *confidence*, but this is just a (reasonable) technical convention, not the real use and meaning of these words.[50] However, it is true that the most theoretically and practically relevant and the most typical notion of trust *is the social one.*

Social trust means *trust towards another autonomous agent perceived (conceived) as such.* That is, towards a purposive, self-governed system, not only with its own resources and

---

[50] Moreover, 'confidence' is very close to 'trust' in a non-technical meaning, it just seems to contain some reliance, and be quite social. It also seems to be based just on learning and experience.

causal-processes, but with its own internal control and *choices*. This is why social trust is there at all; or better, why social interaction requires trust.

The trustor cannot and does not fully 'control' (monitor and guide) Y's activity towards the goal (X's expected result). X passes to Y (part of) the *control* needed for realizing the expected result. X relies precisely on this. On the one side, this is precisely one of the main advantages of *delegating* (task assignment and counting on): *delegating also the control and governance of the activity* (even if X was able to perform it herself). But, on the other side, this is precisely *the specific risk of social trust*: not just possible external (environmental) interferences, but 'internal' interferences (due to Y's nature and functioning).

Y is selecting the right action, employing resources, planning, persisting, executing; he might be defective on this. Moreover, since he has his own control system for purposive behavior, usually he has his own internal goals. Now, those *individual* goals may interfere; taking precedence, being in conflict and prevailing, etc.

If X decides to trust Y, to count on him, X expects (predicts and wishes) that $X's$ goal – adopted by Y – will prevail on Y's autonomous goals, and will be pursued. This is the typical bet of social trust. *There is a peculiar relation between (social) trust and autonomy: we trust in autonomous systems and this is our specific (social) risk: possible indifference, or hostility, or changing of mind, or profiting and exploitation, up to a 'betrayal', which presupposes a specific or general, explicit or implicit, 'pact'.*

Since social trust is directed towards another *autonomous* agent, *considered as* an autonomous agent, with its attitudes, motivations (including the social ones), and some freedom of choice, it requires an *intentional stance* towards a social entity (with its own intentional stance towards us).

However, this is not yet enough to capture the most *typical* social notion of trust; what many authors (like Baier, Hardin, Holton, Tuomela) would like to call *genuine trust*. *Genuine* (social) trust, the basic, natural form of social trust, is based on Y's *adoptive* attitude. That is, *X trusts Y's adoption of her interest/goal, and counts on this.* Y is perceived as taking into account X's goals/interests; and possibly giving priority to them (in case of conflicts). This is true trust in a social agent 'as a social agent'.

Social goal-adoption, is the idea that another agent takes into account in his *mind* – in order to satisfy them– my goals (needs, desires, interests, projects, etc.); he 'adopts' them as goals of himself, since he is an 'autonomous agent', i.e. self-driven and self-motivated (but not necessarily 'selfish'!), and is not an hetero-directed agent, and can only act in view, be driven by, some internal purposive representation (Conte and Castelfranchi, 1995). So – if such an (internally represented) goal will be preferred to others– he will be regulated by my goal; for some motive he will act in order to realize my goal.

A very important case of goal-adoption (relevant for trust theory) is *goal–adhesion*, where X wants and expects that Y adopts her goal, communicates (implicitly or explicitly) this expectation or request to Y; Y knows that X has such an expectation and adopts X's goal not unilaterally and spontaneously, but also because X wants it to be so. Thus not only does Y adopt X's goal that p, but he also adopts X's goal: 'that Y adopts her goal p'. In social trust frequently Y's adoption (cooperation) is precisely due to X's expectation and trust in Y's adoption; and X relies on this response and adhesion.

We agree with Hardin ((Hardin, 2002); Chapter 1) that there is a restrict notion of social trust which is based on *the expectation of adoption* (or even *adhesion*), not just on the prediction of a favorable behavior of Y. When X trusts Y in the strict social sense and counts on him, she expects that Y will *adopt* her goal and this goal will prevail – in case of conflict with other active goals. That is, *X* not only expects an *adoptive goal* by *Y* but an *adoptive decision and intention*. A simple *regularity* based prediction or an expectation simply based on some role or norm prescribing behavior to *Y*, are not enough – we agree with Hardin – for characterizing what he calls 'trust in strong sense', the 'central nature of trust', what we call 'genuine social trust'.

However, in our view, Hardin is not able to realize the broad theory of goal-adoption, and provides us – with his notion of *encapsulated interests* – a restricted and reductive view of it.

The various authors searching for a socially focused and more strict notion of trust go in this direction, but using a non general and not well defined notion, like: *benevolence, good-will, other-regarding attitude, benignity* (Hart, 1988), *altruism, social-preferences, reciprocity, participant stance* (Holton, 1994).

And even the strange and unacceptable notion proposed by Deutsch (Deutsch, 1985) (we discuss this in Chapter 1 and Chapters 8) and repeated several times (for example, Bernard Barber 'to place the others' interests before their own') where in order to trust *Y* one should assume that he is altruistic or even irrational.

What *X* has to believe about *Y* is that:

i) *Y* has *some* motive for *adopting X*'s goal (for doing that action *for X*; for taking care of *X*'s interest); and that he will actually adopt the goal.
ii) Not only *Y* will adopt *X*'s goal (that is, he will formulate in his mind the goal of *X*, because it is the goal of *X*) but also that this goal will become an *intention*, so that *Y* will actually do as desired.

If (i) and (ii) are both true we can say that the adopted goal will prevail against other possible active goals of *Y*, including non-adopted goals (selfish).

More precisely we can claim that the motives *X* ascribes to *Y* while adopting *X*'s goal are assumed to prevail on the other possible motives (goals) of *Y*. Thus, what *X* is really relying on in genuine trust, are *Y's motives for an adoptive intention*.

The fact that a genuine social trust is based/relies on *Y*'s adoption should not be misinterpreted. One should not confuse *goal-adoption* with *specific motives* for adopting. Claiming that *X* counts on *Y*'s adoptive intention is not to claim that she counts on *Y*'s altruism, benevolence, good will, social preferences, respect, reciprocity, or moral norms. These are just specific sub-cases of the reasons and motives *Y* is supposed to adopt *X*'s goal. *X* might count on *Y*'s willingness to be well reputed (for future exchanges), or on his desire to receive gratitude or approval, or of avoiding blame or sanctions, or for his own approval, etc. In other words: *Y can be fully self-interested*.

To realize this it is necessary to keep in mind that the usual structures of goals are means-end chains: not all goals are *final goals*; they can be *instrumental* goals, simple means for higher goals. Thus, on the top of an adoptive and adopted goal there can be other goals, which *motivate* the goal-adoption. For example, I can do something *for* you, just in order to receive what I want for me, what you promise me.

'*It is not from the benevolence of the butcher, the brewer, or the baker that we expect our dinner, but from their regard to their own interest. We address ourselves, not to their humanity but to their self-love, and never talk to them of our own necessities but of their advantages*' (Smith, 1776); however, when I ask the brewer to send me a box of beer and I send the money, I definitely *trust* him to give me the beer.

As seen in Section 1.5.7, we have three kinds of *social goal-adoption* (Conte and Castel-franchi, 1995): *Instrumental*, *Cooperative* and *Terminal*.

X can trust *Y*, and trusts that *Y* will do as expected, *for any kind of adoption*, also (or better, usually) *instrumental* (with both external or internal incentives). Trust in *Y* doesn't presuppose that *Y* is *'generous'* or that he will make *'sacrifices'* for *X*; he can strictly be selfish.

Now, we can formulate in a more reasonable way Deutsch's claim and definition, without giving the impression of trust as counting on *Y*'s altruism or even irrationality.

Y can be self-motivated or interested (autonomous, guided by his own goals) and can even be selfish or egoistic; what matters is that the intention to adopt *X*'s goal (and thus the adopted goal and the consequent intention to do α) will prevail on other non-adoptive, private (and perhaps selfish) goals of *Y*. But this only means that:

Y's (selfish) motives for adopting X's goal will prevail on Y's (selfish) motives for not doing so and giving precedence to other goals.

So, *X* can count on *Y* doing as expected, in *X*'s interest (and perhaps for *Y*'s interest). Trustworthiness is a social 'virtue' but not necessarily an altruistic one. This also makes it clear that *not all 'genuine' trust is 'normative'* (based on norms) (for example, the generous impulse of helping somebody who is in serious danger is not motivated by the respect of a moral/social norm, even if this behavior (later) is socially/morally approved).

Moreover, *not all 'normative' trust is 'genuine'*. We can trust somebody for doing (or not doing α) just because we know that he has to do so (for a specific law or role), independently on his realizing or not and *adopting* or not our goal. For example, I trust a policeman for blocking and arresting some guy who was being aggressive to me, not because he has to respond to my desire, but just because he is a policeman at the scene of a crime (he can even ignore me).[51]

In sum, in genuine trust X just counts upon the fact that Y will understand and care of her (delegated) goal, Y will adopt her goal and possibly prefer it against conflicting goal (for example selfish ones), and this for whatever reason: from selfish advantages to altruism, from duty and obligations to cooperation, from love to identification, and so on.

In addition, May Tuomela (Tuomela, 2003) introduces and defines an interesting notion of 'genuine' social trust. But in our view this notion is too limited and specific. We disagree with her constraint that there is *genuine* trust only when it is symmetrical and reciprocal (for us this is counterintuitive and restrictive). In addition, her conditions (to be respected, the fact that the other will care about my rights, etc.) look quite peculiar in terms of specific – important – social relationships where there is 'genuine' trust, but which exclude other typical situations of trust (like child-mother) that must be covered.[52]

---

[51] It is also important to not mix up 'genuine' adoption-based trust with trust in 'strong delegation': delegation based on *X*'s request and *Y*'s acceptance. 'Genuine' trust can also be there in weak and in mild delegation/reliance: when *Y* ignores *X*'s reliance and acts on his own account, or when *Y*'s behavior is elicited by *X* (but without *Y*'s understanding). In fact, *Y* might have spontaneous reasons for adopting *X*'s interests (and *X* might count on and exploit this), or *X* might elicit in *Y* adoptive motives and attitudes by manipulating *Y*, without *Y* knowing that *X* is expecting and counting upon his adoption.

[52] Paradoxically, sometimes we trust Y precisely for his selfishness, which makes him trustworthy and reliable for that task/mission.

2.11.1 Concern

A very important notion in goal-adoption is the notion of *concern*. How much the goal of X is important for Y; how much Y is concerned with/by X's interest. That is, which is for Y the *value* of X's goal g_X, or best way of X achieving her goal. This value is determined by:

(i) the reasons (higher motivations) that Y has for adopting X's goal, and their value for him; how much and why Y cares about X's welfare;
(ii) $X's$ opinion about the subjective value of g_X for Y.

It is precisely on this basis that the adopted goal will prevail or not against possible costs, against other private conflicting goals of Y, and thus will possibly become/produce an adoptive *intention* of Y; and will also – as intention – persist against possible new interferences and temptations.

It is precisely on Y's *concern* for X's goal (not be confused with benevolence, good will, benignity, and so on) that X relies while betting on Y's adoptive intention and persistence. She also has some 'theory' about the reasons Y should be concerned with her welfare and wish to adopt her goal.

2.11.2 How Expectations Generate (Entitled) Prescriptions: Towards 'Betrayal'

It is characteristic of the most typical/genuine forms of social trust that – in case of failure – X is not only surprised and disappointed (Miceli and Castefranchi, 2002; Castelfranchi and Giardini, 2003), but feels *betrayed*. Where does this affective reaction come from? On which beliefs and goals (present in the trust attitude) is it based?

Social expectation can be entitled, can be based on Y's 'commitment' and thus obligation towards X (Castelfranchi, 1995). What X expects from Y can be 'due'. The violation of this kind of expectations involves not only disappointment but stronger and social emotions, like anger, indignation, etc. In particular it is different if this entitlement, this duty of Y towards X comes from legal norms or from interpersonal relations and merely social norms, like in a promise or like in friendship where fairness and adoption are presupposed.

In these forms of 'genuine' trust, where the expectation of Y's adopting/caring of my needs, requests, wishes (goals), is based on an assumption of *a moral duty* towards me, if Y disappoints this expectation I feel *betrayed* by Y in my trust and reliance on him.

This commitment – and the consequent moral duty, social norm –is not necessarily established in an explicit way; for example by a promise. Not only – as we said – can it be presupposed in the very relationship between us: friends, same family, same group, shared identity (which create some solidarity). It can be established by tacit consent, implicit behavioral communication (Castelfranchi, 2006; Tummolini and Castelfranchi, 2006). See Table 2.5 for an example.

In general, this mechanism is responsible for the tendency of shared social *expectations* (expectations about the behavior of the other agents, which are common knowledge) to become *prescriptions*: not only I predict that you will do something, but I wish so; I want you to behave in such a way (expectation). Moreover, I know that you know (etc.), and you did not disconfirm this (etc.), so you get some obligation of not violating my expectations. And I

Table 2.5   Example of Tacit Consent

IF	*X* decides to count on *Y*, AND *Y* is aware of such an expectation, AND *X* is aware that *Y* is aware, and *Y* knows this, AND *Y* would be able and in condition of rejecting such a delegation, to refuse to 'help' *X*, and to inform *X* about this (a very relevant information for her), and *Y* knows that *X* knows this; AND *Y* says nothing; doesn't provide any sign of his refusal
THEN	*Y* 'tacitly consent': *Y* takes a commitment, a tacit obligation to do as expected, to not disappoint *X*; AND *X* gets a *soft* right towards *Y*: she is entitled to ask and claim for *Y*'s action, and to complain and protest for his not doing as 'committed'

want you to do as expected also for this very reason: because you have a duty and you know (recognize) that I want you to do this simply for this reason. So my expectation becomes a true *prescription* (Castelfranchi and Giardini, 2003). This is how common expectations become social 'conventions' and 'norms'.

In sum, also on the bases of such tacit 'promises' and interpersonal norms, or of those obligations implicit in the relationship, *X* can feel betrayed by *Y*, since she was trusting *Y* on such a specific basis.

2.11.3   Super-Trust or Tutorial Trust

There are very extreme forms of trust, where *X* 'puts herself in *Y*'s hands' in a radical sense; in the sense that she believes and accepts that *Y* will care about her welfare better than her, beyond what she intends, asks, desires. One case is *over-trust*: trust in *Y*'s *over-help*. As we saw, we might confidently expect that, if needed, *Y* will do more than requested or will even violate our request in order to realize the adopted goal of ours. However, we can even go

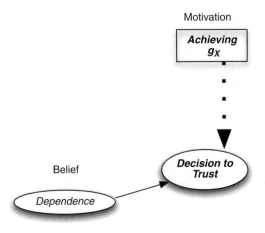

Figure 2.14   Dependence Belief and Goal: First step towards the decision of trusting

beyond this. In over-help Y is supposed to take care of our actual goals, desires, to seek out what we want; but there are forms of trust where we accept that Y goes against our current desires and objectives, while pursuing our (non understood) interests. This is *super-trust* or *tutorial trust*: trust in the 'tutorial' role of Y towards me. I feel so confident in Y that I am convinced that Y is pursuing my good, and is helpful, even when Y is acting against my current goals and I do not understand how he is taking care of me.

In other words, I assume that Y is taking care of my wellness, of doing the best for me, of my (possibly not so clear to me) interests, not just of my actual and explicit goals and desires (and may be against them). He does that for my good.

This presupposes that I feel/believe that I ignore part of my *interests*, of what is good for me (now or in the future), and I assume that, on the contrary, Y is *able* to understand better than me what is good for me (my interests) and *cares* about this, and *wants* – even against me – to protect my interests or oblige me to realize them.

We have modeled (Conte and Castelfranchi, 1995) this kind of social relationship between Y and X (when Y claims to know better than X what is better for X, and care about this, and

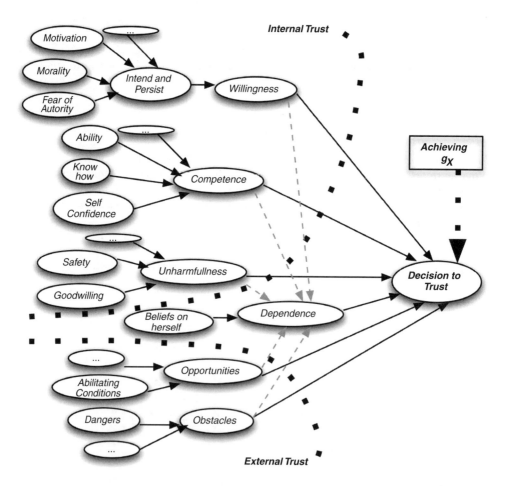

Figure 2.15 The complex set of beliefs converging towards the decision of trusting

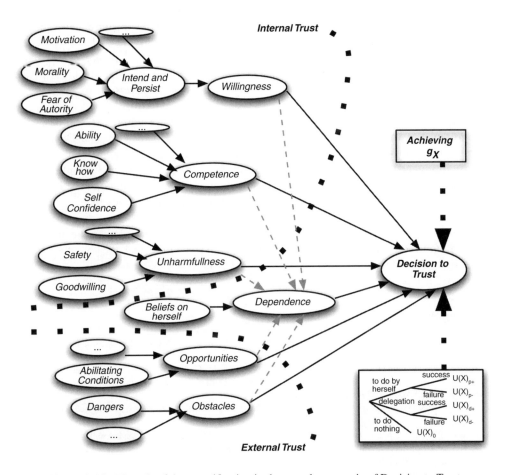

Figure 2.16   The role of the quantification in the complex scenario of Decision to Trust

try to influence X to do what is better for X); and we have labeled this *tutorial* relation of Y towards X. It also exists when X doesn't recognize or even contests it (like between parents and adolescents, or between psychiatrists and patients, etc.).[53]

In *super-trust*, X presumes a *tutorial* attitude and relation from Y, and relies on this, since he feels/believes that Y is really capable of understanding and will care about what is better for X.

2.12   Resuming the Model

Let us resume in a schematic and synthetic way how in our cognitive model of trust the different elements, playing a role in the trust concept, are composed and ordered for producing the trusting behavior of an agent. As we have seen a main role is played by the goal of the trustor that has to be achieved through the trustee (without this motivational component there

---

[53] However, sometimes this is just an arrogant and arbitrary claim, hiding Y's power and advantages, or 'paternalism'.

is no trust). In fact, in addition to the goal, it is also necessary that the trustor believes himself to be (strongly or weakly) dependent from the trustee himself see Figure 2.14.

On the basis of the goal, of her (potential) dependence beliefs,[54] of her beliefs about the trustee attributes (internal trust), of her beliefs about the context in which the trustee performance will come, the trustor (potentially) arrives at the decision to trust or not (Figure 2.15).

As explained in Section 2.2.1, all these possible beliefs are not simply external bases and supports of X's trust in Y (reduced to the Willingness and Competence and Dependence beliefs, and to the Decision and Act), but they are possible internal sub-components and forms of trust, in a recursive trust-structure. The frame looks quite complicated and complex, but, in fact, it is only a potential frame: not all these sub-components (for example, the beliefs about X's morality, or fear of authority, or self-esteem) are necessarily and already there or explicitly represented.

Moreover, as we will see in detail in Chapter 3, a relevant role is played by the quantification of the different elements: the weight of the beliefs, the value of the goal, the potential utilities resulting from a delegation and so on (see Figure 2.16).

References

Bacharach, M. and Gambetta, D. (2000) Trust in signs. In K. Cook (ed.), *Trust and Social Structure*. New York: Russel Sage Foundation.

Bacharach, M. and Gambetta, D. (2001) Trust as type detection, in: C. Castelfranchi, Y.-H. Tan (eds.), *Trust and Deception in Virtual Societies*, Kluwer Academic Publishing, Dordrecht, The Netherlands, pp. 1–26.

Baier, A. Trust and Antitrust, *Ethics* 96: 231–260, 1986.

Bandura, A. (1986) *Social Foundations of Thought and Action: A social cognitive theory*. Englewood Cliffs, NJ: Prentice-Hall.

Butz, M.V. (2002) *Anticipatory Learning Classifier System*. Boston, MA: Kluwer Academic Publisher.

Butz, M.V. and Hoffman, J. Anticipations control behaviour: animal behavior in an anticipatory learning classifier system. *Adaptive Behavior*, 10: 75–96, 2002.

Castaldo, S. (2002) *Fiducia e relazioni di mercato*. Bologna: Il Mulino.

Castelfranchi, C. Social Commitment: from individual intentions to groups and organizations. In ICMAS'95 *First International Conference on Multi-Agent Systems*, AAAI-MIT Press, 1995, 41–49 (versione preliminare in *AAAI Workshop on 'AI and Theory of Group and Organization'*, Washington, DC, May 1993).

Castelfranchi, C. (1996) Reasons: belief support and goal dynamics. *Mathware & Soft Computing*, 3: 233–247.

Castelfranchi, C. (1997) Representation and integration of multiple knowledge sources: issues and questions. In Cantoni, Di Gesu', Setti e Tegolo (eds), *Human & Machine Perception: Information Fusion*. Plenum Press.

Castelfranchi, C. (1998) Modelling social action for AI agents. *Artificial Intelligence*, 103: 157–182, 1998.

Castelfranchi, C. Towards an agent ontology: autonomy, delegation, adaptivity. *AI*IA Notizie*. 11 (3), 1998; Special issue on 'Autonomous intelligent agents', Italian Association for Artificial Intelligence, Roma, 45–50.

Castelfranchi, C. (2000a) Again on agents' autonomy: a homage to Alan Turing. *ATAL 2000:* 339–342.

Castelfranchi, C. (2000) Affective appraisal vs cognitive evaluation in social emotions and interactions. In A. Paiva (ed.) *Affective Interactions. Towards a New Generation of Computer Interfaces*. Heidelberg, Springer, LNAI 1814, 76–106.

Castelfranchi, C. Through the agents' minds: cognitive mediators of social action. *Mind and Society*. Torino, Rosembergh, 2000, pp. 109–140.

---

[54] To be true, the dependence belief already implies some belief about *Y*'s skills or resources, that are useful for *X*'s goal.

Castelfranchi, C. The micro-macro constitution of power, protosociology, an international journal of interdisciplinary research double vol. 18–19, 2003, *Understanding the Social II – Philosophy of Sociality*, Edited by Raimo Tuomela, Gerhard Preyer, and Georg Peter.

Castelfranchi, C. Mind as an anticipatory device: for a theory of expectations. In 'Brain, vision, and artificial intelligence' Proceedings of the First international symposium, BVAI 2005, Naples, Italy, October 19–21, 2005. Lecture notes in computer science ISSN 0302–9743 vol. 3704, pp. 258–276.

Castelfranchi, C. SILENT AGENTS: From Observation to Tacit Communication. *IBERAMIA-SBIA* 2006: 98–107.

Castelfranchi, C. and Falcone, R. (1998) Principles of trust for MAS: cognitive anatomy, social importance, and quantification, *Proceedings of the International Conference on Multi-Agent Systems (ICMAS'98)*, Paris, July, pp. 72–79.

Castelfranchi, C. Falcone R. (1997) Delegation Conflicts, Proceedings of the 8th European Workshop on Modelling Autonomous Agents in a Multi-Agent World: Multi-Agent Rationality, LNAI, Vol. 1237, pp. 234–254.

Castelfranchi, C. and Falcone, R. (1998) Towards a theory of delegation for agent-based systems, robotics and autonomous systems, special issue on multi-agent rationality, *Elsevier Editor*, 24 (3–4):.141–157.

Castelfranchi, C., Giardini, F., Lorini, E., Tummolini, L. (2003) The prescriptive destiny of predictive attitudes: from expectations to norms via conventions, in R. Alterman, D. Kirsh (eds) *Proceedings of the 25th Annual Meeting of the Cognitive Science Society*, Boston, MA.

Castelfranchi, C. and Lorini, E. (2003) Cognitive anatomy and functions of expectations. In *Proceedings of IJCAI'03 Workshop on Cognitive Modeling of Agents and Multi-Agent Interactions*, Acapulco, Mexico, August 9–11, 2003.

Castelfranchi, C. and Paglieri, F. (2007) The role of beliefs in goal dynamics: Prolegomena to a constructive theory of intentions. *Synthese*, 155, 237–263.

Castelfranchi, C., Tummolini, L. and Pezzulo, G. (2005) From reaction to goals - AAAI workshop on From Reaction to Anticipation.

Cohen, J. (1992) *An Essay on Belief and Acceptance*, Oxford University Press, Oxford.

Conte, R. and Castelfranchi, C. (1995) *Cognitive and Social Action*. London: UCL Press.

Cook (ed.) *Trust and Social Structure*, New York: Russel Sage Foundation, forthcoming.

Cooper, J. and Fazio, R.H. A new look at dissonance theory. *Advances in Experimental Social Psychology*, 17: 229–265, 1984.

Dennett, D.C. (1989) *The Intentional Stance*. Cambridge, MA: MIT Press.

Deutsch, M. (1985) *The Resolution of Conflict: Constructive and destructive processes*. New Haven, CT: Yale University Press.

Drescher, G. (1991) *Made-up Minds: a constructivist approach to artificial intelligence*. Cambridge, MA: MIT Press.

Elster, J. (1979) *Ulysses and the Sirens*. Cambridge: Cambridge University Press.

Engel, P. Believing, holding true, and accepting, *Philosophical Explorations* 1, 140–151, 1998.

Falcone, R. and Castelfranchi, C. (1999) Founding agents adjustable autonomy on delegation rheory, *Atti delSesto Congresso dell'Associazione Italiana di Intelligenza Artificiale (AI*IA-99)*, Bologna, 14–17 settembre, pp. 1–10.

Falcone, R. and Castelfranchi, C. (2001) Social trust: a cognitive approach, in *Trust and Deception in Virtual Societies* by Castelfranchi, C. and Yao-Hua, Tan (eds.) Kluwer Academic Publishers, pp. 55–90.

Falcone, R., Singh, M., Tan, Y.H. (eds.) (2001) Trust in cyber-societies. Lecture Notes on Artificial Intelligence, n°2246, Springer.

Falcone, R., Singh, M., and Tan, Y. (2001) Bringing together humans and artificial agents in cyber-societies: a new field of trust research; in *Trust in Cyber-societies: Integrating the Human and Artificial Perspectives* R. Falcone, M. Singh, and Y. Tan (eds.), LNAI 2246 Springer. pp. 1– 7.

Falcone, R. (2001) Autonomy: theory, dimensions and regulation, in C. Castelfranchi and Y. Lesperance (eds.) *Intelligent Agents VII, Agent Theories Architectures and Languages*, Springer, pp. 346–348.

Falcone, R., Barber, S., Korba, L., Singh, M. (eds.) (2003) Trust reputation, and security: theories and practice. Lecture Notes on Artificial Intelligence, n°2631, Springer.

Festinger, L. (1957) *A Theory of Cognitive Dissonance*. Stanford, CA: Stanford University Press.

Gambetta, D. 'Can we trust trust?' (1998) In *Trust: Making and Breaking Cooperative Relations*, edited by D. Gambetta. Oxford: Blackwell.

Good, D. (2000) Individuals, interpersonal relations, and trust. In Gambetta, D. (ed.) *Trust: Making and Breaking Cooperative Relations*, electronic edition, Department of Sociology, University of Oxford, pp. vii-x, <http://www.sociology.ox.ac.uk/papers/gambettavii-x.doc>.

Grosz, B. and Kraus, S. Collaborative plans for complex group action, *Artificial Intelligence*, 86:269–358.

Hardin, R. (2002) *Trust and Trustworthiness*, New York: Russel Sage Foundation.

Hart, K. (1988) Kinship, contract and trust: economic organization of migrants in an African city slum. In *Trust: Making and Breaking Cooperative Relations*, edited by Diego Gambetta. Oxford: Blackwell.

Hertzberg, L. (1988) On the Attitude of Trust. *Inquiry* 31 (3): 307–322.

Holton, R. Deciding to trust, coming to believe. *Australian Journal of Philosophy* 72: 63–76, 1994.

Jennings, N. R. Commitments and conventions: The foundation of coordination in multi-agent systems. *The Knowledge Engineering Review*, 3: 223–250, 1993.

Jones, A. J. On the concept of trust, *Decision Support Systems*, 33 (3): 225-232, 2002, Special issue: *Formal modeling and electronic commerce.*

Johnson-Laird, P. N. (1983) *Mental models: towards a cognitive science of language, inference and consciousness.* Cambridge, UK: Cambridge University Press.

Levesque, H. J. (1984) A logic of implicit and explicit belief. In *Proceedings of the Fourth National Conference on Artificial Intelligence (AAAI-84)*, pages 198–202, Austin, TX.

Lorini, E. and Castelfranchi, C. (2007) The cognitive structure of Surprise: looking for basic principles. Topoi: An International Review of Philosophy, 26(1), 133–149.

Luhmann, N. (1979) *Trust and Power*, Wiley, New York.

Marsh, S.P. (1994) Formalising trust as a computational concept. PhD thesis, University of Stirling. Available at: http://www.nr.no/abie/papers/TR133.pdf.

Mayer, R. C., Davis, J. H., and Schoorman, F. D. An integrative model of organizational trust. *Academy of Management Review*, 20 (3): 709–734, 1995.

McKnight, D. H. and Chervany, N. L. Trust and distrust definitions: one bite at a time. In *Trust in Cyber-societies*, Volume 2246 of Lecture Notes in Computer Science, pages 27–54, Springer, 2001.

Meyer, J.J. Ch. and Van Der Hoek, W. (1992) A modal logic for non monotonic reasoning. In W. van der Hoek, J.J. Ch. Meyer, Y. H. Tan and C. Witteveen, editors, *Non-Monotonic Reasoning and Partial Semantics*, pages 37–77. Ellis Horwood, Chichester, 1992.

Miceli, M., Castelfranchi, C. and Parisi, D. Verso una etnosemantica delle funzioni e dei funzionamenti: 'Rotto', 'guasto', 'non funziona', 'malato'. *Quaderni di semantica*, 8 (IV), 1983, 179–208.

Miceli, M. and Castelfranchi, C. (1997) Basic principles of psychic suffering: A preliminary account. *Theory & Psychology*, 7, 769–798.

Miceli, M. and Castelfranchi, C. (2000) The role of evaluation in cognition and social interaction. In K. Dautenhahn (ed.), *Human Cognition and Agent Technology* (pp. 225–261). Amsterdam: Benjamins.

Miceli, M. and Castelfranchi, C. (2002) The mind and the future: The (negative) power of expectations. *Theory & Psychology*. 12 (3): 335–366.

Miceli, M. and Castelfranchi, C. Anxiety as an 'epistemic' emotion: an uncertainty theory of anxiety. *Anxiety, Stress, and Coping*, 18: 291–319.

Miceli, M. and Castelfranchi, C. *Hope: the power of wish and possibility. Thoery & Psychology*, in press.

Miller, G. A., Galanter, E., and Pribram, K. A. (1960) *Plans and the Structure of Behavior*. New York: Holt, Rhinehart, & Winston.

Pezzulo, G., Butz, M., Castelfranchi, C. and Falcone, R. (Editors) *The Challenge of Anticipation*. LNAI State-of-the-Art Survey N°5225, pp. 3–22.

Searle, John R. (1995) *The Construction of Social Reality*, Free Press: NY.

Sichman, J. R., Conte, C., Castelfranchi and Y. Demazeau. A social reasoning mechanism based on dependence networks. In *Proceedings of the 11th ECAI*, 1994.

Smith, A. (1776) *An Inquiry into the Nature and Causes of the Wealth of Nations*, London: Methuen & Co., Ltd.

Snijders, C. (1996) Determinants of Trust, *Proceedings of the workshop in honor of Amnon Rapoport, University of North Carolina at Chapel Hill*, USA, 6–7 August.

Tummolini, L. and Castelfranchi, C. Trace Signals: The Meanings of Stigmergy. E4MAS 2006: 141–156, 2006.

Tuomela, M. (2003) A collective's rational trust in a collective's action. In *Understanding the Social II: Philosophy of Sociality, Protosociology*. 18–19: 87–126.

van Linder, B. (1996) Modal Logics for Rational Agents, PhD thesis, Department of Computing Science, University of Utrecht.

3

Socio-Cognitive Model of Trust: Quantitative Aspects

So far in this book, we have analyzed trust from the *qualitative* point of view: we have carefully discussed the cognitive ingredients of trust, their relationships and formalizations. Before going on (describing trust dynamics, trust sources, trust generalization, etc.) we have to evaluate, understand, describe and formalize the *quantitative* nature of the trust concept.

It is true that when *Xania* decides to trust *Yody*, she has considered the different aspects of trustworthiness, like ability, willingness, context, and so on, and the various reasons and causes these aspects are based on. But, at the same time, she has also evaluated their amount: if the weight of each element is enough, if the quantity of their complete composition (also considering potential overlapping and interferences) can be evaluated as sufficient for trusting *Yody*. Every day we participate in discussions where judgments like: 'John is really trustworthy, because he knows his work and is very competent and serious with the customers', are expressed, where quantitative evaluations are expressed in an approximate, colloquial way: what does it mean 'really', 'very'? How much does know about John his work? How much is serious? How much are we sure about this? In fact, directly or indirectly, there is always a judgment of quantification over these properties. And we always test these quantifications: 'How much do you trust him?', 'Is your trust in him sufficient?', 'Is he so trustworthy?' 'Are you sure?', and so on.

So although the qualitative analysis of the trust components is fundamental for getting the real sense of trust concept, the quantification of its ingredients and an adequate composition of them will permit the results of its application to be effectively evaluated and simulated.

3.1 Degrees of Trust: a Principled Quantification of Trust

The idea that trust is measurable is usual (in common sense, in social sciences, in AI (Snijders, 1996), (Marsh, 1994)). In fact, in the majority of the approaches to the trust study the quantification aspect emerges and prevails over the qualitative and more analytic aspects (that are considered less relevant and sometimes useless). Because of this, in these approaches

Trust Theory   Cristiano Castelfranchi and Rino Falcone
© 2010 John Wiley & Sons, Ltd

no real definition and cognitive characterization of trust is given; so, the quantification of trust is quite *ad hoc* and arbitrary, and the introduction of this notion or predicate results in being semantically empty.[1]

On the contrary, in our studies we try to understand and define the relationships between the cognitive definition of trust, its mental ingredients, and, on the one hand, its value. On the other hand, its social functions and its affective aspects (Chapter 5). More precisely the latter are based on the former.

In this chapter we will show our efforts to ground the degree of trust of X in Y in the cognitive components of X's mental state of trust.[2] In particular, given our belief and evaluation based model, we predict and claim that *the degree of trust is a function:*

- *on the one hand, of the estimated degree of the ascribed 'quality' of Y on which the positive expectation is based;*
- *on the other hand, it is a function of the subjective certainty of the pertinent beliefs.*

Let us be more specific: the first component describes the quantitative level of Y's quality under analysis: for example, if X is evaluating Y's *ability* (about a given task τ) she has to select among different discrete or continuous values the one (or ones) she considers the more adequate to attribute to Y. These values could be either directly numerical or described by linguistic categories referable to a set of numerical attributions ('very good', 'good', 'sufficient', 'poor', just to give some examples). X could have, for example, a main (prevalent) belief that Y is either with ability 0.7 or 0.8 (in the scale $(0,1)$) and a secondary (less relevant) belief that Y is not so able (ability included in the 0.2-0.3 interval). See Figure 3.1 as an example.

At the same time, X also has a *meta-belief*,[3] about the subjective certainty of these beliefs (the second component indicated above): how much is X sure of her evalutative beliefs about Y's quality?

These meta-beliefs in fact translate the strength of the reasons that produced the first-level beliefs. Are they based on a consistent (or just superficial) set of experiences, reasoning, facts, deductions, a priori judgments, and so on? There is, of course, a correlation between the construction of the first kind of belief and the building of the second kind, but we can distinguish (and it is useful to do this for analytical reasons) between the different semantic and functional roles of the two categories. In any case, for simplicity, in the following part of the chapter we will consider the integration of beliefs and meta-beliefs.

We will use the degree of trust to formalize a rational basis for the decision of relying and betting on Y. We will also consider – for the 'decision' to trust – the quantitative aspect of another basic cognitive ingredient: *the value or importance or utility of the goal g_X.*

As we said, trust always implies risks, and frequently 'perceived' (evaluated) risks (Section 2.8.2). So we will also introduce *the evaluation of the risk* (depending on the potential

---

[1] As reported also in other parts of this book Williamson (Williamson, 1993) claims that 'trust' is an empty and superfluous notion – used by sociologists jn a rhetorical way – since it is simply reducible to subjective probability/risk (Chapter 8).

[2] Precisely the 'affective' trust in part represents an exception to that; since its 'degree' is not based on arguable 'reasons' but it is due to the 'intensity' of the feeling or of the evoked 'somatic markers'.

[3] An 'explicit' belief about other beliefs, or an 'implicit' one; that is, just some index of the belief strength or certainty; like the one we have introduced in the analysis of 'expectations' and of their 'strength' (Chapter 2).

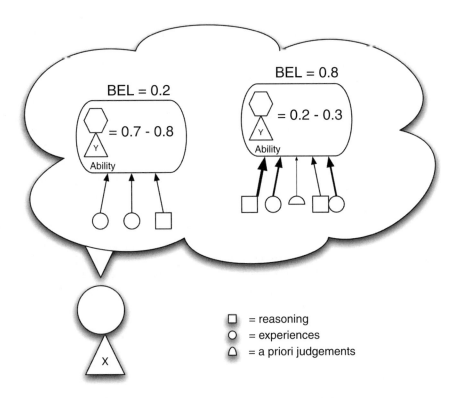

Figure 3.1   Quantifying *X*'s Beliefs about *Y*'s ability

failures of the various components of trust). Finally, we will introduce different quantitative thresholds linked with the different quantitative dimensions included in our cognitive model.

In sum, *the quantitative dimensions of trust are based on the quantitative dimensions of its cognitive constituents*. For us, trust is not an arbitrary index just with an operational importance, without a real content, or a mere statistical result; but it is derived from the quantitative dimensions of the pertinent beliefs.

3.2   Relationships between Trust in Beliefs and Trust in Action and Delegation

The solution we propose is not an *ad hoc* solution, just to ground some degree of trust. It instanciates a general claim. Pears (Pears, 1971) points out the relation between the *level of confidence* in a belief and the likelihood of a person *taking action based on the belief*: 'Think of the person who makes a true statement based on adequate reasons, but does not feel confident that it is true. Obviously, *he is much less likely to act on it, and, in the extreme case of lack of confidence, would not act on it*' (p. 15) (We stressed the terms clearly related to theory of trust).

'It is commonly accepted that people behave in accordance with their knowledge' (notice that this is precisely our definition of a *cognitive agent* but it would be better to use the word 'beliefs' instead of 'knowledge'). '*The more certain the knowledge then the more likely, more rapid and more reliable is the response.* If a person strongly believes something to be correct which is, in fact, incorrect, then the performance of the tasks which <u>rely on</u> this erroneous belief or misinformation will likewise be in error – even though the response may be executed rapidly and with confidence.' (p. 8). (Hunt and Hassmen, 1997). [4]

Thus under our foundation of the degree of trust there is a general principle: Agents act depending on what they believe, i.e. *relying on* their beliefs. And they act on the basis of the degree of reliability and certainty they attribute to their beliefs. In other words, trust/confidence in an action or plan (reasons to choose it and expectations of success) is grounded on and derives from trust/confidence in the related beliefs.

The case of trust in delegated tools or agents is just a consequence of this general principle in cognitive agents. Also, beliefs are something one bets and risks on, when one decides to base one's action on them, although, frequently, without an explicit deliberation about this, but with a procedural and implicit assumption about their reliability. Chosen actions too are something one bets, relies, counts on and depends upon. We trust our beliefs, we trust our actions, we trust delegated tools and agents. In an uncertain world any single action would be impossible without some form of trust (Luhmann, 1990).

3.3   A Belief-Based *Degree of Trust*

Let's call the degree of trust of X in Y about τ:

$$DoT_{XY\tau} \tag{3.1}$$

with $0 \leq DoT_{XY\tau} \leq 1$, where $DoT_{XY\tau} = 0$ means absolutely no trust, and $DoT_{XY\tau} = 1$ means full trust (in fact, a sort of *faith*): these two values are in fact two asymptotic limits (they are contradictory to the definition of trust as always including some risk:[5] although, strictly subjectively speaking, the risk might be ignored).

As described in Section 3.1, we can distinguish between evaluative beliefs (about the qualities of the trustee or its contextual environment) and meta-beliefs (how much the trustor is sure about that evaluative belief). Suppose X is considering Y's ability (Ab_Y, with $0 \leq Ab_Y \leq 1$) about the task τ. For different reasons (direct experiences, reasoning about categories in which Y is included, and so on) X could have several values to attribute to Ab_Y ($Ab_{1Y} = 0.2$, $Ab_{2Y} = 0.4$, $Ab_{3Y} = 0.8$). Each of these possibilities has different strengths (suppose $Bel_X(Ab_{1Y}) = 0.7$, $Bel_X(Ab_{2Y}) = 0.5$, $Bel_X(Ab_{3Y}) = 0.4$, respectively) (see Figure 3.2).

Imagine, for example (see Figure 3.2), the case in which *Xania* observed *Yody* carrying out a specific task (many years ago: so the strength of the source is not that high (0.40)) performing rather well (0.80), in addition, someone not so reliable as source (0.50) informed her about a

---

[4] This correct view, is just incomplete; it ignores the dialectic, circular, relationships between action and beliefs: a successful action – based on certain assumptions – automatically and unconsciously reinforces, 'confirms' those beliefs; an unsuccessful action, a failure, arouses our attention (surprise) about the implicit assumptions, and casts some doubt over some of the grounding beliefs.

[5] In the case of $DoT_{XY\tau} = 1$ we would have no risks, the full certainty of Y's success. In the case of $DoT_{XY\tau} = 0$ we would have absolute certainty of Y's failure (analogously no risks, that subsume uncertainty).

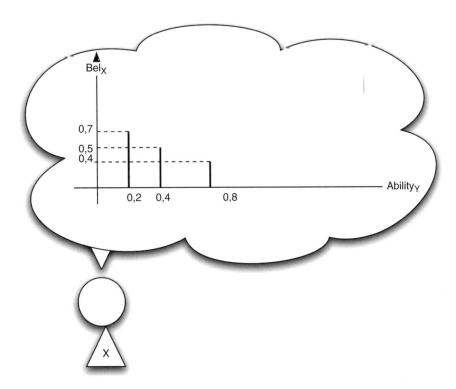

Figure 3.2   Relationships among different evalutative Beliefs and their strenghts

modest ability of *Yody* (0.40) and at the same time *Xania* believes (with high certainty (0.70)) *Yody* belongs to a category which, generally (for its own features), performs very badly (0.20) on those kinds of task.

At this stage of the analysis it is also possible to have concurrent hypotheses with similar values (the sum should not necessarily be equal to *1*: again, they are not coherently integrated but just singularly evaluated). The meta-belief (we can also consider it the strength of this belief) on *Y*'s ability has to consider the different beliefs about this quality[6] and, on the basis of a set of constraints (see later), define the degree of credibility *DoC* of the specific quality.

In general, we could write that:

$$DoC_X(Qual\text{-}i_{(s1,...sn),Y}(\tau)) = F_{X,Y,\tau}(Bel_X(Str_1 Qual\text{-}i_{s1Y}(\tau)),$$
$$Bel_X(Str_2 Qual\text{-}i_{s2Y}(\tau)), \ldots, Bel_X(Str_n Qual\text{-}i_{snY}(\tau))) \tag{3.2}$$

In other words, the degree of credibility for *X* (*DoC_X*) about the *i*-th *Quality of Y* on the task τ (*Qual_{iY}(\tau)*) on the basis of the *n* belief sources *(S_1, ..., S_n)* is *a function (depending on X, Y*

---

[6] Here we are considering *Y*'s ability as a quality of *Y*. We have used the term quality in Chapter 2 at a more detailed level: the set of features characterizing an entity. In fact *Y*'s ability might be considered as a meta-quality (composed by a set of more specific qualities).

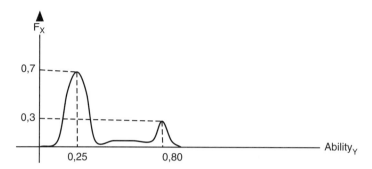

Figure 3.3 Probabilities for two different values of *Y*'s ability to perform the task τ (as output of the function *F*)

and τ) *of the different strengths of the belief sources (in X's point of view: Str₁, Str₂,. . ., Strₙ) that weigh the different contents (Qual-i_{s1Y}(τ), Qual-i_{s2Y}(τ), . . ., Qual-i_{snY}(τ)).*

$F_{X,Y,\tau}$ is a *selective and normalizing function* that associates couples *quality-values/ strengths-of-beliefs* with a probability curve (with all the constraints the probability model introduces).

In clearer terms, $F_{X,Y,\tau}$ should produce a matrix with *n* rows and two columns, where the number of rows corresponds with the *number* of *quality-values* that have to be taken into consideration by the selective process (not necessarily all the values reported by the belief sources[7]) while the two columns represent the *contents* of the *quality-values* and their *normalized probabilities* respectively (the sum of the probabilities in the second column must be equal to *1*).

As an example we can show how, starting from Figure 3.2, $F_{X,Y,\tau}$ by selection (for example the values *0.2* and *0.4* are collapsed in *0.25*) and by normalization (producing *0.7* and *0.3* as probabilities values) gives the output of the matrix:

$$\begin{pmatrix} 0.25 & 0.7 \\ 0.80 & 0.3 \end{pmatrix}$$

or, to see this in a clear visual form see the graph in Figure 3.3

In the following, for the sake of simplicity, $DoC_X(Qual-i_{(s1,...sn),Y}(\tau))$ will give as a result the more probable value of that quality weighted on the basis of its strength and of the probability of the other results and strengths. In other words, *we reduce the function F to a single number averaging the complexity of the different values, sources and strengths.*

Given that we postulate that the degree of trust basically is a function of the 'strength' of the trusting beliefs, i.e. of their *credibility* (expressing both the subjective probability of the fact and trust in the belief): the stronger *X*'s belief in *Y*'s competence and performance, the stronger *X*'s trust in *Y*.

---

[7] In some cases maybe there is an integration of some of these values.

Assuming that the various credibility degrees are independent from each other, we can say (for simplicity we consider $p=g_X$):

$$DoT_{XY_\tau} = C_{opp} DoC_X[Opp_Y(\alpha, p)]^* C_{ab} \ DoC_X[Ability_Y(\alpha)]^*$$
$$\times C_{will} \ DoC_X[WillDo_Y(\alpha, p)] \qquad (3.3)$$

where:

- C_{opp}, C_{ab}, and C_{will} are constant values and represent the weights of the different credibility terms[8]: they take into account the variable *relevance* or *importance* of the different components of Y's trustworthiness. Depending on the kind of task, on X's personality, etc, Y's *competence* or Y's *reliability* do not have equal impact on his global trustworthiness for task τ, and on X's decision.

If, for example, τ is quite a technical task (like repairing an engine, or a surgical intervention) Y's competence is more important, and its *weight* in the evaluation and decision is more determinant; if τ is not technically demanding but its deadline is very important then Y's punctuality or reliability is more relevant than his competence or skills.

- $DoC_X[Opp_Y(\alpha,p)]$, is the degree of credibility of X's beliefs (for X herself) about Y's opportunity of performing α to realize p; in more simple words, it takes into account all the contextual factors in which Y is considered to act.
- $DoC_X[Ability_Y(\alpha)]$, the degree of credibility of X's beliefs (for X herself) about Y's ability/ competence to perform α;
- $DoC_X[WillDo_Y(\alpha,p)]$, the degree of credibility of X's beliefs (for X herself) about Y's actual performance.

In a case in which Y is a cognitive agent, the last degree $(DoC_X[WillDo_Y(\alpha,p)])$ will become:

$$DoC_X[WillDo_Y(\alpha, p)] = DoC_X[Intend_Y(\alpha, p)]^* DoC_X[Persist_Y(\alpha, p)]$$

and can be interpreted as the degree of credibility of X's beliefs (for X herself) about Y's willingness to actually perform α to realize p; where the willingness can be split in the composition of intention and persistence.

Finally, of course:

$$0 \leq DoC_X[Opp_Y(\alpha, p)] \leq 1; \ 0 \leq DoC_X[Ability_Y(\alpha)] \leq 1;$$
$$0 \leq DoC_X[WillDo_Y(\alpha, p)] \leq 1.$$

3.4 To Trust or Not to Trust: Degrees of Trust and Decision to Trust

In this paragraph we analyze the complex process of taking a real (reason-based) decision about trusting or not, on the basis of the mental ingredients described in Chapter 2 and of their quantitative values.

---

[8] In fact the role of these factors would be more complex than simple constant values, they should represent the set of non linear phenomena like saturation effects, possible interference among the different credibility degrees, and so on.

Resuming the trustor's mental state we have two main subsets:

(a) A set of mental states $(MS\text{-}CT_{X,Y})$ -called *Core Trust-* with these components:
   - a set of X's goals and, in particular, one specific of them (g_X) in order to trust Y;
   - a set of X's competence beliefs $(B\text{-}Com_{X,Y})$ on Y about τ;
   - a set of X's disposition beliefs $(B\text{-}Dis_{X,Y})$ on Y about τ and
   - a set of X's practical opportunities beliefs $(B\text{-}PrOp_{X,Y})$ on Y about τ at that given moment (time) and site (space).
(b) A set of mental states $(MS\text{-}REL_{X,Y})$ -called *Reliance-* that must be added to the 'core trust' ones and that are strictly linked with the decision to trust; in particular:
   - a set of X's dependence beliefs $(B\text{-}Dep_{X,Y})$ (it is needed or it is better to delegate than not delegate to Y ((Sichman *et al.*, 1994), (Jennings, 1993)) and
   - a set of X's preference beliefs $(B\text{-}Pref_{X,Y})$ for delegating to Y (in fact, although this notion is related to the dependence notion, we like to mark it).

We can imagine that each one of the above listed beliefs will have a specific value.

In order that X trusts Y about τ, and thus delegates that task to Y, it is not only necessary that the $DoT_{XY\tau}$ exceeds a given threshold (depending on X, Y and the task τ), but also that *it constitutes the best solution* (compared with other possible and practicable solutions).

In any circumstance, an agent X endowed with a given goal, has three main choices:[9]

i) to try to achieve the goal by itself;
ii) to delegate the achievement of that goal to another agent Y;
iii) to do nothing (relative to this goal).

So we should consider the following abstract scenario (Figure 3.4)
where we call:

$U(X)$, the agent X's utility function, and specifically:

$U(X)_{p^+}$, the utility of X's success performance (directly realized by agent X);

$U(X)_{p^-}$, the utility of X's failure performance (directly realized by agent X);

$U(X)_{d^+}$, the utility of a successful delegation (utility due to the success of the delegated action to Y);

$U(X)_{d^-}$ the utility of a failure delegation (damage due to the failure of the delegated action to Y);

$U(X)_0$ the utility of doing nothing.

---

[9] The choice of collaborating on a given goal implies the agreed delegation of a subgoal (of a subpart of the main goal) to Y: that means to apply the choice number (ii).

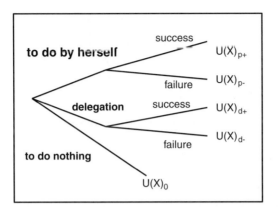

Figure 3.4 Decision Tree for Trust-based Delegation. (Reproduced with kind permission of Springer Science+Business Media © 2001)

One should also consider that trust (the attitude and/or the act) may have a value *per se*, independently on the achieved results of the 'delegated' task ((McLeod, 2006) *Stanford Encyclopedia*). This value is taken into account in the decision. For example, if to have/show/put trust in Y is a positive thing (for the subject or for the social environment X cares about) this 'value' (the satisfaction of this additional goal) should be included among the results (outcomes) of the decision to trust Y; and it might be determinant for the choice, even winning against worries and doubts. Vice versa, if trust is a negative fact (for example, a sign of naivety, of weakness of character, etc., of stupidity) this effect too will be taken into account in the decision.

Among the positive results of the act of trusting (successful or unsuccessful; see Figure 3.4) X will put the value of trusting *in se* and *per se*; in the opposite case, among the positive results of the act of trusting (successful or unsuccessful) X will put the negative value (cost, harm) of trusting *in se* and *per se*.

However, for the sake of simplicity, we will consider the following scenario (Figure 3.5): In the simplified scenario, in order to delegate we must have (using the Expected Utility Theory approach by Bernoulli and von Neumann and Morgenstern (von Neumann and Morgenstern, 1944)):

$$DoT_{XY\tau}\,^*U(X)_{d+} + (1\text{-}DoT_{XY\tau})U(X)_{d-} > DoT_{XX\tau}\,^*U(X)_{p+} + (1\text{-}DoT_{XX\tau})\,U(X)_{p-}$$
$$(3.4)$$

where $DoT_{XX\tau}$ is the *selftrust* of X about τ.

Analyzing more carefully the different kinds of utilities we can say that:

$U(X)_{p+} = Value(g) + Cost\ [Performance(X\ \tau)],$

$U(X)_{p-} = Cost\ [Performance(X\ \tau)] + Additional\ Damage\ for\ failure$

$U(X)_{d+} = Value(g) + Cost\ [Delegation(X\ Y\ \tau)],$

$U(X)_{d-} = Cost\ [Delegation(X\ Y\ \tau)] + Additional\ Damage\ for\ failure$

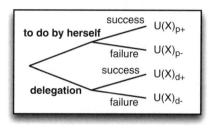

Figure 3.5   Simplified Decision Tree for Trust-based Delegation

where it is supposed that it is possible to attribute a quantitative value (importance) to the goals and where the values of the actions (delegation and performance) are supposed to be negative (having costs: energies and resources committed in the enterprise).

We have called *Additional Damage for Failure* the negative results of both the failures (direct performance and delegation): they represent not only the frustration of the missed achievement of the goal but also the potential additional damages coming from these failures (in terms of compromising other goals, interests, resources, and so on): in particular, in the case of trusting Y there are all the risks of reduced control in the situation following the delegation act.

From the formula (3.4) we obtain:

$$DoT_{XY\tau} > DoT_{XX\tau} \, {}^{*}A + B \qquad\qquad (3.5)$$

where:

$A = (U(X)_{p^+} - U(X)_{p^-}) / (U(X)_{d^+} - U(X)_{d^-})$
$B = (U(X)_{p^-} - U(X)_{d^-}) / (U(X)_{d^+} - U(X)_{d^-})$

Let us consider now, the two terms A and B separately.

As for term A, (considering $B=0$, that means that the trustor has the same damage in terms of cost in the case of the failure of the direct performance and in the case of the failure of action delegated to Y) if:

- $A=1$ (the differences between success utility and failure utility in the case of X's direct performance and in the case of delegation to Y are the same):

then the formula (3.5) becames, $DoT_{XY\tau} > DoT_{XX\tau}$ in practice, *in this case (B=0 and A=1) it is necessary more trust in Y than in herself for X's delegation.*
This (A=1, B=0) is a clear and rational result: in the case in which:

1) *the utility's difference between success and failure is the same (in performing the task by themselves or delegating the task to the trustee)(A=1); and*
2) *the failure of X's performance has the same damage of the failure of the Y's delegated performance; then*
   X has to trust Y more than herself for delegating to him that task: for X is rational to delegate to Y for any value greater than zero between these two trust values ($DoT_{XY\tau}$- $DoT_{XX\tau}$ >0).

- case $A>1$:

$$U(X)_{p^+} - U(X)_{p^-} > U(X)_{d^+} - U(X)_d$$

then $A^*DoT_{XX\tau} > DoT_{XX\tau}$

i.e. if the difference between the utility of the success and the utility of the failure in delegation is smaller than the difference between the utility of the success and the utility of the failure in the direct performance of X, then (for the term A) in order to delegate return *the trust of X in Y must be bigger than the selftrust of X* (about τ).

More precisely, suppose two subcases:

(i) In the case of $DoT_{XY\tau} > (DoT_{XX\tau} * a + B)$ *(with $B=0$ and $1<a<A$, $A>1$) even if the X's degree of trust in Y is greater than the selftrust, it is not sufficient for delegation.*
(ii) In the case of $DoT_{XY\tau} > (DoT_{XX\tau} * a + B)$ *(with $B=0$ and $a \geq A$, $A>1$), the X's degree of trust in Y is sufficient for delegating.*

In other words, for delegating X's trust in Y has to be greater thnt X's selftrust at least of the value given from A.

Also this ($A>1$, $B=0$) is a clear and rational result: in the case in which:

1) *the difference between the utility of the success and the utility of the failure in direct performance of X is greater than the difference between the utility of the success and the utility of the failure in delegation to Y, and*
2) *the failure of X's performance has the same damage of the failure of Y's delegated performance; then*
   X has to trust Y more than herself for delegating to him that task: the difference in the trust values has to be at least of a factor given from A.

Vice versa, if

- $A<1$:

$$U(X)_{p^+} - U(X)_{p^-} < U(X)_{d^+} - U(X)_{d^-}$$

then $A^*DoT_{XX\tau} < DoT_{XX\tau}$

i.e., if the difference between the utility of the success and the utility of the failure in delegation is bigger than the difference between the utility of the success and the utility of the failure in X's direct performance, then (given $B=0$) in order to delegate

the trust of X in Y should be smaller (of A factor) than the X's selftrust (about τ).

Even in this case we have to distinguish two alternative subcases:

(i) In the case of $DoT_{XY\tau} > (DoT_{XX\tau} * a + B)$ *(with $B=0$ and $0<a \leq A$, $A<1$) X's degree of trust in Y is sufficient for delegating.*

(ii) In the case of $DoT_{XY\tau} > (DoT_{XX\tau} * a + B)$ *(with B=0 and* A<a<1, A<1) *X's degree of trust in Y is not sufficient for delegation.*

Note that even if in both the cases *X's trust in Y is smaller than X's selftrust, only in one of them is* it possible to delegate: so we can say that (for utilities reasons) *it is possible to delegate to agents which X trusts less than herself*.
  Considering now also the term B $(B \neq 0)$,

- If $U(X)_{p^-} - U(X)_{d^-} > 0$, then a positive term is added to the A: $A + B > A$,
  i.e., *if the utility of the failure in case of X's direct performance is bigger than the utility of the failure in case of delegation, then - in order to delegate - the trust of X in Y about τ must be greater than in the case in which the right part of (3.5) is constituted by A alone.*
  Vice versa,
- If $U(X)_{p^-} - U(X)_{d^-} < 0$, then $A + B < A$,
  i.e., *if the utility of the failure in the case of non-delegating is smaller than the utility of the failure in the case of delegation, then – in order to delegate – the trust of X in Y about τ must be smaller than in the case in which the right part of the formula (3.5) is constituted by just A alone.*[10]

Since $DoT_{XY\tau} \leq 1$, from the formula (3.5) we can obtain (starting from $1 > DoT_{XX\tau}*A+B$):

$$DoT_{XX\tau} < (U(X)_{d^+} - U(X)_{p^-})/(U(X)_{p^+} - U(X)_{p^-}) \qquad (3.6)$$

From the formula (3.6) we have two consequences in the dynamics of trust; to delegate X to Y the task τ, as the selftrust $(DoT_{XX\tau})$ grows either:

1) the difference between the utility of the success in delegation and the utility of the failure in the direct performance increases; or
2) it reduces the difference between the utility of the success and of the failure in direct performance.

Because $DoT_{XX\tau} \geq 0$, from (3.6) we obtain (starting from $0 < (U(X)_{d^+} - U(X)_{p^-})/(U(X)_{p^+} - U(X)_{p^-})$:

$$U(X)_{d^+} > U(X)_{p^-} \qquad (3.7)$$

(consider that for definition we have $U(X)_{p^+} > U(X)_{p^-}$).
  In practice, for delegating, a necessary (but not sufficient) condition is that the utility of the success in delegation is greater than the utility of the failure in direct performance *(as intuitively rational).*
  Let us conclude this section by underlining the fact that (in our model and in real life) *we do not necessarily delegate to the most trustworthy agent*; we do not necessarily choose the alternative where trust is greater. We might prefer to choose a partner or to rely on a device that is not the most reliable one, simply because there are other parameters involved in our

---

[10] Both for A and B there is a normalization factor $(U(X)_d{}^+ - U(X)_d{}^-)$: the more its value increases, the more the importance of the terms is reduced.

decision to delegate (to 'trust', as action): costs, risks, utility and so on. For example, the most competent and trustworthy doctor might be the most expensive or not immediately available (because very busy). In this case, we could delegate a less competent and cheaper one.

Another important thing to be underlined is that the complete scenario of Figure 3.4, with all its branches and precise pros and cons of each alternative, is an ideal situation just rarely effectively evaluated (by humans) in real life (and often also in artificially simulated scenarios): it is more a normative model of a trust decision. However, increasing the importance of the goal to be achieved, the risks of potential damages, and the time for the decision, the choice of enquiring the potential alternative branches becomes a necessity beyond the available and achievable information: it is the paradigmatic scenario upon which trust reasoning must be based.

3.5   Positive Trust is not Enough: a Variable Threshold for Risk Acceptance/Avoidance

As we saw, *the decision to trust is based on some positive trust*, i.e. on some evaluation and expectation about the capability and willingness of the trustee and the probability of success. And on the necessity/opportunity/preference of this delegation act.

First of all, those beliefs can be well justified, warranted and based on reasons. This represents the 'rational' (reasons based) part of the trust in Y. But they can also be unwarranted, not based on evidence, even quite irrational, or intuitive (based on sensations or feelings), faithful. We call this part of the trust in Y: 'faith'.[11]

Notice that irrationality in trust decision can derive from these unjustified beliefs, i.e. on the ratio of mere faith.

Second, *positive trust is not enough* to account for the decision to trust/delegate. We do not distinguish in this book the different role or impact of the rational and irrational part of our trust or positive expectations about Y's action: the entire positive trust (reason-based + faithful) is necessary and contributes to the *Degree of Trust*: its sum should be greater than discouraging factors. We do not go deeply in this distinction (a part of the problem of rational Vs irrational trust) also because we are interested here in the additional fact that this (grounded or ungrounded) positive expectation can not be enough to explain the *decision/act* of trusting. In fact, another aspect is necessarily involved in this decision: the decision to trust/delegate necessarily implies *the acceptance of some risk*. A trusting agent is a risk-acceptant agent, either consciously or unconsciously. Trust is never certainty: always it retains some uncertainty (ignorance)[12] and some probability of failure, and the agent must accept this and be willing to run such a risk (see Chapter 2) with both positive and negative expectations, and the fact that they can remain just 'implicit' or 'potential'.

Thus a fundamental component of our decision to trust Y, is our acceptance and felt exposition to a risk. Risk is represented in previous quantification of *DoT* and in the criteria

---

[11] To be more precise, non-rational blind trust is close to faith. Faith is more than trust without evidence, it is trust without the need for and the search for evidence, and even against evidence.

[12] We do not want to introduce here a more sophisticated model where 'ignorance' and 'uncertainty' are explicitly represented and distinct from probability; like the Dempster & Shafer model ((Dempster, 1968), (Shafer, 1976)) and the theory of 'plausibility'. We use (less formally) this more sophisticated model in other chapters to explain trust 'optimism' and other aspects.

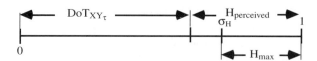

Figure 3.6 Degree of Trust and Hazard Threshold: the case of the Delegation Branch. (Reproduced with kind permission of Springer Science+Business Media © 2001)

for decision. However, we believe that this is not enough. A specific risk policy seems necessary for the trust decision and bet; and we should aim to capture this aspect explicitly.

The equation (3.4) – that basically follows classical decision theory – introduces the degree of trust instead of a simple probability factor. In this way, it permits one to evaluate when to delegate rather than to do it herself in a rigid, rational way. The importance of this equation is to establish what decision branch is the best on the basis of both the relative (success and failure) utilities for each branch and the probability (trust based) of each of them. In this equation no factor can play a role independently from the others. Unfortunately, in several situations and contexts, not just for the human decision makers but – we think – also for good artificial decision makers, it is important to consider the absolute values of some parameter independently from the values of the others. This fact suggests that some saturation-based mechanism, or threshold, by which to influence the decision, needs to be introduced.

For example, it is possible that the value of the damage *per se* (in case of failure) is too high to choose a given decision branch, and this is independent either from the probability of the failure (even if it is very low) or from the possible payoff (even if it is very high). In other words, that danger might seem to the agent an intolerable risk. In this paragraph we analyze (just in a qualitative way) different possible threshold factors that must play an additional role when choosing between alternatives like in Figure 3.5.

First, let us assume that each choice implies a given failure probability as perceived by X (and let's call this: 'hazard' or 'danger'), and a given 'threat' or 'damage': i.e. a negative utility due to both the failure (the cost of a wasted activity and a missed reward) and the possible additional damages.[13]

Second, we assume that X is disposed to accept a maximum hazard (*Hmax*) in its choices, in a given domain and situation. In other words, *there is a 'hazard' threshold over which X is not disposed to pursue that choice.*

We are considering the case of delegation branch ($DoT_{XY\tau}$, $U(X)_{d^-}$, $U(X)_{d^+}$), but the same concepts are valid in the case of X's performance (substituting $DoT_{XX\tau}$, $U(X)_{p^-}$, $U(X)_{p^+}$). In Figure 3.6 we have:

$H_{perceived}$ is the failure hazard perceived by X;

H_{max} is the maximum failure hazard acceptable by X;

σ_H is the hazard threshold.

---

[13] Thus here we will use the term 'risk' as the result of the entity of losses (damage or threat) and of its probability (hazard or danger). Risk theory (Kaplan and Garrik, 1980) calculates the risk as the product of uncertainty (subjective probability) and damage; other authors propose – for the objective risk – the product of frequency and magnitude of the danger. We are interested in the subjective dimension, so risk should be in our terminology hazard * damage. (Common sense would prefer to call 'risk' the probability, and 'danger' the global result of probability and damage).

To choose a given path it is necessary that:

$$DoT_{XY\tau} \geq \sigma_H = (1 - H_{max})$$

We claim that such a threshold can vary, not only from one agent to another (*personality*) but also depending on several factors in the same agent. In particular, we claim that the acceptable hazard varies with the importance of the *threat-damage* and with the expected reward. In other words, σ_H (where $0 \leq \sigma_H \leq 1$) is a function of both $(U(X)_{d-})$ and $(U(X)_{d+})$: $\sigma_H = f(U(X)_{d-}, U(X)_{d+})$.

More precisely: the greater the damage $(U(X)_{d-})$ the more it grows σ_H; while the greater the utility of the potential achievements $(U(X)_{d+})$ the more σ_H is reduced.

Moreover, we may also introduce an *'acceptable damage'* threshold σ_d: it fixes the limit of the damage X can endure. Under this value the choice would be regarded as unacceptable.

We have also introduced a *minimal acceptable value* for $U(X)_{d+}$ (σ_a, *payoff threshold*): under this value the choice would be considered inconvenient.

The function σ_H is such that when $U(X)_{d-}$ is equal (or lesser) than σ_d then σ_H is equal to 1 (in practice, that choice is impossible).

At the same time we can say that when $U(X)_{d+}$ is equal (or lesser) than σ_a then σ_H is equal to 1. For each agent both σ_d and σ_a can assume different values.

One might also have one single dimension and threshold for *risk* (by using the formula 'damage * hazard'). However, we claim that there could be different heuristics for coping with risk (this is certainly true for human agents). For us, a great damage with a small probability and a small damage with a high probability do not necessarily represent two equivalent risks. They can lead to different decisions, they can pass or not pass the threshold.

To go back to the case of delegation branch (it is sufficient to substitute $U(X)_{d-}$ with $U(X)_{p-}$, $U(X)_{d+}$ with $U(X)_{p+}$, to obtain the case of X's performance branch) we have:

$$\sigma_H = f(U(X)_{d-}, \ U(X)_{d+}) \text{ and in particular}$$
$$\sigma_H = 1 \text{ when } (U(X)_{d-} \leq \sigma_d) \text{ or } (U(X)_{d+} \leq \sigma_a).$$

In other words, we assume that *there is a risk threshold* – more precisely a *hazard* threshold depending also on a *damage* threshold – under which the agent refuses a given choice even if the equation (3.4) suggests that choice is the best. It might be that a choice is convenient (and the best) for the ratio between possible payoff, costs and risk, but that the risk *per se* is too high for that agent in that situation.

Let us consider an example (Figure 3.7):

Given $U(X)_{p+} = 10$, $U(X)_{p-} = 1$, $U(X)_{d+} = 50$, $U(X)_{d-} = 5$, and $Dot_{XY\tau} = Dot_{XX\tau} = 0.7$, the equation (3.5) is satisfied: $0.70 > (0.70 * 9/45) + (4/45) = 0.23$. So on the basis of this equation agent X should delegate the task τ to agent Y. However, suppose that the maximum acceptable damage for X is $\sigma_d = 4$ (the damage grows as the $U(X)_{d-}$ is reduced) then the choice to delegate is stopped from the saturation effect.

Vice versa, considering the example in Figure 3.8, with $Dot_{XY\tau} = 0.7$ and $Dot_{XX\tau} = 0.1$, the equation (3.4) is also satisfied: $0.7 > (0.1 * 10/11) + (6/11) = 0.63$. Also, again on the basis of this equation, agent X should delegate the task τ to agent Y. But if the *minimal acceptable value* for delegation is $\sigma_a = 18$, then the choice to delegate is stopped from the saturation effect because it is considered unconvenient.

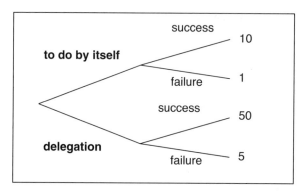

Figure 3.7   An Example in which the maximum acceptable damage is over the threshold. (Reproduced with kind permission of Springer Science+Business Media © 2001)

It is possible that all the branches in the decision scenario would be in a situation of saturation ($\sigma_H=1$). What choice does the agent have to make? In these cases there could be several different possibilities.

Let us consider the scenario in Figure 3.5. There could be at least four possibilities:

1) Saturation due to σ_d for branch 'to do by itself' (the potential damage is too high); saturation due to σ_a for branch 'delegation' (the payoff is too low).
2) Saturation due to σ_a for branch 'to do by itself' (the payoff is too low); saturation due to σ_d for branch 'delegation' (the potential damage is too high).
3) Saturation due to σ_a for branch 'to do by itself' (the payoff is too low); saturation due to σ_a for branch 'delegation' (the payoff is too low).
4) Saturation due to σ_d for branch 'to do by itself' (the potential damage is too high); saturation due to σ_d for branch 'delegation' (the potential damage is too high).
    *   In the cases (1) and (2) the choice could be the minimum damage (better a minor payoff than a high damage).
    *   In the case (3) if $(\sigma_{a^-}\, U(X)_{d^+}) < (\sigma_{a^-}\, U(X)_{p^+})$ then the choice will be 'to do by itself' and vice versa in the opposite case. In other words, given a payoff threshold greater than both the utilities (performance and delegation) the action with greater utility (although insufficient) will be better.

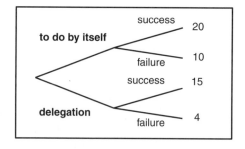

Figure 3.8   An Example in which the *minimal acceptable value* is under the threshold

- In the case (4) if $(U(X)_{d^+} - \sigma_d) > (U(X)_{p^+} - \sigma_d)$ then the choice will be 'to do by itself' and vice versa in the opposite case (the choice is the minor damage).
- In the cases (3) and (4) if $(\sigma_{a^-} U(X)_{d^+}) = (\sigma_{a^-} U(X)_{p^+})$ and $(U(X)_{d^+} - \sigma_d) = (U(X)_{p^+} - \sigma_d)$ then will the equation (3.5) be the one that will decide what is the right choice.

3.6 Generalizing the Trust Decision to a Set of Agents

It is possible to determine a trust choice starting from each combination of credibility degrees $- \{DoT_{Ag1, Agi, \tau}\}$ with $Ag_i \in \{Ag_1, \ldots, Ag_n\}$ – of the main beliefs included in Core-Trust and Reliance of Ag_1, and from a set of Ag_1's utilities $\{U_{p^+}, U_{p^-}, U_{di^+}, U_{di^-}, U_0\} = U(Ag_1)$, with $i \in \{2, \ldots, n\}$.

It is possible that – once fixed the set of utilities and the kind and degree of control – different combinations of credibility degrees of the main beliefs produce the same choice. However, in general, *changing the credibility degree of some beliefs should change the final choice about the delegation* (and the same holds for the utilities and for the control).

So, if we suppose we have a set constituted by: $\{DoT_{Ag1,Agi,\tau}\}$ and $U(Ag_1)$, we will have, as a consequence, the delegation $Kind_0Delegates(Ag_1 \ Ag_i \ \tau_0)$ with $Ag_i \in \{Ag_2, \ldots, Ag_n\}$, and $Kind_0Delegates \in \{Performs, Weak\text{-}Delegates, Mild\text{-}Delegates, Strong\text{-}Delegates, Nothing\}$[14].

At a different time we might have a new set $\{DoT_{Ag1,Agi,\tau}\}$ and/or a new set of utilities.

In Chapter 7 (about the delegation adjustments) we will see how, in order to adjust a given delegation/adoption, it is necessary for the agent to have specific reasons, that is new beliefs and goals. What, in fact, this means, is simply that:

- the delegator's mental state has changed in at least one of its components in such a way that the action to choose is different from the previous one; or
- the delegee's level of self-trust or the delegee's trust in the environment has changed, and there is some disagreement with *the delegator* about this.

At the same time the new sets of beliefs and utilities might suggest various possible strategies of recovery of the trust situation: i.e. given $Kind_0Delegates(Ag_1 \ Ag_i \ \tau)$, $DoT_{Ag1,Agi,\tau}$ and $U(Ag_1)$ we might have an adjustment of $Kind_0Delegates$ (for example from *Weak-Delegates* to *Strong-Delegates*).

This adjustment reflects a modification in the mental ingredients. More precisely, the trustor/delegator either *updates or revises* their delegation beliefs and goals, i.e.:

- either she revises its *core trust beliefs* about *the trustee/delegee* (the latter's goals, capabilities, opportunities, willingness);
- or she revises its *reliance beliefs* about: i) her dependence on *the trustee/delegee*, or ii) her preference to delegate to *trustee/delegee* than to do it herself, or to delegate to Ag_3 (a third agent) or to renounce the goal;
- or she changes her risk *policy* and more or less likely she accepts the estimated risk (this means that the trustor changes either her set of utilities ($U(Ag_1)$) or her set of thresholds.

---

[14] For the meaning of *Weak, Mild* and *Strong Delegation* see Chapter 2. *Perform* means that the trustor does not delegate but personally performs the task.

In other words, either Ag_1's trust of Ag_2 is the same but her preferences have changed (including her attitude towards risk), or Ag_1 has changed her evaluations and predictions about relying on Ag_2. Another important role is played by the control (see Chapter 7) that can allow delegation also to be given to a not very trusted agent. For an analysis on this relationship see (Castelfranchi and Falcone, 2000)).

3.7 When Trust Is Too Few or Too Much

Trust is not always rational or adaptive and profitable. Let's see when it is rational or irrational, and when it is not useful, although well grounded.

3.7.1 Rational Trust

In our view trust can be rational and can support rational decisions. *Trust as attitude* (*core Trust*) is epistemically rational when it is reason-based. When it is based on well motivated evidence and on good inferences, when its constitutive beliefs are well grounded (their credibility is correctly based on external and internal credible sources); when the evaluation is realistic and the esteem is justified, not mere faith.

The *decision/action of trusting* is rational when it is based on an epistemically rational attitude and on a sufficient degree relative to the perceived risk. If my expectation is well grounded and the degree of trust exceeds the perceived risk, my decision to trust is subjectively rational.[15]

To trust is indeed irrational either when the accepted risk is too high (relative to the degree of trust), or when trust is not based on good evidence, is not well supported. Either the faith component (unwarranted expectations) or the risk acceptance (blind trust) are too high.[16]

3.7.2 Over-Confidence and Over-Diffidence

Trust is not always good – also in cooperation and organization. It can be dangerous both for the individual and for the organization. In fact the consequences of over-confidence (the excess of trust) at the individual level are: reduced control actions; additional risks; non careful and non accurate action; distraction; delay in repair; possible partial or total failure, or additional cost for recovering.

The same is true in collective activity. But, what does 'over-confidence' i.e. excess of trust actually mean? In our model it means that X accepts too much risk or too much ignorance, or is not accurate in her evaluations. Noticed that there cannot be too much positive trust, or esteem of Y. It can be not well grounded and then badly placed: the actual risk is greater than the subjective one. Positive evaluation on Y (trust in Y) can be too much only in the sense that it is more than that reasonably needed for delegating to Y. In this case, X is too prudent and has searched for too much evidence and information. Since also knowledge has costs and

---

[15] For a more detailed discussion about rational and irrational motive for trust see Chapter 8, Section 8.1.

[16] Rational trust can be based not only on reasons and reasoning, on explicit evaluations and beliefs, but also on simple learning and experience. For example the prediction of the event or result can be based not on some understanding of the process or some model of it, but just on repeated experiences and associations.

utility, in this case the cost of the additional knowledge about Y exceeds its utility: X already has enough evidence to delegate. Only, in this case, the well-grounded trust in Y is 'too much'. But notice that we cannot call it 'over-confidence'.

In sum, there are three cases of 'too much trust':

- More positive trust in Y than necessary for delegating. It is not true that 'I trust Y too much' but it is the case that I need too much security and information than effectively necessary.
- I have more trust in Y than he deserves; part of my evaluations and expectations are faithful and unwarranted; I do not see or do not take into account the actual risk. This is a case of *over-confidence*. This is dangerous and irrational trust.
- My evaluation of Y is correct but I'm too risk prone; I accept too much ignorance and uncertainty, or I bet too much on a low probability. This is another case of *over-confidence*, and of dangerous and irrational trust.

Which are the consequences of over-confidence in delegation?

- Delegating to an unreliable or incompetent Y.
- Lack of control over Y (Y does not provide his service, or provides a bad service, etc.).
- Delegation which is too 'open': unchecked misunderstandings, Y's inability to plan or to choose, etc.

Which, on the contrary, are the consequences of insufficient confidence, of an excess of diffidence in delegation?

- We do not delegate and rely on good potential partners; we miss good opportunities; there is a reduction of exchanges and cooperation.
- We search and wait for too many evidences and proofs.
- We make too many controls, losing time and resources and creating interferences and conflicts.
- We specify the task/role too much without exploiting Y's competence, intelligence, or local information; we create too many rules and norms that interfere with a flexible and opportunistic solution.

So, some diffidence, some lack of trust, prudence and the awareness of being ignorant are obviously useful; but, also, trusting is at the same time useful. Which is the right ratio between trust and diffidence? Which is the right degree of trust?

- The right level of positive trust in Y (esteem) is when the marginal utility of the additional evidence on Y (its contribution for a rational decision) seems inferior to the cost for acquiring it (including time).
- The right degree of trust for delegating (betting) is when the risk that we accept in the case of failure is inferior to the expected subjective utility in the case of success (the equation is more complex since we have also to take into account alternative possible delegations or actions).

3.8   Conclusions

What is to be noticed in this chapter is how one can derive a precise model of the 'degree' of trust simply from independently postulated beliefs, expectations, evaluations, and their properties (like the 'certainty' of the belief, or the quantity of the quality or virtue). In general, in our model of cognition, the pursued goals, the intentions, are based on beliefs, on 'reasons'.

This is also why trust and trust decision can be the object of argumentation and persuasion: I can provide you with reasons for trusting or not trusting Y; I can convince you. Of course, trust can also be the result of mere suggestion, of manipulation, of attraction, and other affective maneuvers (see Chapter 5); but here we were modeling explicit and arguable trust.

It is also important to notice that the impact of such a trust degree in decision making is not just due to the 'expected utility'; the process is more complex: there are specific thresholds, there are differences between high probability and low value versus low probability and high value.

It is also important not to have simplistic models of trust degree in terms of mere statistics or reinforcement learning; or of trust decision in terms of delegating to the most trustful guy. An important additional sophistication we should have introduced – at least for modeling human trust – would be the asymmetric evaluation of gains (and missed gains) and of losses (and avoided losses), as explained by 'Prospect Theory': the same amount of money (for example) does not have a comparable impact on our decision when considered as acquisition and when considered as loss; and, as for losses, we are risk prone (we prefer uncertain losses to certain losses), while for winnings we are risk averse (prefer certain winnings to uncertain ones) (Allais, 1953).

In sum, trust (as *attitude and disposition*) is *graded* for seven different reasons:

1. Because it is based on explicit beliefs (like 'evaluations') with their degree of subjective *certainty*, recursively due to trust in evidences and sources: on such a basis, X is more or less sure, convinced that, and so on.
2. Because it is based on implicit, felt 'beliefs': sensations, somatic markers, emotional activations, with their *intensity* and affective qualities (safety, worry, etc.); the functional equivalents of 'beliefs' and explicit evaluations.
3. Because those judgments are about Y's *qualities*, virtues, and they can be gradable: Y can be more or less skilled, or competent, or persistent, etc. In other words, trust is graded because *trustworthiness* is graded.
4. Because it is *multi-dimensional* (and trustworthiness too); and the global judgment or feeling is the combination of those dimensions.
5. Because it is relative to some Goal of X's, and goals have a '*value*': they are more or less important.
6. Because it is a prediction about a future event, and thus about a subjective *probability* of such an event.
7. Because it presupposes some *risks* (both failure, costs, and possible dangers), that might be perceived with some tangible amount and threshold.

As *decision and act*, trust can be more or less convinced and sure, but cannot really be graded, since X has to decide or not, given some threshold of risk acceptance and convenience.

References

Allais, M. (1953) Le comportement de l'homme rationnel devant le risque: critique des postulats et axiomes de l'école Américaine, *Econometrica* 21, 503–546.

Castelfranchi, C. and Falcone R. (2000) Trust and control a dialectic link, *Applied Artificial Intelligence Journal*, Special Issue on Trust in Agents. 14 (8).

Dempster, Arthur P. (1968) *A Generalization of Bayesian Inference*, Journal of the Royal Statistical Society, Series B, Vol. 30, pp. 205–247.

Hunt, D. P., and Hassmen, P. (1997) *What it means to know something*. Reports from the Department of Psychology, Stockholm University, N. 835. Stockholm: University of Stockholm.

Jennings N. R. (1993) Commitments and conventions: The foundation of coordination in multi-agent systems. *The Knowledge Engineering Review*, 3: 223–250.

Kaplan, S. and Garrick, J. (1980) On the quantitative definition of risk. In *Risk Analysis*, 1 (1).

Luhmann, N. (1990) Familiarity, confidence, trust: Problems and alternatives. In D. Gambetta (ed.), *Trust* (Chapter 6, pp. 94–107). Oxford: Basil Blackwell.

Marsh, S.P. (1994) Formalising Trust as a computational concept. PhD thesis, University of Stirling. Available at: http://www.nr.no/abie/papers/TR133.pdf.

Pears, H. (1971) *What is Knowledge ?* New York, Harper and Row.

Sichman, J. R., Conte, C., Castelfranchi, Y., Demazeau. A social reasoning mechanism based on dependence networks. In *Proceedings of the 11th ECAI*, 1994.

Shafer, G. (1976) *A Mathematical Theory of Evidence*, Princeton University Press.

Snijders, C. and Keren, G. (1996) Determinants of Trust, Proceedings of the workshop in honor of Amnon Rapport, University of North Carolina at Chapel Hill, USA, 6-7 August.

von Neumann, John and Morgenstern, Oskar (1944) *Theory of Games and Economic Behavior*, Princeton University Press.

Williamson, O. E. Calculativeness, trust, and economic organization, *Journal of Law and Economics*, XXXVI: 453–486, April, 1993.

4

The Negative Side:
Lack of Trust, Implicit Trust,
Mistrust, Doubts and Diffidence

In this chapter we analyze the theory of lack of trust, mistrust, diffidence, and the nature of pessimism and optimism in 'giving credit' and entrusting, and also the implicit forms of trust, not based on explicit evaluations and specific beliefs (putting aside trust as feeling and affect – see Chapter 5). These are fundamental issues, not to be theoretically simplified or just reduced to a trivial technical trick.

4.1 From Lack of Trust to Diffidence: Not Simply a Matter of Degree

Between full trust and absolute mistrust there is not just a difference of quantity and a continuum from *0* (*complete lack of trust*) to *1* (*full trust*) (see Figure 4.1); like for scholars just considering 'trust' as subjective probability of a favorable event.

Neither is it a value between *−1* (negative 'trust', the complete mistrust) and *+1*; where *0* would be the simple lack of trust. As already argued by Ulmann-Margalit (Ulmann-Margalit, 2001), however, without a complete and formal systematization of the issue – *trust and its negative counterparts are qualitatively different mental states*.

There exist diverse forms and kinds of lack of trust, which are not just a matter of 'degree' or 'intensity', but must be analyzed in their specific ingredients.[1]

Actually we have to characterize *five* different states (*seven*, if we consider also 'diffidence' and '(not)giving credit') and complicated relations of conceptual or extensional inclusion or exclusion among them.

---

[1] Also because trust is in part based on the judgments about features and 'qualities', on evaluations. As beliefs – as we saw - they can have a 'degree' (the strength of subjective certainty), but as qualities or 'signals' not all of them are 'gradable': for example, 'PhD' or 'married' or 'nurse'.

Trust Theory   Cristiano Castelfranchi and Rino Falcone
© 2010 John Wiley & Sons, Ltd

Figure 4.1   From Mistrust to Full Trust

Let us put aside – for the moment – 'quantities' and the additional problem of 'uncertainty' and of giving or not 'credit' (see below). Let us consider the issue from the point of view of the theory of evaluation (given that – as we have shown – trust basically is a matter of 'evaluation') and of the theory of beliefs (given that explicit trust is grounded on a set of beliefs). In a belief-based approach 'mistrust' is not simply 0 trust or negative (-1) trust (Marsh, 2005), but it is the concrete presence of a negative evaluation about Y, with its possible supports, and/or of a negative expectation.

4.1.1   Mistrust as a Negative Evaluation

Let's start from the belief that 'Y is NOT reliable; one cannot trust him':

$$Bel_X \left(Not \left(Trustworthy \ Y \right) \right) \tag{4.1}$$

We will call this belief 'mistrust' (and its consequences): a specific judgment about Y, a relevant negative evaluation of Y, not simply a lack of relevant positive evaluations. If this negative evaluation/expectation is specific (for example: 'If we delegate him without controlling him, everything will be really indecent!') and possibly based on specific 'qualities' (for example: 'He is a dirty and untidy guy'), it is even clearer that we are not simply speaking of degrees of certainty and/or of a value in a continuum.

There are two kinds of negative evaluations (Miceli, 2000) (Castelfranchi, 2000).

- *Inadequacy evaluation*

  'Y is *not* able to, is *not* good, apt, useful, adequate, . . . for my goal';

  $$Bel_X \left(Not \left(Good\text{-}For \ Y \ g \right) \right) \wedge \left(Goal_X g \right)^2 \tag{4.2}$$

- *Nocivity/dangerosity evaluation*

  'Y is 'good' (powerful) but for bad things (contrary to my goal), it is noxious, dangerous'

  $$Bel_X \left(Good\text{-}For \ Y \ g' \right) \wedge \left(Goal_X \ Not(g') \right) \tag{4.3}$$

---

[2] *GoodFor* is the evaluative form/use of the concept of 'Power of' (Chapter 2).

Correspondingly, there are two kinds of 'negative trust' or better of 'mistrust'; two opposites of trust: mistrust in Y as for being good/able/reliable/..; suspicion, worry, and diffidence towards Y:

- **Negative Trust-1** or **Mistrust** it is when X believes that Y is not competent or able, or that his behavior is not predictable and reliable. X isn't afraid of Y being malevolent, indifferent, irresponsible, or with other defects; she is just convinced that Y cannot/will not realize her goal g effectively. He doesn't have the needed qualities and virtues (*powers of*).
- **Negative Trust-2** or **Diffidence/Suspicion** it is more than this. At the explicit level it is some sort of 'paradoxical trust': X believes that Y has powers and abilities, and also that he can intend and realize something; but something bad (for X)! X is not simply predicting a failure (X cannot rely on Y), but probable harms from Y, because of Y's bad powers or dispositions.[3]

4.2   Lack of Trust[4]

Both the previous forms of negative trust clearly imply 'lack of trust', since the belief that *'Y is NOT reliable for g; one cannot trust him (as for. . .)'* (*formula (4.1)* is true) – which logically excludes that *'Y is reliable'* – is just a sub-case of the mental state: *'X does NOT believe that Y is reliable/trustworthy'* (where *'Not $(Bel_X q)$'* denotes the absence of such Belief in X's mind):

$$Not\,(Bel_X\,(Trustworthy\;Y)) \tag{4.1a}$$

We actually call this mental state 'lack of trust'.

This belief – per se – is clearly compatible with and can also cover another idea: *'X does NOT believe that Y is reliable/trustworthy' Not (Bel_X (Not (Trustworthy Y)))*. This is when X doesn't know, has no idea: the case of ignorance or pending judgment. When *'X does NOT believe that Y is trustworthy'*, either she believes that Y is not trustworthy, is unreliable; or she doesn't know how to evaluate Y. The 'lack of trust' covers both *mistrust* and *doubt*:

$$Not\,(Bel_X\,(Trustworthy\;Y)) \wedge Not(Bel_X\,(Not\,(Trustworthy\;Y))) \tag{4.1b}$$

What we have just said does not imply that there is trust only when it is enough for entrusting ('delegating' and rely on)[5] or that there is 'lack of trust' or even 'mistrust' when the agent decides not to entrust Y. Logical relations should not be mixed up with quantitative criteria, which are necessary for the decision. These logical relations of inclusion or incompatibility define some 'spaces'. Within the space of *'X trusts Y'* (with some degree) then X has no mistrust or doubts about Y; however, this does not entail that X trusts Y enough to entrust him.

---

[3] Perhaps Y might even realize X's goal g; but by exposing X to serious risks and dangers.

[4] One might prefer – for this case – the term 'no trust' and to limit the expression 'lack of trust' just for case 4.1, where trust is not 'sufficient'. However, actually this does not correspond to the current use of the expression (we try to be as coherent as possible). It is in fact normal to say something like: *'But this is lack of trust!'* when for example X has delegated Y but is full of doubts and worries, and would like to continuously check Y's work.

[5] It is a typical, correct, non-contradictory expression to say: *'I trust Y, but not enough'*. This is because 'I trust Y' can mean two different things: (i) 'I have a certain amount of trust (positive evaluations) about Y'; or, (ii) 'I have a sufficient amount of trust in Y to entrust him, rely on him'. This thanks to the pragmatic implications of the sentence 'I trust Y' and to the meaning of 'to trust' not just as disposition but as decision and action (that presupposes the 'sufficiency' of the evaluation).

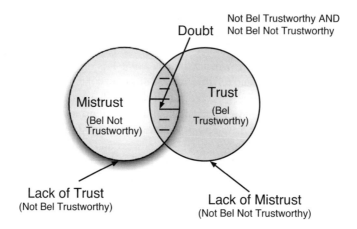

Figure 4.2   Trust, Mistrust, Lack of Trust, Lack of Mistrust, Suspect, Doubt

Correspondingly, within the 'space' of lack of trust, clearly there is no trust. However, it is perfectly possible and usual that X 'in part' trusts Y (and perhaps enough to delegate to Y) and 'in part' doesn't trust or even mistrusts Y.[6]

More than this: *any trust implies some degree of mistrust or at least a lack of trust and doubt*. Since we do not admit that trust is the total certainty and evidence, and we claim that in any case trust entails some (at least *implicit*) risk of failure or some perceived ignorance. Vice versa, any mistrust too is not certainty, and presupposes some (although minimal) doubt in favor. This is a general theory about beliefs with a degree of certainty, and applies to evaluation, to expectations, etc. For example, any hope – by definition – implies some worry; and any worry (implicitly) contains some hope; although the complementary mental state is not necessarily derived (it can remain just implicit) or focused on. (See Chapter 2).

4.3   The Complete Picture

We have also to add: *'X believes that Y is reliable'* ('trust'), which is the only state incompatible with 'lack of trust': *formula (4.1)*. However, this case also implies another belief and is a sub-case of its extension: *'X does NOT believe that Y is NOT reliable/trustworthy'* (*Not (Bel$_X$ (Not (Trustworthy Y))))*.

The complete picture (including also 'suspicion') is shown in Figure 4.2.

Here we can see that – obviously – 'trust' and 'mistrust' exclude each other, while 'lack of trust' and 'lack of mistrust' are not fully incompatible; they partially overlap in an area of 'doubt'. 'Diffidence' is not explicitly represented; it would be a specification within the entire area of 'lack of trust' or of 'mistrust', that is, of the doubt or 'suspicion' that there might be

---

[6] This does not deny that when we use the expression 'I trust' usually we in practice intend 'I trust enough', while when we use the expression 'I do not have trust' we mean 'I do not have enough trust'. Similarly, when we say 'tall' we actually mean 'taller than the norm/average'. But this is not logically necessary; it is just a pragmatic implication. This is why there is no logical contradiction while saying: 'I distrust, I'm diffident, but not enough for not entrusting him'.

some betrayal or danger (the second kind of negative evaluation). It is an active suspicion, that is, to be alerted, to be vigilant for promptly detecting signs of wrong or dangerous acts of Y. This is an *epistemic* attitude which implies epistemic actions (Section 2.3.1), at least to give attention to, to monitor.

As we can see it is a complex picture of various well-defined states. As we said, this becomes even more clear and strong if we assume – like in our model – that our trust-evaluations are based on and supported by specific positive or negative evaluations about Y's characteristics (which actually are other pieces of the trust in Y, sub-components of the whole trust judgment, see Section 2.2.1): *Y is expert; Y is persistent; Y is a good friend of mine;* etc. This makes it even more evident that trust and its negative counterparts cannot be reduced at all to a unique dimension and to a simple 'quantity'.

Of course, while reintroducing quantities and degrees the borders and transitions from one case/state to the other (in Figure 4.2) become more fuzzy. What actually does it mean by 'X does not believe. . . .'? Which degree of certainty is needed for not-believing? 70%? In other words, 'do not believe' because there is an evidence of 70% will be different from 'do not believe' based on an evidence of 60% or 50%.

Moreover, in the intermediate state of the *formula (4.1b)* we can quantify the two components: how much I believe that Y is reliable (although believing that he is unreliable) versus how much I believe that he is unreliable (although not believing that he is so). There can be different degrees and thus various kinds of 'doubt' more or less favorable to Y.

The stronger it is *Not (Bel$_X$ (Not (Trustworthy Y)))* the closer X is to 'trust' *(Bel$_X$ (Trustworthy Y))*. In sum, the discrete *YES/NO* character of the 'decision' to believe and do not believe makes discontinuous the spectrum of our evaluation of trustworthiness.

4.4   In Sum

With the conceptual approach we can characterize *four* different basic mental states relative to perceive Y's trustworthiness, but not a continuum:

$$(\textbf{\textit{Mistrust}}) \quad Bel_X (Not (Trustworthy \; Y)) \tag{4.1}$$

which also includes *diffidence* if not *suspicion* (that mainly is about Y's intentions, motivations, good disposition).

$$(\textbf{\textit{Lack of trust or no-trust}}) \quad Not (Bel_X (Trustworthy \; Y)) \tag{4.1a}$$

$$(\textbf{\textit{Lack of mistrust or No Mistrust}}) \quad Not (Bel_X (Not (Trustworthy \; Y))) \tag{4.1c}$$

$$(\textbf{\textit{Trust}}) \quad Bel_X (Trustworthy \; Y) \tag{4.1d}$$

(Ulman-Margalit, 2001) is right about asymmetries and implications of those different states, but actually they are merely pragmatic; rules and presuppositions of communication and its implications. On the conceptual and logical plan:

(4.1) is incompatible with (4.1d) and (4.1c), while it is compatible with (4.1a): (4.1a) covers (4.1) and (4.1) logically entails (4.1a). In order to characterize state

(4.1a) without (4.1) one should say: (4.1a) + (4.1c) = (4.1b); that is, pending judgment: neither trust nor mistrust.

(4.1a) is incompatible with (4.1d), but compatible with (4.1) and (4.1c);

(4.1c) is incompatible with (4.1) and with (4.1d), but compatible with (4.1a)

(4.1d) is incompatible with (4.1c), that is, (4.1c) can cover (4.1d) too, while (4.1d) logically entails (4.1c) (thanks to the principle of 'non-contradiction').

4.5   Trust and Fear

'Fear' is not just the expectation of a danger or harm, neither is it the emotion due to (elicited by) such an expectation: by the simple explicit representation of a possible harm. To feel fear the danger should be 'significant', not just explicitly taken into account: the subjective expected (dis)utility must be greater than the personal and contextual threshold of acceptable 'risk'. A perceived risk that cannot be coped with and is unbearable, elicits fear (Lazarus, 1991). In fact, every moment we consider (view) possible harms or risks in our situations and decisions, but not all of them elicit 'fear' as a specific emotion.

'Fear' is the extreme lack of trust, is a possible part and reason of 'distrust': X not only believes that Y is unwilling or unable to do as needed, and one cannot rely on him; but, in particular, X believes that Y is untrustworthy because he is dangerous: either out lack of attention or through a bad disposition, character, or out of hostility, envy, etc. Y can produce some harm to X (even intentionally), especially if X is exposed and unprotected because of her trusting attitude; some harm going beyond the simple unattainment of the delegated task (failure).

So 'fear' usually is one basis for 'distrust' and is difficult (but not impossible) when it coexists with a decision to trust. Sometimes X – although perceiving her exposition and vulnerability, and considering a risk and even feeling some fear – either has no alternative for τ or believes Y to have the willingness and competence to do τ, that she decide to trust Y and rely on him, in spite of having some worry and fear. As we said, trust is a bet, necessarily entails some risk, and frequently implies some perceived risk.

4.6   Implicit and by Default Forms of Trust

Given the two different opposites of trust (the two kinds of negative evaluations), one can identify an important implicit form of trust where X neither has specific and explicit positive evaluations about Y, nor has she explicit suspects and worries. She is just *without doubts, suspect and worries*, she naively relies upon Y: not because of an explicit and reason based evaluation of Y. She trusts by default and because she has no reason to be cautious and to suspect, does so without any explicit examination of whether Y is able, willing, or dangerous. One could represent this attitude as the absence of mistrust, of suspicion, but also of explicit positive evaluation and trust.

This implicit, passive, and spontaneous or naive form of trust consists of not having the typical trust beliefs, but also in *not having* negative ones, negative expectations: to be without any alarm and suspicion. Consider in fact that not having a given belief (*I do not believe that it is raining*) is a very different cognitive state than having the opposite belief (*I believe that it is not raining*).

As not trusting *Y* for *g* is not the same as expecting harm from *Y*, analogously not dis/mistrusting, not worrying, is not the same as positively believing that *Y* is capable and willing. However, this lack of dis/mistrust can be sufficient for relying upon *Y*. It depends on the agent's rule. If the agent has a default which in order to delegate requires specific positive evaluations, specific trustworthiness beliefs, the absence of these beliefs actually is distrust. On the contrary, if the agent's default rule is '*except you have specific reasons for not relying on Y, specific negative evaluations, then trust Y*', the lack of mistrust is factually an (implicit) form of trust.

In this weak form the behavior/action of trust consists of the absence of cautions, of controls, of any search for evidence for evaluation, and in the absence of a true 'decision' about trusting or not *Y*. Only after some negative unexpected experience, this kind of trust is damaged. Whoever uses explicit, evaluation-based trust, based on evidence, is no longer naive: she has already considered the situation as problematic; she has some doubt. There is, on the contrary, a form of trust without and before any question like: 'Can/should I trust *Y*?' (See also Chapter 5).

It is important also to distinguish between *uncertainty* (the fact that we do not have complete evidence of our positive evaluation of (*trust in*) *Y*, we are not 100% sure of our beliefs), that make *Y*'s behavior (and results) not completely subjectively predictable; from the actual presence of contrasting, negative evaluations and expectations. The absence of a belief is a mental state significantly different from the presence of the negative belief, with completely different consequences at the reasoning and at the pragmatic level. When *X* has positive evaluations of *Y*, and *does not have any negative (pertinent) evaluation,* although this positive evaluation leaves some room for ignorance and uncertainty, this is very different from a situation where *X* has negative beliefs about *Y* which make *Y* 'ambivalent' (attractive and repulsive, positive and negative, at the same time) and destroys *X*'s 'trust in' *Y*, his trustworthiness. Non-ambivalent although uncertain evaluation is very different from ambivalent evaluation. Thus, we have to distinguish between two types of 'unharmfulness': 'safety' and 'there is nothing to worry' etc.: the implicit and the explicit.

Implicit un-harmfulness simply consists of the absence of suspicions, doubts, reasons to worry, diffidence, no perceived threats; some sort of 'by default' naive and non-arguable confidence. I do not have reasons to doubt *Y*'s pro-attitude (active or passive adoption), I do not have negative beliefs about this.

Explicit un-harmfulness consists of explicit beliefs about the fact that 'I have nothing to worry from *Y*'.

Both, the implicit or explicit un-harmfulness can be based on other beliefs about *Y*, like '*He is a friend of mine*', '*I'm likeable*', '*I feel his positive emotional disposition*' (empathy), '*He is honest and respectful of norms and promises*', '*He fears me enough*', . . . and also '*He trusts me and relies on me*'.[7]

Another important kind of 'implicit trust' is the procedural, automatic trust, or better 'confidence', based on perceived regularities, learning, and confirmation of practices.

For example, in motor behavior there are a lot of implicit 'expectations' about objects, movements, etc. and their 'reliability'. And when we (as expected) successfully perform a

---

[7] This unharmfulness perception and then trust in *Y* based on *Y*'s trust in *X*, is important for the circular dynamics of trust and to explain how trust can create trust (Chapter 6).

given action we unconsciously confirm a lot of implicit assumptions and predictions. We are 'confident' in the ground, in a chair, in a key, etc. Only when something goes wrong, we become suspicious and worry about something. But this form of trust is fundamental also in social life and for the maintenance of social order, which is based on trust and is the basis for trust (Chapter 9).

In sum, the lack of *explicit* trust covers three different mental states:

- *insufficient trust* (X does not estimate enough Y to count on him, she has some negative evaluation on Y) (see below);
- *mistrust* (X worries about Y);
- *implicit trust*, be it either spontaneous, naive and by default (lack of suspect), or be it automatic and procedural, just based on previous positive experience and learning.

4.6.1   Social by-Default Trust

By-default trust is very relevant in social life, especially in communication: we – for example – ask for information from people who we have never met before and we will never meet again, which might deceive us without any external sanction. In general our claim is that Grice's principles about linguistic communication are in fact two default rules. It is obviously true that people can lie, but they should have some reason for doing so; it is true that people can not believe, reject information, but they have to have some reason for this. The speaker's and the hearer's default rules in linguistic communication are as follows; and the hearer rule is a 'trust' rule.

The speaker's default rule:

'except you have specific reasons for deceiving, say the useful truth'

The hearer's default rule:

'except you have specific reasons for being diffident, believe the speaker'

We justify the existence of this default-rule (to say the relevant truth and to ask for information and believe in people) with a form of 'reciprocal altruism' in humans about exchanging and circulating knowledge (Conte, 1995). Knowledge is such a crucial resource for human beings and is made more valid by social circulation and sharing, that passing on in a reciprocate fashion becomes a serious advantage.

An important form of by-default trust is also in 'generalized trust' (see Chapter 6):

'Except you have specific reasons for diffidence towards a given agent, trust everybody in this context/community/category'

The reason why we are so sensitive to a trust atmosphere is that this is the 'signal' of a diffuse default-rule, and we can adopt it while greatly reducing the cost of search and monitoring of information, and our stress and worries.

In a sense 'by-default' trust is another form of 'conditional' trust. It is a belief of trustworthiness relative to another belief:

$$IF\& \ UNTIL\,(Not\,(Bel_X\,(Not\,(Trustworthy\ Y)) \rightarrow Bel_X\,(Trustworthy\ Y) \qquad (4.4)$$

where $\rightarrow$ means 'implies'.

Or it is some sort of conditional rule about the trust act:

$$IF\& \ UNTIL\,(Not(Bel_X\,(Not\,(Trustworthy\ Y)) \Rightarrow Trust\,(X\ Y) \qquad (4.5)$$

where $\Rightarrow$ means 'produces'.

This means that X will control the validity of her assumption; and its confirmation.

Of course, this default rule is very different (in feeling and behavior) from the possible opposite one; some sort of *conditional distrust*:

$$IF\& \ UNTIL\,(Bel_X\,(Trustworthy\ Y)) \rightarrow Not\,(Trust(X\ Y)) \qquad (4.6)$$

4.7   Insufficient Trust

Trust insufficiency is when the trust that X nevertheless has in Y does not exceed the threshold (adopted by X in that context for that task) necessary for the decision to delegate, entrust Y.

Notice that:

- The amount of trust can be even greater than the amount of distrust or of lack of trust; but, nevertheless, insufficient (see Figure 4.3).
- when trust is not sufficient, lack of trust is too much, but it is not necessarily all 'distrust': there can be a lot of ignorance or doubt (see below).

While introducing more sophisticated and quantitative models, we get other notions that are very interesting, like 'evidence/reason-based' trust/distrust *versus* trust/distrust just based on giving/not-giving credit.

There is another form of trust 'insufficiency': a relative not an absolute one.

Negative evaluations can in fact be 'absolute' or 'comparative' and 'relative' to a standard, threshold, etc. (Chapter 2 and 3). If the evaluation of Y is 'inferior' to the needed threshold or standard, or inferior to the evaluation of Z, it becomes 'negative' (Y is insufficient, is inferior). In a sense here we are at a meta-evaluative level; there is a sort of meta-evaluation: an evaluation of the evaluation: '*Is this evaluation sufficient?*' '*No; it is not a really good evaluation*'. In particular, it is not really a 'lack of trust' a trust which is or would be 'enough'

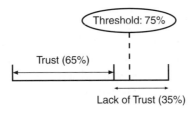

Figure 4.3   For the Delegation, Trust must be over the threshold

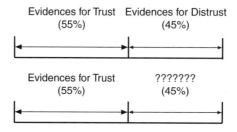

Figure 4.4   Evidence-based Trust matched with. a) Evidence-based Distrust and b) lack of knowledge

for trusting, but being inferior to another Agent (*Z*) is not winning. *X* trusts *Y*, she finds *Y* even reliable and might entrust him, but – since there is also *Z*, who is even better – she will chose *Z* and (en)trust him. This is an 'insufficient' evaluation of *Y* but not really a 'lack of trust' in *Y*. Or better, it is a 'relative lack of trust' but not an 'absolute lack of trust'.

4.8   Trust on Credit: The Game of Ignorance

Trust (usually) goes *beyond* the evidences and certainty of well-grounded beliefs; in this it consists of 'giving credit' (or not), that is, to believe even beyond the evidence.

Let us adopt a model *à la* Dempster and Shafer,[8] where, between the evidence in favor of a given eventuality that *P* (thus of the well established probability of *P*) (say 55%) and the evidence in favor of *Not-P* (thus its estimated probability) (say 15%), there is a gap of 'ignorance' (say 30%), a lack of proof in favor or against *P*. First of all, let us observe that with a given level of 'evidence-based' trust (say 55%) it is very different whether the rest, the complement, is supported distrust or just lack of knowledge, just possibility, not probability of *Not-P* (see Figure 4.4).

These two scenarios are psychologically very different: they induce opposite behaviors. For example, in the second scenario the subject might suspend her judgment and decision while waiting or searching for new data; she might perceive a strong uncertainty and be anxious. While in the first situation she might feel able to take a decision (positive or negative).

This is additional confirmation of the fact that it is not sufficient to have just an index, a number (say 55%, or 0.5) to represent trust.

It is important to stress that *with just one and the same level/degree of trust there may be completely different feelings, decisions and behaviors*. Moreover, an 'insufficient' trust can be due to quite different cases: factual distrust, negative evidence, or too much uncertainty and ignorance.

4.8.1   Control and Uncertainty

As we just said, lack-of-trust because of insufficient information is different from lack-of-trust due to supported low evaluations. In the former case, I can wait or search for additional information. Sometimes, it is possible to adopt such a strategy as 'run time', 'work in progress', and then start to delegate (trust is enough!) with the reserve of *monitoring*.

---

[8] See for example: http://en.wikipedia.org/wiki/Dempster–Shafer_theory

Control is useful and can be a remedy in both cases, but with different roles. Control means in fact (see Chapter 7 for a more detailed analysis):

a) The possibility of improving Y's performance, or of intervening in time – on the basis of new evidence – for remedying, revoking the delegation, etc. Control is a remedy to low expectations about Y.
b) The possibility of acquiring, through *monitoring,* additional evidence and information run-time.

In sum, control (monitoring + intervention) is also a remedy against uncertainty-based lack-of-trust: '*In case something is bad I will know, and I will be able to do something*'. I use monitoring in order to *update* and complete my beliefs, since I know that my prediction of the future and my knowledge is incomplete. This is different from a low probability-based lack-of-trust: '*In case some prediction is wrong I will know, and I will be able to do something*'. I use monitoring in order to *revise* my beliefs that were wrong.

Of course, the two problems and the two functions of monitoring can co-occur.

4.8.2   Conditional Trust

Searching or waiting for additional and determinant evidence also means having to formulate a form of 'conditional' trust. X trusts Y but if and only if/after Y has performed a given action, provided a given assurance, or proof. '*Only if he swears on his sons*'; '*Only if I can check*', '*Only if has this documented experience*', and so on.

$$IF \, (Bel_X \, (Predicate \ Y)) \rightarrow (Trust(X \ Y)) \tag{4.7}$$

Where, *Predicate* represents either an act or a feature of Y, or a *sign* of such a feature.

The difference between this 'conditional' trust and normal 'evidence-based' trust is just that X is waiting for such evidence; but in a sense she has already decided to trust Y, provided that the expected evidence will be true.

4.8.3   To Give or Not to Give Credit

The second remark is that it is extremely important for there to be the possibility to mentally ascribe that gap of ignorance (that part without evidence; what is 'possible' but not grounded) in favor or against P or *Not-P*.

In fact, given this model, the part that Dempster and Shafer call 'plausibility', the empty part, is in fact ascribable to P or to *Not-P*. P is probable 55% but *possible* and plausible (not against evidence) up to the 85%! While *Not-P* is probable 15% but plausible up to the 45%. In our opinion, applied to the prediction of the behavior of $Y X$ counts on, this gap represents what we call to 'give credit'. If X trusts Y beyond supported evidences and grounded probabilities, and gives in favor of Y all the 'plausibility' space (till the limit of the opposite evidences), then she is 'giving credit' to Y; she is trusting (believing) beyond evidence (see Figure 4.5 and Section 4.2.1).

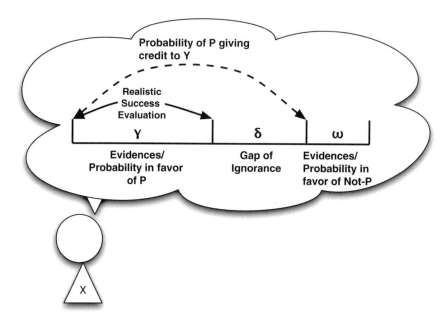

Figure 4.5   Giving credit

If, on the contrary, X has a pessimistic or prudential attitude and decides to trust somebody only after due consideration, only on the basis of evidence and certain data, then the entire space β of 'possibility' is assumed in favor of Not-P, as lack of trust (in this frame we will call that space b, see Figure 4.5).

Some scholars would like to call 'trust' only this 'credit', only reliance not based on evidences and good reasons; but, this is a prescriptive attitude, in contrast with the meaning of the term in various languages, and with the concrete social use of this notion. Also because even proofs, certainty, reasons, are just a subjective fact; a matter of believing, and of trusting sources (there is no way out from such a recursion of trust).

The truth is that *there exist two faces and components of trust* (and of lack of trust): the one '*after due considerations*', that is, given in return for received evidences; the other, given '*on credit*', for free, in view of future confirmations. In both cases, X is exposed to risks and has not full certainty; in fact, the 'bet' present in any act of trust is not in trust *per se*, but in counting on the other, in entrusting, in accepting the possibility of a failure or harm.

To 'give trust' in the sense of 'giving credit' is – as stressed by many authors – an *optimistic* attitude[9] not fully prudent and reasonable, but frequently quite effective, and some sort of self-fulfilling prophecy (see Chapter 8).

The inclusion of the theory of ignorance and uncertainty within the theory of trust – which is an additional confirmation that the simple 'subjective probability' of the favorable event is really unsuitable – has interesting consequences.

---

[9] This is not necessarily due to a personality or stable disposition. It can be due to circumstances, to Y and his 'cues' that inspire trust; or to a given context or transitory mood.

In sum, there are three cases, three 'games' played by our ignorance: (i) to play in favor of trust; (ii) to remain indifferent, in between; (iii) to play against trust. (ii) and (iii) are both 'lack of trust', but with two distinguishable faces: unsolved doubts *versus* suspect and diffidence.

On the one hand, X's trust in Y acquires two 'components': the *positive evaluation part* (γ), and the *simply credited part* (δ). And there can be quite different proportions between them, for example:

- $\gamma = 0$, that is, pure *'faith'* in Y, without any evidence (see note 11 in Chapter 3);
- $\delta = 0$, that is, a stingy trust (perhaps sufficient), mere esteem based on evidences and 'after due considerations', without any additional credit.

It is important to note that δ trust can be necessary (in some decisions) in order to pass the decision threshold; but probably with a greater perception of *uncertainty*. In our model, in fact, the subjective/perceived 'uncertainty' is the function of two independent dimensions: the amount of ignorance, of lack of data; the balance between the pros and the cons. The greater the lack of data, the *perceived ignorance*, the greater the perceived 'uncertainty'; but also: the smaller the difference between the data in favor of P and those in favor of *Not-P*, the greater the perceived 'uncertainty'.

For example, given the same gap of ignorance ('plausibility'), say 60%, our perceived uncertainty is greater if the probability of P is 20% and the probability of *Not-P* is 20%, rather than if the probability of P is 35% while *Not-P* is just 5%.

These two kinds of uncertainty have a different nature: the first (*due to lack of data*) is about the evidence; the second (*equilibrium between evidence in favor of P and Not-P*) is about the decision to take. However, there is a relationship between them: for example, the first implies the second one.

Our previous claim was simply that the feeling of trust (given a decision and delegation, that is, when $\gamma + \delta$ are higher than the required threshold) will be quite different if γ is very consistent or even sufficient and δ is just additional and reassuring, compared with the situation where α is weak or insufficient and β is necessary for the decision. This also means that the *risk perceived* (taken into account) and accepted in the moment of the decision to trust, has two faces: on the one hand, it is the part representing true distrust, *bad evaluations and prediction* (ω); on the other hand, it can just be the *lack of positive evidence*, perceived ignorance, but perhaps given as 'plausibly' in favor.

Not only it is a matter of how much risk X is disposed to take (subjectively), but also how blindly X accept it.

4.8.4   Distrust as Not Giving Credit

If there is an γ (evidence-based) trust and a δ trust ('there is no evidence but I give it in favor of P'), then there also exists a *distrust* of the first kind (ω) and another form of distrust: a 'lack of trust' assumed in a pessimistic attitude as against P (λ).[10] As we said, the 'lack of trust' can cover both 'distrust' in the strict sense (negative evaluations and expectations), and lack

---

[10] Notice that this operation of 'taking as favorable/good' or 'taking as unfavorable/bad' more precisely is an operation of assumption ('acceptance') than of 'belief'.

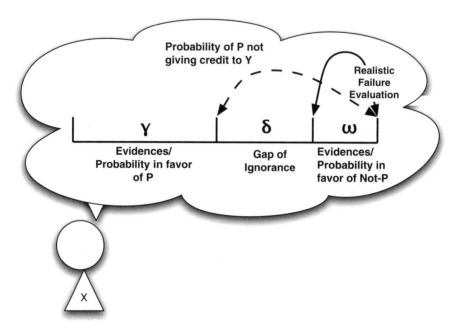

Figure 4.6   Be prudent

of beliefs and evidence (pros or cons); but also this lack of evidence, assumed as unfavorable, as suspect.

A 'diffident' attitude consists, on the one hand, of being vigilant and alarmed (worrying about possible harms), on the other hand, precisely in being prudent, *not giving credit*, not believing without evidence, being pessimistic and taking as potentially negative all the unclear possibilities (see Figure 4.6).

As we said, the perceived ignorance must not necessarily be ascribed in favor of P or *Not-P*. It can be just neutral, as perceived lack of information; and perhaps the decision will be suspended. This entails – as we said – that when trust (γ) is not sufficient for entrusting, its complement, is mistrust; it can just be perceived ignorance (Figure 4.7).

Notice that in a sense, ω represents the estimated probability of a failure (of *Not-P*) and thus a crucial component of the perceived 'risk' that might have a specific threshold (Chapter 3). X might not accept a given level of ω, and thus renounces on betting on Y if the bad evidence on Y is too great. Clearly, this is independent of the positive trust and on its acceptance threshold.

For example, if ω is at the 15% – like in our previous example – it might be acceptable; while if it was at 30% (even still being γ at the 55%) the risk might be unacceptable. As we said, X may decide not to trust Y because trust is not enough, not because distrust is greater than trust, or because all the rest is distrust.

We will develop in Chapter 8 a more complete analysis of 'Optimism' and also the fact that both trust and distrust are actually 'self-fulfilling prophecies', with important consequences on 'rationality' of trust bet, on trust dynamics (Chapter 6), on trust spreading and self-organization.

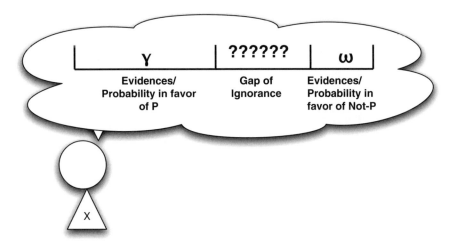

Figure 4.7   Perceived ignorance and doubt

References

Castelfranchi, C. (2000) Affective appraisal vs cognitive evaluation in social emotions and interactions. In A. Paiva (ed.) *Affective Interactions. Towards a new generation of computer interfaces* (pp. 76–106). Heidelberg: Springer.

Conte, R. and Castelfranchi, C. (1995) *Cognitive and Social Action*. UCL Press.

Lazarus, R. S. (1991) *Emotion and Adaptation*. New York: Oxford University Press.

Lorini, E. and Castelfranchi, C. (2004) The role of epistemic actions in expectations. In Proceedings of the Second Workshop of Anticipatory Behavior in Adaptive Learning Systems (ABIALS 2004), Los Angeles, 17 July 2004.

Marsh, S., Dibben, M. R. Trust, untrust, distrust and mistrust - an exploration of the dark(er) side, Proceedings of Third *iTrust* International Conference, 2005.

Miceli, M., and Castelfranchi, C. (2000) The role of evaluation in cognition and social interaction. In K. Dautenhahn (ed.), *Human Cognition and Agent Technology* (pp. 225–261). Amsterdam: Benjamins.

Edna Ullmann-Margalit TRUST, DISTRUST, AND IN BETWEEN. Overview1 © Draft September 2001. 'http://www.law.nyu.edu/clppt/program2001/readings/ullman_margalit/Trust,%20distrust%20and%20in%20between.pd'

5

The Affective and Intuitive Forms of Trust: The Confidence We *Inspire*

In this chapter[1] we analyze an aspect of trust which is in a sense 'marginal' relative to our systematic and analytic theory,[2] but not marginal at all from a complete and adequate account of trust and of its real significance and functioning in human interaction: the *affective version or components of trust*. This aspect is also very crucial for social, economic, and moral theories. Is it not actually too 'cold' and reductive to treat trust as a judgment and a reason-based decision? Is not trust something that we just feel and cannot explain?

We have just analyzed the cognitive *explicit* facet of trust as beliefs and goals about something, and a consequent decision of relying upon it. We have completely put aside the affective side: the trust that we 'inspire', the merely intuitive, emotional facet. *It is true that trust can also be this or just this: no judgment, no reasons, but simply attraction and sympathy.* This is an automatic, associative, unconscious form of appraisal: *we do not know why* we prefer Y and are attracted by Y.

There are beautiful experiments by Bargh's group on this form of affective appraisal. One should also account for the personality aspects of trust as disposition or as default attitude. Some emotions are based on and elicited by true evaluations (beliefs), and also, trust as affective disposition can be based on trust as esteem and good expectations. And the affective aspect of trust can play a role by modifying the belief process, source, and 'decision'. But, on the other hand, trust can be a non-belief-based emotional reaction, an affective attitude simply activated by unconscious perception of signs or associations, by 'somatic markers' (Damasio, 1994), (Castelfranchi, 1998).

---

[1] This chapter does build heavily on the work with Maria Miceli on the theory of emotions, evaluation, intuitive appraisal.

[2] It is also marginal in part for applications in some domains like ICT (but not 'affective interaction' or web communities), but not at all in other domains like marketing or politics.

Trust Theory   Cristiano Castelfranchi and Rino Falcone
© 2010 John Wiley & Sons, Ltd

5.1   Two Forms of 'Evaluation'

As we said trust is a form of appraisal (of the trustee by the trustor); it is an *attitude* based on
or implying an evaluation (and an *act* signaling it). However, there are two different forms of
evaluation/appraisal in cognitive agents: *explicit evaluative judgments (beliefs)* versus *implicit,
affective appraisal*. The first kind has been systematically explained in Chapter 2 (with the
theory of 'qualities', 'standards', two kinds of 'negative evaluations' and so on). The other
kind deserves some more attention (Miceli, 2000), (Castelfranchi, 2000).

5.2   The Dual Nature of Valence: Cognitive Evaluations Versus
         Intuitive Appraisal

There are at least two kinds of appreciation of the valence of events, situations, and entities;
two kinds of 'evaluation' in a broad sense.

a) A *declarative or explicit form of evaluation*, that contains a judgment of a means-end
   link, frequently supported by some reason for this judgment, relative to some 'quality' or
   standard satisfaction.
   This is a *reason-based evaluation* that can be discussed, explained, argued upon. Also
   the goal of having/using the well-evaluated entity (which is the declarative equivalent of
   'attraction') can be 'justified'. This is the classical approach to values (moral or ethical)
   that is synthesized by the 'motto' (of Aristotelian spirit):

      'it is pleasant/ we like it, because it is good/beautiful'

b) A *non-'rational' (or better non-'reasons-based') but adaptive evaluation*, not based on
   justifiable arguments; a mere 'appraisal',[3] which is just based on associative learning and
   memory.

   In our view, in the psychological literature on emotions, in particular in the very important
and rich literature on emotions as based on a cognitive appraisal of the situation (Frijda, 1986),
(Frijda, 1988), (Arnold, 1960), (Scherer, 1986), (Scherer, 1999), there is a systematic and
dangerous confusion between these two kinds of 'evaluation' (also in Damasio). Incoherent
terms and properties are attributed indifferently to the term 'appraisal' or 'evaluation'. This
fundamental forerunner and component of the emotion is characterized – at the same time – as
'cognitive', 'intuitive', 'immediate', 'unconscious', implying also inferences and predictions,
etc. We propose (see also (Miceli, 2000), (Castelfranchi, 2000) (Castelfranchi, 2009)) *to
distinguish between 'appraisal' - that should be the unconscious or automatic, implicit,
intuitive orientation towards what is good an what is bad for the organism- and 'evaluation'*.
We reserve this last term (evaluation) for the cognitive judgments relative to what is good or
bad for the goal (and why).

---

[3] Although the English term 'appraisal' is basically a synonym of 'evaluation', let's use it – for the sake of simplicity –
for characterizing the second form of evaluation: the intuitive, implicit, affective, somatic, ... appraisal.

5.3   Evaluations

Let us assume there is a good understanding of what 'evaluations' are (Chapter 2, in particular Section 2.2.7), and look in more detail now at the relationships between evaluations, goals, and emotions.

Evaluations imply goals by definition, in that the latter are a necessary component of evaluations, namely, the second argument of the GOOD-FOR predicate. From a more 'substantialist' perspective, evaluations imply goals in the sense that they *originate* from them: it is the existence of some goal g (either X's or someone else's) that makes the words good or bad, justifies and motivates both the search for a means m to achieve it, and the belief that m is (not) GOOD-FOR g. Goals and evaluations endow objects and people with 'qualities' and 'faults'.

The relationship between evaluations and goals is even closer, because *evaluations not only implies goals, but can also generate them.* In fact, if X believes m is good for some goal, and X has that goal, X is also likely to want (possess, use) m. So there is a rule of 'goal generation' which might be expressed as follows: if X believes something m to be a means for X's goal g, X comes to have the goal of exploiting/using the means m.

Evaluations, that is, knowledge about 'what is good for what', and 'why', play a crucial role in all the cognitive activities that are based upon symbolic and explicit representations, reasoning and deliberation. For example, in problem solving and decision making, the particular advantage offered by evaluative knowledge is precisely a preliminary relationship established between descriptive knowledge and goals, in terms of beliefs about 'what is good for what', derived from either one's experience about problems solved in the past, or one's reasoning and inferences (think for instance of evaluation by standards), or others' communication.

Evaluations make such a relationship explicit; they fill the gap between knowledge and goals, by 'reinterpreting' the properties, qualities, and characteristics of objects and situations in terms of means for the system's (potential or actual) goals.

The cognitive network ceases to be neutral and becomes 'polarized' toward goals, that is ready for problem solving and decision-making.

In a cognitive agent preferences can be internally represented both at the *procedural* and at the *declarative* (propositional) level.

- Having a *procedural preference* means that, at a given level of their processing, a system's goals present different degrees or indexes of activation, priority, weight, value, importance (or whatever), that in fact create some rank order among them, which will be followed by some choice/selection procedure.
- Having a *declarative preference* means that the system is endowed with an explicit belief such as: 'm is better than n (for goal g)'. In particular, three types of beliefs are relevant for preferences: (a) simple evaluations, that is beliefs about how good/useful/apt/powerful are certain entities relative to a given goal ('m is very useful for g'; 'n is quite insufficient for g'); (b) comparative evaluations like 'm is better than n for g'; (c) reflective preference statements, of the kind 'I prefer m to n (for g)'. Generally, *(b)* are based on *(a)*; while *(c)* are based on *(b)*.

Both procedural and declarative preferences can coexist in a human mind (and would be of some use in artificial minds too), and each level of preference representation – though having its own mechanisms of reasoning – *is translatable into the other.* One can derive a 'weight' from the declarative evaluations and their arguments, and vice versa, one can explicitly express (as beliefs) some priority of attractiveness, urgency, activation, or whatever.

However, being able to <u>deliberate</u>, that is, to choose an alternative on the grounds of explicit evaluations concerning the 'goodness' of the various options, and being capable of reasoning aimed at supporting such judgments will add further advantages to the mere fact of making choices. In these cases, in fact, the system can *justify* its choices, as well as *modify* the 'values' at stake through reasoning. Moreover, it is liable to persuasion, that is, it can modify its preferences on the grounds of the evaluations conveyed by others (argumentation).

We interact with people on the basis of the image and trust we have of them, i.e. on the basis of our evaluations of them: this defines their 'value' and reputation. And also, social hierarchies are just the resultant of the evaluations that the individuals and the groups receive from others.

5.3.1 Evaluations and Emotions

Given this 'cold' view of evaluation ('cold' if compared with others', e.g., (Mandler, 1975)), what is the relationship between evaluation and emotion? As we claim in (Castelfranchi, 2009):

- *Evaluations do not necessarily imply emotions*

No doubt many evaluations show some emotional feature. For instance, if I believe a certain food, book, person, and so on, to be 'good', I will be likely to feel attracted to it (or him or her). But evaluations and emotions are not necessarily associated with each other, because not any belief about the goodness or badness of something necessarily implies or induces an emotion or an attraction/rejection with regard to that 'something'. There also exist 'cold' evaluations: if, for instance, I believe that John is a good typist, I will not necessarily feel attracted to him. This is especially true because X (for example a neutral consultant or expert) can formulate evaluations relative to Y's goals: what would be good or bad for Y.

Evaluations luckily have emotional consequences if they simultaneously:

  i) are about our own goals (the evaluator is the goal owner);
 ii) these goals are currently active;
iii) they are important goals.

- *Emotions do not necessarily imply evaluations*

One may view attraction or rejection for some m as a (possible) consequence of an evaluation; so, in this case the emotion 'implies' an evaluation in the sense we have just considered. On the other hand, however, one may view attraction or rejection *per se* as forms of evaluation of the 'attractive' or 'repulsive' object. In the latter case, we are dealing with a supposed identification: to say that an emotion implies an evaluation means to claim that the two actually coincide, which is still to be proved.

In fact, we view attraction and rejection as <u>pre-cognitive implicit</u> evaluation, that we call 'appraisal'. In a sense, any emotion implies and signals an 'appraisal' of its object.

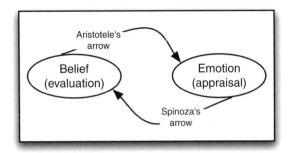

Figure 5.1   From Evaluation to Appraisal and vice-versa

5.4   Appraisal

We assume that a positive or negative emotional response can be associated with some stimulus. The automatic activation of this associated internal response (in Damasio's terms, a 'somatic marker'; (Damasio, 1994)) is the 'appraisal' of the stimulus postulated by several theories of emotions (Arnold, 1960), (Frijda, 1986), (Frijda, 1988), (Lazarus *et al.*, 1970) (Ortony, 1990). The associated negative or positive affective response makes the situation bad or good, unpleasant or pleasant, and we dislike or we like it.

'Appraisal' consists of an automatic association (conscious or unconscious) of an internal affective response/state either pleasant or unpleasant, either attractive or repulsive, etc., to the appraised stimulus or representation.

It does not consist of a judgment of appropriateness or capability – possibly supported by additional justifications; on the contrary, it just consists of a subjective positive or negative experience/feeling associated with the stimulus or with the mental representation, usually previously conditioned to it in similar circumstances, and now retrieved.

This gives us a completely different 'philosophy' of valence and value: now the 'motto' is the other way around – in Spinoza's spirit – (see Figure 5.1)

'It is good/beautiful what we like/what is pleasant'

As a *cognitive evaluation of m is likely to give rise to some goal*: if the evaluator X believes something m to be a means for X's goal g, X comes to have the goal 'of acquiring and using the means m' (we call q this instrumental goal); also *the emotional appraisal of m gives rise to a goal*: it activates a very general goal linked to the emotional reaction. This is the *cognitive* aspect of emotional appraisal. Positive appraisal activates an 'approach goal' ('to be close to m; to have m'), while negative appraisal activates a generic avoidance goal ('not to be close to m; to avoid m '). We consider *these sub-symbolic, implicit forms of 'evaluation' as evolutionary and developmental forerunners of cognitive evaluations*. Thus we believe the answer to the question 'do emotions imply evaluations?' depends on the level of analysis addressed.

In sum, in our view:

(1) *Appraisal is an associated, conditioned somatic response* that has a central component and involves pleasure/displeasure, attraction/repulsion. Here attraction/repulsion is not a

motor behavior but just the preliminary, central and preparatory part of a motor response. And pleasure/displeasure simply is the activation of neural centers.

(2) This associated response can be *merely central*, because the somatic-emotional component can also be reduced to its central trace (Damasio's somatic markers) and because emotions have a central response component which is fundamental. But of course this response can also be more complete involving peripheral overt motor or muscle responses or visceral emotional reactions.

(3) This associated response is *automatic, and frequently unconscious*.
   • Appraisal is a way of *'feeling' something*, thanks to its somatic (although central) nature.
   • Appraisal *gives 'valence'* to the stimulus because it makes it attractive or repulsive, good or bad, pleasant or disagreeable.
   • Appraisal has *'intentionality'* i.e. the association/activation makes what we feel 'about' the stimulus, makes it nice or bad, fearful or attractive. It gives the stimulus the character that Wertheimer called 'physiognomic'. (How this happens, how the associated response is 'ascribed to', 'attributed to', and 'characterizes and colors' the stimulus; how it does not remain concurrent, but dissociated, is not so clear – at least to us – and probably just the effect of a neural mechanism).

(4) When it is a response just to the stimulus it is *very fast, primary*. It anticipates high level processing of the stimulus (like meaning retrieval) and even its recognition (it can be subliminal). In this sense the old Zajonc's slogan 'preferences need no inferences' proves to be right (although not exclusive: there are preferences which are based on reasoning and inferences; and also emotions based on this).

(5) There can be an analogous associative, conditioned, automatic *response to high level representations*: to beliefs, to hypothetical scenarios and decisions (Damasio, 1994), to mental images, to goals, etc.

We have to change our usual view of cognitive 'layers', where association and conditioning are only relative to stimuli and behaviors, not to cognitive explicit mental representations.

Any emotion as a response implies an appraisal in the above mentioned sense.

It implies the elicitation of a central affective response involving pleasure/displeasure, attraction/repulsion, and central somatic markers if not peripheral reactions and sensations. This is what gives emotions their 'felt' character. (While not all emotions presuppose or imply a cognitive evaluation of the circumstances).

5.5 Relationships Between Appraisal and Evaluation

Evaluation and affective appraisal have much in common: in particular, their function (Castelfranchi, 2009). Evaluations favor the acquisition of adequate means for one's goals, and the avoidance of useless or dangerous means, and precisely the same function can be attributed to emotions.

More than that: emotions – though they have traditionally been attributed the negative role of clouding and altering rational thought – seem to help at least some kind of reasoning. In fact, they provide 'non-conscious biases' that support processes of cognitive evaluation and reasoning (Bechara, 1997), enabling one for instance to choose an advantageous alternative before being able to explicitly evaluate it as advantageous.[4]

However, all this should not prevent one from acknowledging the differences between emotional appraisal and cognitive evaluation, addressing the latter in their own right, and trying to establish their specific functions. For instance, in some context emotional appraisal by itself might prove insufficient for assuring adaptive responses, in that, the *more changeable and complex the world becomes* (because of the increasing number of goals and situations to deal with, and the complex relations among such goals and contexts), *the more one is in need of analytical and flexible judgments about objects and events*, rather than (or in addition to) more global and automatic reactions. In fact, evaluations allow one to make subtle distinctions between similar (but not identical) goals and means, and to find out the right means for some new goal, never pursued in the past.

Moreover, evaluations allow one to reason about means and goals, and to construct and transmit theories for explaining or predicting the outcome of behavior. Therefore, though emotional appraisal can be conceived of as an evolutionary forerunner to cognitive evaluation (as well as a valuable 'support' for it), being an evolutionary 'heir' does not imply maintaining the same nature as the forerunner; on the contrary, one might suppose that the same function has favored the development of different means, at different levels of complexity.

It is also important to consider that evaluation and appraisal about the same entity/event can *co-occur*, and give rise to *convergence* and enhancement of the valence, or to *conflicts*; in fact, either:

- the means that we are rationally considering for our ends are associated to previous or imagined positive experiences; or
- what I believe to be the right thing to do frightens me; what I believe to be wrong to do attracts me.[5]

Evaluation and appraisal can also derive one from the other.

It is possible to verbalize, to translate a merely affective reaction towards *m* into a declarative appreciation. This is for example what happens to the subjects in the experiment by Bargh and Chartrand, 1999. They do not realize that their evaluation is just a post-hoc rationalization of some arbitrary association (conditioning) they are not aware of.

---

[4] A number of studies conducted by Damasio and his collaborators (e.g. Bechara, Damasio, Damasio, and Anderson 1994) have pointed to the crucial role of emotion in cognitive evaluation and decision making. Their patients with lesions of the ventromedial prefrontal cortex show emotional blunting as well as difficulties in making decisions, especially in real-life contexts. If compared with normal subjects, they do not show stress reactions (as measured, for instance, by skin conductance response) when trying to make choices in uncertain and risky contexts (e.g. a gambling task). The interesting fact is that such emotional reactions, displayed by the normal subjects especially before making a wrong choice (i.e. a kind of choice previously associated with some punishment), help them to avoid it, and to opt for a less risky alternative. Such a choice is made before reasoning over the pertinent beliefs, including cognitive evaluations about the game, its options, and the possible strategies of decision making.

[5] On this, Damasio's model of the role of the somatic markers in decision-making looks rather simplistic: somatic markers do not 'prune' the tree of possible choices, but just add some weight or value to them; it is always possible that we decide to pursue a goal that actually disgust us, or that we do not pursue a goal that was very exciting and attracting.

It is also possible to go down the opposite path – from a cold evaluation to a hot appraisal; especially for personal, active, important goals, and in particular for felt kinds of goals like needs, desires, etc. (Castelfranchi, 1998).

This possible translation from one form to the other is very important, because it also helps to explain a very well known vicious and irrational circle of our emotional life (Elster, 1999). We mean the fact that we feel our emotional activation, what we feel towards *m*, as a possible evidence, *confirmation* of our beliefs that give rise to that emotion itself. So, for example, we start with a belief that *m* can be dangerous, we predict possible harm, on such a basis we feel some fear, and then this fear (as an appraisal of *m*) 'feeds back' on the beliefs and increases their certainty, i.e. confirms them; something like: 'Since I'm afraid, actually there is some danger here' (which is not such a rational evidence; it is a case of self-fulfilling prophecy and also the origin of 'motivated reasoning', (Kunda, 1990) (see Figure 5.1).

Applied to trust this means that there are two possible paths:

(i) On the one side, it is possible to formulate a judgment, and explicit evaluation ('John is honest; John is a serious guy; John is really expert in this') and then *feeling* a positive trust disposition towards John;

(ii) Feeling – for some unconscious and unclear analogy and evocation of previous affective experiences – a positive affective disposition of safety, reliability, other's benevolence, towards John, and on such a basis to formulate real beliefs and explicit evaluations on him.

In fact, why do we need to spend so much time on the theory of implicit, affective appraisal? Because there are forms of trust just based on this, not on explicit beliefs about *Y*. And also because trust usually has an affective component, is some sort of weak 'emotion', or at least a 'feeling', an affective disposition.

5.6 Trust as Feeling

Trust is also a 'feeling', something that the agent 'feels' towards another agent, something one 'inspires' to the others. It can be just *confidence* (similar to self-confidence) not a judgment. It can be not arguable and based on reasons or explicit experiences; it can be just 'dispositional' or just 'intuitive' and based on tacit knowledge and implicit learning.

At a primitive level (consider a baby) trust is something not express/ed/ible in words, not made of explicit beliefs about *Y*'s competence or reliability. It is a spontaneous, non reasonable or reasoned upon (non rational) reliance, and a feeling of confidence in a given environment or in a person.

What is this kind or this facet of trust?

Trust as a feeling is characterized by a sensation of 'letting oneself go', of relaxing, a sort of confident surrendering; there is an attenuation of the alert and defensive attitude (consider the trust/confidence of a baby towards her mother).

Affective components of trust result in a felt freedom from anxiety and worried; *X* feels safe or even protected; there is no suspicion or hostility towards *Y*, which is appraised/felt as benevolent and reliable ('S/he will take care of...'). Towards a benevolent *Y*, we are benevolent, good-willing; towards a good/skilled *Y*, we are not aroused, alerted, cautious, worried (*X* could say: 'I am in your hands'; while *Y* would say 'Let yourself go, do not resist, *trust* me').

It is very important to note – coherently with our model of trust – that it is not true that in this kind of trust we (implicitly) evaluate/perceive only the good disposition, the *good will* of the other; we also appraise his *power* (ability, competence): for example, for a feeling of protection and safety, and for relying on the other as for being protected against bad guys with a sense of safety, also the perception of his physical strength and character is crucial.

Notice how these possible affective components of trust are coherent and compatible with our cognitive analysis of trust. However, they can also be independent of any judgment; they can be just the affective, dispositional consequence of an intuitive appraisal (cf. Section 5.2) and of learning. They can even be by default or just the result of lack of (bad) experiences, lack of negative evaluations (Chapter 4). In fact, a bad experience with *Y* would be a bad 'surprise' for *X*; something one was not suspecting at all. There are no really *subjectively* 'rational' justifications for that attitude, but it can be 'rational' in relation to a repeated experience or to evolutionary adaptive functions.

5.7   Trust Disposition as an Emotion and Trust Action as an Impulse

An emotion or feeling is 'caused' by, elicited by, it is spontaneously arousing on the basis of given beliefs (not only of 'stimuli', like for more simple and primitive emotions, like a very fast reaction of startle and fear due to a terrible noise, before any understanding of it). *The emotion is the 'response' of our mind-body to a given (mental) event (the internal configuration of our representation of the situation).* Those beliefs are not 'reasons' for the affective reaction, like they are 'reasons' for a believing or for a decision ('arguments').

This is our model of emotions in terms of 'cognitive anatomies', i.e. in terms of the specific beliefs and goals that are needed for eliciting and entertaining that emotion, which are necessary constituents, ingredients, but also bases for it. We would say that certain beliefs are the causes, the triggers of a feeling of shame, guilt, etc. and from an external point of view, they are also the 'reasons' for that emotion, but not from the internal, subjective point of view: emotions have no 'reasons' in the strict sense.

Feelings and emotions usually activate a goal (a behavioral tendency or a more abstract desire to be translated into actions). For example, fear activates the goal of escaping, of being safe; shame, the goal of disappearing; pity, the goal of being of help; guilt, the goals of repairing, atoning, and not doing the same again.

We can accept trust as feelings, for the same kind of analysis:

- a feeling, an affective response arousing from a given more or less explicit perception and appraisal of the world;
- an activated goal on the basis of this feeling and mental configuration.

We have also to remind our reader that in our model (as in several others), there is the possibility that the relation and path from assumptions to feeling can be reversed: instead of feeling fear because one sees or thinks that there is some danger, one can assume that there is some danger just because one is feeling fear; using the sensation as a sort of (non 'rational') 'evidence'.

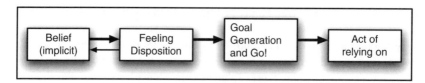

Figure 5.2 From Assumptions to Feeling

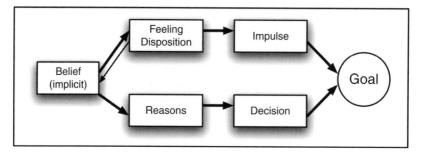

Figure 5.3 Two Kinds of Trust

So the model of the process would be something like that shown in Figure 5.2.

What matters here is the fact that a purposive behavior is generated not on the basis of a true 'decision' based on reasons, but is affectively activated, and has a strength not due to its supports and their credibility but just to the *intensity* of the affective arousal (Castelfranchi, 1998), and it not necessarily acted-out on the basis of a true 'deliberation' process.[6]

We might say that there are *two kinds of trust and two parallel paths of trust* (following in a sense a 'dual' model à la Kahneman) (see Figure 5.3).

In both cases, *trust* is not only the disposition towards the other (appraisal, evaluation, expectations), but is a winning goal of relying upon the other and the act of making oneself relied upon (many authors miss this point) and thus *making* oneself vulnerable and non-defended to the other.

One way or the other, complete trust arrives in both cases at a goal (the goal of counting on Y as for g) and an act (some sort of intention in action, even without any 'reason'), and then to a 'relation' which is not only evaluative, affective, or dispositional, but is actual, practical: of a concrete and exploited 'dependence' (relying/counting upon).

5.8 Basing Trust on the Emotions of the Other

The relationships between trust and affect/feelings are more complex. Not only can trust be just a feeling, an affective attitude and move towards Y; but affects from Y's side are also very important.

---

[6] M. Tuomela's (Tuomela, 2003) analysis of trust is correct under this respect (as 'affective Trust not implying a true 'decision' and 'intention') but remains too restricted: only based on a perceived relation of mutual respect, of shared values, of recognition of my rights. This kind of trust can also be elicited just by love and protective attitude of the other towards me, or by pity, etc.

Y's perceived trustworthiness/reliability can just be based on the perception or ascription of affects and emotions in *Y*. We ascribe to the other not only goals and beliefs but also emotions and moods, and our positive (or negative) evaluation as for trusting *Y*, or our prediction can just be based on the perceived/ascribed emotion.

For example, I can rely on the fact that *Y* will carry out a given action (for example, helping a poor guy sleeping in the streets) because I see that he is really feeling pity or guilt, and he is pushed to care about that guy. Or I expect that *Y* will interpose between *Z* and me (*Z* is aggressing me) just because *Y* is very angry with *Z* and aggressively reacting to *Z*'s aggression. Or I expect help from *Y* because I realize that he has very nice sentiments and positive emotions towards me.

X's emotions and *Y*'s emotions can be strongly related with each other. On the one hand, *X*'s perception of *Y*'s emotional attitude can be based on *X*'s own emphatic emotions. On the other hand, *X*'s emotion of trust can be just the affective response and reciprocation of *Y*'s perceived positive emotions towards *X*. Moreover, *Y*'s perceived emotions can be elicited by *X*'s trustful and affective attitude (Chapter 8).

But again the affective-based and the reason-based attitude of trust can be (and usually are) mixed up with each other. Even an evaluation or a decision to trust *Y*, based on reasons and evidence, can elicit not only an affective disposition in *X* towards *Y*, but also an affective response in *Y*. Or vice versa, an affective disposition of *Y*, can be the base for a well-argued evaluation of *Y* as for a given task or relation (for example, as a babysitter).

5.9 The Possible Affective Base of 'Generalized Trust' and 'Trust Atmosphere'

We deny that the theory of generalized or diffuse trust (the so called 'social capital') is a separate and different model from the theory of interpersonal trust. In our view, the former must be built upon and derived from the latter (Chapter 6).

On the one hand, *trust capital is a macro, emerging phenomenon; but it must be understood also in terms of its micro foundations*.

On the other hand, conceptually speaking, the *notion of trust is just one and the same notion*,[7] at different levels of abstraction and generalization.

It is true that social trust relationships and networks (Chapter 9) are more than the simple summation of *X*'s trust towards *Y* (and *Z*, and *W*) and *Y*'s (etc.) trust towards *X* (etc.). Social trust builds richer 'relationships': *'I trust you; in so doing our relationship moves a little further forward.'* ((Holton, 1994) p. 68). But, why is it so?

Because *X*'s trust is also a *signal*, it exhibits an internal evaluation, etc.; because it gives or shows to the other the opportunity to harm *X* and thus *the opportunity to choose not to harm X*, of favoring *X*; because it elicits gratitude, relax, trust improvement; because it creates reciprocity of both attitude and behavior.

5.10 Layers and Paths

In conclusion, *in our model there are three layers or stages and two paths or families of trust* (see Figure 5.4).

---

[7] Both, as emotional disposition, feeling, and action tendency; and as an explicit evaluation, belief, and grounded expectation and decision.

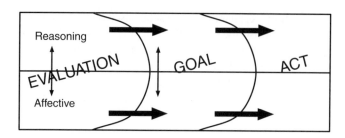

Figure 5.4   Layers and Paths of Trust

When discussing the layers we have:

1) Appraisal and disposition towards Y;
2) Goal generation and selection;
3) Action and the consequent practical relation (*counting on*).

About the paths/families we have:

1) Reason-based trust; and
2) Affective or feeling-based trust.

In Figure 5.4, the phase of evaluation corresponds with the double potential question: What I think/feel about Y (as for my depending for τ).

The two roots are not completely alternative and independent: they can coexist and influence each other. This is why we put some crossing arrow between them; in order to signal possible influences of the feeling on the goal, although this is part of a real deliberation, and is based on explicit reasons and evaluations.

As we have already said, in this book we are mainly focused on reasons and decision-based trust, not on the merely affective and 'impulsive' one; but they are 'equifinal' (respond to similar 'functions') and the former is an evolved form of the latter.

5.11   Conclusions About Trust and Emotions

Here we are just analyzing an affective-based form of trust, not the relationships between trust and emotions in general. However, notice that our socio-cognitive analysis of trust as psychological state and behavioral attitude in terms of cognitive and motivational components, provides a systematic and grounded basis for predicting and explaining the relationships between trust and emotions.

On the one hand, some trust constituents are also constituents of emotional reaction, which – thus – can co-occur with trust. This is the relationship between trust and hope, or between trust and worries/fear; or trust in Y and sympathy/benevolence towards Y; or trust and relaxation, not to feel exposed or in danger with Y.

Moreover, the trust attitude and decision exposes to new affective states and reactions, like 'disappointment' (since trust implies positive expectations), or to 'feel betrayed' (when trust relies on *Y*'s adoption and commitment).

A good componential analysis (cognitive-motivational anatomy) of a mental state should be able to account for the relationship between different kinds and levels of that phenomenon, of the relationships between it and close or possibly co-occurring mental states, and conceptual families.

We do not propose to provide a systematic treatment of that here, but in different parts of this book we stress the justified relations between trust and other affective or cognitive states: expectations, evaluations, uncertainty, anxiety, fear, surprise, hope, faith, optimism, sympathy, and so on.

References

Arnold M.B. (1969) *The Nature of Emotion*. Penguin Books.

Bargh, J. A., and Chartrand, T. L. The unbearable automaticity of being. *American Psychologist*, 54: 462–479, 1999.

Bechara, A., Damasio, A.R., Damasio, H. and Anderson, S.W. Insensitivity to future consequences following damage to human prefrontal cortex. *Cognition*, 50: 7–15, 1994.

Bechara, A., Damasio, H., Tranel, D. and Damasio, A.R. Deciding advantageously before knowing the advantageous strategy. *Science*, 275: 1293–1295, 1997.

Castelfranchi, C. (1998) To believe and to feel: The case of 'needs'. In D. Canamero (ed.) *Proceedings of AAAI Fall Symposium 'Emotional and Intelligent: The Tangled Knot of Cognition'*, AAAI Press, 55–60.

Castelfranchi, C. (2000) Affective appraisal vs cognitive evaluation in social emotions and interactions. In A. Paiva (ed.) *Affective Interactions. Towards a New Generation of Computer Interfaces*. Heidelberg, Springer, LNAI 1814, 76–106.

Castelfranchi, C., Miceli, M. (2009) The cognitive-motivational compound of emotional experience. *Emotion Review*, 1 (3): 221–228.

Damasio, A. R. (1994) *Descartes' Error*. New York, Putnam's Sons.

Elster, J. (1999) *Alchemies of the Mind: Rationality and the Emotions*, Cambridge University, Press Cambridge UK.

Frijda, N. H. (1986) *The Emotions*, Cambridge University Press.

Frijda, N. H. (1988) *Cognitive Perspectives on Emotion and Motivation*, Springer.

Holton, R. Deciding to trust, coming to believe. *Australian Journal of Philosophy* 72: 63–76, 1994.

Kunda, Z. (1990) *The Case for Motivated Reasoning*. Princeton University.

Lazarus, R. S., Averill, J. R., and Opton, E. M. (1970) Towards a cognitive theory of emotion. In M. B. Arnold (ed.), *Feelings and emotions, The Loyola Symposium*. New York: Academic Press. (pp. 207–232).

Mandler, G. (1975) *Mind and Emotions*, New York: Wiley.

Miceli, M., and Castelfranchi, C. (2000) The role of evaluation in cognition and social interaction. In K. Dautenhahn (ed.), *Human Cognition and Agent Technology* (pp. 225–261). Amsterdam: Benjamins.

Ortony, A., Clore, G. L., Collins, A. (1990) *The Cognitive Structure of Emotions*.

Scherer, K. R. (1986) *Experiencing Emotion*. Cambridge University Press.

Scherer, K. R. (1999) Appraisal theories. In T. Dalgleish, and M. Power (eds.). *Handbook of Cognition and Emotion* (pp. 637–663). Chichester: Wiley.

Tuomela, M. (2003) A Collective's Rational Trust in a Collective's Action. In Understanding the Social II: Philosophy of Sociality, *Protosociology*. 18–19: 87–126.

6

Dynamics of Trust

Trust in its intrinsic nature is a dynamic phenomenon. Trust has to change on time, because all the entities participating in the trust relationship are potentially modifiable. In real interactions, we never have exactly the same interactive situation in different time instants.

Trust changes with experience, with the modification of the different sources it is based on, with the emotional or rational state of the trustor, with the modification of the environment in which the trustee is supposed to perform, and so on. In other words, being trusted is an attitude depending on dynamic phenomena, as a consequence it is itself a dynamic entity.

In fact, trust is in part a socially emergent phenomenon; it is a mental state, but in socially situated agents and based on social context. In particular, trust is a very dynamic phenomenon; not only because it is based on *the trustor*'s previous experiences, but because it is not simply an external observer's prediction or expectation about a matter of fact.

There are many studies in literature dealing with the dynamics of trust ((Jonker and Treur, 1999), (Barber and Kim, 2000), (Birk, 2000), (Falcone and Castelfranchi, 2001)). We are interested in analyzing four main basic aspects of this phenomenon:

i) The traditional problem of how trust changes on the basis of the trustor's experiences (both positive and negative ones).
ii) The fact that in the same situation *trust is influenced by trust* in several rather complex ways.
iii) How diffuse trust diffuses trust (*trust atmosphere*), that is how X's trusting Y can influence Z trusting Y or W, and so on.
iv) The fact that it is possible to predict how/when an agent who trusts something/someone will therefore trust something/someone else, before and without a direct experience (*trust through generalization reasoning*).

The first case (i) considers the well known phenomenon about the fact that trust evolves in time and has a history, that is X's trust in Y depends on X's previous experience and learning with Y himself or with other (similar) entities. In the following sections, we will analyze this case showing that it is true that in general a successful performance of Y increases X's trust in him (and vice versa a failing performance drastically decreases X's trust) but we will also

Trust Theory Cristiano Castelfranchi and Rino Falcone
© 2010 John Wiley & Sons, Ltd

consider some not so easily predictable results in which trust in the trustee decreases with positive experiences (when the trustee realizes the delegated task) and increases with negative experiences (when the trustee does not realize the delegated task).

The dynamic nature of trust is also described by the second case (ii) where we will study the fact that in one and the same situation *trust is influenced by trust* in several rather complex ways. In particular, we will analyze two main crucial aspects of trust dynamics.

How trust creates a reciprocal trust, and distrust elicits distrust; but also vice versa: how *X*'s trust in *Y* could induce lack of trust or distrust in *Y* towards *X*, while *X*'s diffidence can make *Y* more trustful in *X*. In this chapter we will examine also an interesting aspect of trust dynamics: *How the fact that X trusts Y and relies on him in situation* Ω *can actually (objectively) influence Y's trustworthiness in the* Ω *situation*. Either trust is a self-fulfilling prophecy that modifies the probability of the predicted event; or it is a self-defeating strategy by negatively influencing the events. And also how *X* can be aware of (and takes into account) the effect of its own decision in the very moment of that decision (see also Section 8.9). We will also analyze the trust atmosphere. This is a macro-level phenomenon, and the individual agent does not calculate it. Finally, we will consider the power of the trust cognitive model for analyzing and modeling the crucial phenomenon of a trust transfer from one agent to another or from one task to another.

As we have argued in Chapters 2 and 3, we will resume in the following, that trust and reliance/delegation are strictly connected phenomena: trust could be considered as the set of mental components on which a delegation action is based. In the analysis of trust dynamic, we have also to consider the role of delegation (*weak*, *mild* and *strong* delegation) (Castelfranchi and Falcone, 1998).

6.1   Mental Ingredients in Trust Dynamics

From the point of view of the dynamic studies of trust, it is relevant to underline how the basic beliefs, described in Chapter 2, might change during the interaction between the trustor and the trustee: for example, they could change the abilities of the trustee or his reasons/motives for willing (and/or the trustor's beliefs on them); or again it might change the dependence relationships between the trustor and the trustee (and so on).

Another important characteristic of the socio-cognitive model of trust is the distinction (see also Section 2.7.2 and Section 8.3.3) between trust 'in' someone or something that on the basis of its *internal characteristics* can realize a useful action or performance, and the global trust in the global event or process and its result which is also affected by *external factors* (to the trustee) like opportunities and interferences.

Trust in *Y* (for example, 'social trust' in the strict sense) seems to consist of the two first prototypical beliefs/evaluations identified as the basis for reliance: *ability/competence* (that with cognitive agents includes *knowledge* and *self-confidence*), and *disposition/motivation* (that with cognitive agents is based on *willingness*, *persistence*, *engagement*, etc.).

Evaluation about external opportunities is not really an evaluation about *Y* (at most the belief about its ability to recognize, exploit and create opportunities is part of our trust 'in' *Y*). We should also add an evaluation about the probability and consistence of obstacles, adversities, and interferences.

Let us now introduce some formal constructs. We define $Act=\{\alpha_1, . . ,\alpha_n\}$ be a finite set of *actions*, and $Agt=\{X, Y, A, B,..\}$ a finite set of *agents*. Each agent has an action repertoire, a plan library, resources, goals, beliefs, motives, etc.

As introduced in Chapter 2, the action/goal pair $\tau=(\alpha,g)$ is the real object of delegation, and we called it 'task'.[1] Then by means of τ, we will refer to the action (α), to its resulting world state $(g)^2$, or to both.

Given an agent Y and a situational context Ω (a set of propositions describing a state of the world), we define as trustworthiness of Y about τ in Ω (called *trustworthiness (Y, τ, Ω)*), the objective probability that Y will successfully execute the task τ in context Ω. This objective probability is in terms of our model computed on the basis of some more elementary components:

- An *objective degree of ability* (*OdA*, ranging between 0 and 1, indicating the level of Y's ability about the task τ); we can say that it could be measured as the number of Y's successes (s) on the number of Y's attempts (a): s/a, when a goes to ∞:

$$OdA_Y = lim \, (a{-}{>}\infty) \, s/a \qquad (6.1)$$

and
- An *objective degree of willingness* (*OdW*, ranging between 0 and 1, indicating the level of Y's intentionality/persistence about the task τ); we can say that it could be measured as the number of Y's (successfully or unsuccessfully) performances (p) of that given task on the number of times Y declares to have the intention (i) to perform that task: p/i, when i goes to ∞:

$$OdW_Y = lim \, (i{-}{>}\infty) \, p/i \qquad (6.2)$$

we are considering that an agent declares its intention each time it has got one.

So, in this model we have that:

$$Trustworthiness \, (Y, \, \tau, \, \Omega) = F(OdA_{Y\tau\Omega}, \, OdW_{Y\tau\Omega}) \qquad (6.3)$$

Where F is in general a function that preserves monotonicity, and ranges in $(0,1)$: for the purpose of this work it is not relevant to analyze the various possible models of the function F. We have considered this probability as *objective* (absolute, not from the perspective of another agent) because we hypothesize that it measures the real value of Y's trustworthiness; for example, if *trustworthiness(Y $\tau\Omega$) = 0.80*, we suppose that in a context Ω, *80%* of times Y tries and succeeds in executing τ.

As the reader can see, we have considered the opportunity dimension as included in Ω: the external conditions favoring, allowing or inhibiting, impeding the realization of the task.

---

[1] We assume that *to delegate an action necessarily implies delegating some result of that action*. Conversely, *to delegate a goal state always implies the delegation of at least one action (possibly unknown to Y) that produces such a goal state as result.*

[2] We consider $g = g_X = p$ (see Chapter 2, Section 2.1).

6.2   Experience As an Interpretation Process: Causal Attribution for Trust

It is commonly accepted ((Jonker and Treur, 1999), (Barber and Kim, 2000), (Birk, 2000)) and discussed in another work of ours (Falcone and Castelfranchi, 2001)) that one of the main sources of trust is direct experience. It is generally supposed that, on the basis of the realized experiences, to each success of the trustee (believed by the trustor) there is a significant increment or a confirmation of the amount of the trustor's trust towards him, and that to every trustee's failure (believed by the trustor) there is a corresponding reduction of the trustor's trust towards the trustee itself.

There are several ways in which this qualitative model could be implemented in a representative dynamic function (linearity or not of the function; presence of possible thresholds (under a minimum threshold of the trustworthiness's value there is no trust, or vice versa, over a maximum threshold there is full trust), and so on).

This view is very naïve, neither very explicative for humans and organizations, nor useful for artificial systems, since it is unable to discriminate cases and reasons of failure and success adaptively. However, this primitive view cannot be avoided until trust is modeled just as a simple index, a dimension, an all-inclusive number; for example, reduced to mere subjective probability. We claim that a cognitive attribution process is needed in order to update trust on the basis of an 'interpretation' of the outcome of X's reliance on Y and of Y's performance (failure or success). In doing this, a cognitive model of trust – as we have presented – is crucial. In particular we claim that the effect of both Y's failure or success on X's trust in Y depends on X's 'causal attribution' ((Weiner, 1992)) of the event.

Following 'causal attribution theory', any success or failure can be either ascribed to factors *internal* to the subject, or to environmental, *external* causes, and either to *occasional* facts, or *stable* properties (of the individual or of the environment).

So, there are four possible combinations: *internal* and *occasional*; *internal* and *stable*; *external* and *occasional*; *external* and *stable*.

Is Yody's guilt or merit based on whether he was failing or successful on τ? Or was the real responsibility about the conditions in which he worked? Was his performance the standard performance he was able to realize? Were the environmental conditions the standard ones in which that task is realized?

The cognitive, emotional, and practical consequences of a failure (or success) *strictly depend on this causal interpretation*. For example – psychologically speaking – a failure will impact on the self-esteem of a subject only when attributed to *internal* and *stable* characteristics of the subject itself. Analogously, a failure is not enough for producing a crisis of trust (see Chapter 9); it depends on the causal *interpretation* of that outcome, on its attribution (the same for a success producing a confirmation or improvement of trust). In fact, we can say that a first qualitative result of the causal interpretation can be resumed in the following flow chart (Figure 6.1).

Since in agent-mediated human interaction (like Computer Supported Cooperative Work or Electronic Commerce) and in cooperating autonomous Multi-Agent Systems it is fundamental to have a theory of, and instruments for 'Trust building' we claim that a correct model of this process will be necessary and much more effective. However, this holds also for marketing and its model of the consumer's trust and loyalty towards a brand or a shop or a product/service; and for trust dynamics in interpersonal relations; and so on.

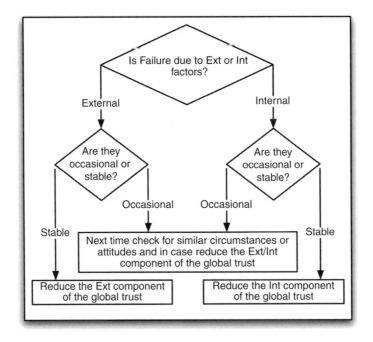

Figure 6.1   Flow-chart to identify the causes of failure/success

Let's first present a basic model (which exploits our cognitive analysis of trust attitude and the 'Causal Attribution Theory' which are rather convergent), and later discuss possible more complex dynamics. The following analysis takes into account the stable facts.

We consider a general function by which the agent X evaluates its own trust (*degree of trust*) in agent Y about the task τ (to be performed) in the environment Ω ($DoT_{X,Y,\tau,\Omega}$):

$$DoT_{X,Y,\tau,\Omega} = f(DoA_{X,Y,\tau},\ DoW_{X,Y,\tau},\ e(\Omega)) \qquad (6.4)$$

Where: f (like F) is a general function that preserves monotonicity. In particular, $DoA_{X,Y,\tau}$ is Y's degree of ability (in X's opinion) about the task τ; $DoW_{X,Y,\tau}$ is Y's degree of motivational disposition (in X's opinion) about the task τ (both $DoA_{X,Y,\tau}$ and $DoW_{X,Y,\tau}$ are evaluated in the case in which Y would try to achieve that task in a standard environment: an environment with the commonly expected and predictable features); $e(\Omega)$ takes into account the part of the task not directly performed by Y (this part cannot be considered as a separated task but as an integrating part of the task and without which the same task cannot be considered as complete) and the hampering or facilitating conditions of the specific environment.

In a simplified analysis of these three sub-constituents ($DoA_{X,Y,\tau}$ (*Abilities*), $DoW_{X,Y,\tau}$ (*Motivations*), and $e(\Omega)$ (*Environment*)) of X's degree of trust, we have to consider the different possible dependencies among these factors:

i) We always consider *Abilities* and *Motivations* as *independent* to each other. We assume this for the sake of simplicity.

ii) Case in which *Abilities* and *Motivations* are both *independent* from the *Environment*.

It is the case in which there is a part of the task performed (activated, supported etc.) from the *Environment* and, at the same time, both *Abilities* and *Motivations* cannot influence this part of the task. Consider for example, the task of urgently delivering a piece of important machinery to a scientific laboratory in another town. Suppose that this apparatus could be sent by using any service of delivery (public, private, fast or normal, and so on) so that a part of the task (to materially bring the apparatus) is independent (once made the choice) from the actions of the trustee.

iii) Case in which *Abilities* and *Environment* are *dependent* on each other.

We have two sub-cases: first, the *Environment* favours or disfavours the *Y*'s *Abilities* (useful for the task achievement); second, the *Y*'s *Abilities* can modify some of the conditions of the *Environment* (both these sub-cases could be known or not before the task assignment).

iv) Case in which *Motivations* and *Environment* are dependent with each other.

Like for case (iii), there are two sub-cases: first, the *Environment* influences *Y*'s *Motivations* (useful for the task achievement); second, *Y*'s *Motivations* can modify some of the conditions of the *Environment* (both these sub-cases could be known or not before the task assignment).

Given this complex set of relationships among the various sub-constituents of trust, a well informed trustor who is supplied with an analytic apparatus (a socio-cognitive agent), could evaluate which ingredients performed well and which failed in each specific experiential event (analyzing and understanding the different role played by each ingredient in the specific performance).

Let us start from the case in which *Abilities* and *Motivations* both are considered as composed of internal properties and independent from the *Environment* (case (ii)). After an experiential event the trustor could verify:

$$Actual(DoA, \ DoW) - Expected(DoA, \ DoW) > 0 \qquad (6.5)$$

$$Actual(DoA, \ DoW) - Expected(DoA, \ DoW) < 0 \qquad (6.6)$$

$$Actual(e(\Omega)) - Expected(e(\Omega)) > 0 \qquad (6.7)$$

$$Actual(e(\Omega)) - Expected(e(\Omega)) < 0 \qquad (6.8)$$

Where the operators *Actual* and *Expected* give the values of the arguments as respective evaluations after the performance of the task and before it.

In (6.5) and (6.7) both the trustee (*internal-trust*) and the environment (*external-trust*) are more trustworthy than expected by the trustor; vice versa, in (6.6) and (6.8) they are both less trustworthy than expected by the trustor.

In Table 6.1 all the possible combinations are shown.

Where: 'More Int-trust' ('Less Int-trust') means that the trustor after the performance considers the trustee more (less) trustworthy than before it (he performed better (worst) than expected); 'More Ext-trust' ('Less Ext-trust') means that the trustor after the performance considers the environment more (less) trustworthy than before it (it performed better (worst) than expected).

Table 6.1 Performances of the trustee and the environment in combination with the success or failure of the global task

	Success of the performance	Failure of the performance
Δ(int-trust) > 0 Δ(ext-trust) > 0	**A** More Int-trust; More ext-trust	**A'** More Int-trust; More ext-trust
Δ(int-trust) > 0 Δ(ext-trust) < 0	**B** More Int-trust; Less ext-trust	**B'** More Int-trust; Less ext-trust
Δ(int-trust) < 0 Δ(ext-trust) > 0	**C** Less Int-trust; More ext-trust	**C'** Less Int-trust; More ext-trust
Δ(int-trust) < 0 Δ(ext-trust) < 0	**D** Less Int-trust; Less ext-trust	**D'** Less Int-trust; Less ext-trust

Cases of particular interest are:

(*B in Table 6.1*) in which even if the environment is less trustworthy than expected, the better performance of the trustee produces a global success performance.

In fact, three factors have to be considered: the trustor over-evaluated the environmental's trustworthiness; she under-evaluated the trustee's trustworthiness; the composition of internal and external factors produced a successful performance.

(*C in Table 6.1*) in which even if the trustee is less trustworthy than expected, the better performance of the environment produces a global success performance.

Also in this case, three factors have to be considered: the trustor under-evaluated the environmental's trustworthiness; she over-evaluated the trustee's trustworthiness; the composition of internal and external factors produced a successful performance. Tom actually made a mess, was really a disater, but was also incredibly 'lucky': accidentally and by circumstantial factors eventually the desired result was there. But not thanks to his ability or willingness!

This is a very interesting case in which the right causal attributions make it possible the trust in the trustee to decrease even in presence of his success. Of course, two main possible consequences follow: the new attributed trustworthiness of the trustee is again (in the trustor's view) sufficiently high (over a certain threshold) for trusting him later; vice versa, the new attributed trustee's trustworthiness is now unsufficient for trusting him later (under the threshold).

(*D and A' in Table 6.1*) In which expectations do not correspond with the real trustworthiness necessary for the task: too high (both in the trustor and in the environment) in D and too low (both in the trustor and in the environment) in A'. These cases are not possible if the trustor has a good perception of the necessary levels of trustworthiness for that task (as we suppose in the other cases in Table 6.1).

(*B' in Table 6.1*) in which even if the trustee is more trustworthy than expected (so increases the trust in him), the unexpected (at least for the trustor) difficulties introduced by the environment produce a failure of the global performance. *This is another interesting case in*

*which the right causal attributions make it possible to increase the trust in the trustee even in
the presence of his failure.*

Consider this situation: 'In the last tour in Italy I saw that the coach driver did his best to
arrive on time at the Opera in Milano, and was very competent and committed: to recognise
his shortcomings, drive quickly, but safely, and do everything he could, but unfortunately,
given that we were already late because of some of the passengers and due to the traffic on the
highways we missed the Opera. In any case my trust in that driver actually increased due to
the respect I had in him and I would definitely use him again'.

Again, the case is made more complex when there is some dependence between the internal
properties and the environment (cases (iii) and (iv). In this case, in addition to the introduced
factors Δ*(int-trust)* and Δ*(ext-trust)*, we have to also take into account the possible influences
between internal and external factors. We consider these influences as not expected from the
trustor in the sense that the expected influences are integrated directly in the internal or external
factors. We can – for example – consider the case of a violinist. We generally trust him for
playing very well; but, suppose he has to do the concert in an open environment and the
weather conditions are particularly bad (very cold): may be these conditions can modify the
specific hand abilities of the violinist and his performance; at the same time, it is possible that
a special distracted, inattentive, noisy audience could modify his willingness and consequently
again his performance.

Concluding with the experience-based trust we have to say that the important thing is not
only the final result of the trustee's performance but in particular *the trustor's causal attribution
to all the factors producing that result.* It is on the basis of these causal attributions that the
trustor updates her beliefs about the trustworthiness of the trustee, of the environment, and of
their reciprocal influences.

So the rational scheme is not the simplified one showed in Figure 6.2 (where there is a
trivial positive or negative feedback to Y on the basis of the global success or global failure),
but the more complex one showed in Figure 6.3 (where in the case of either failure or success
both the components and their specific contributions are considered).

6.3   Changing the Trustee's Trustworthiness

In this paragraph we are going to analyze how a delegation act (corresponding to a decision
making based on trust in a specific situational context) could change (just because of that
delegation action and in reaction to it) the trustworthiness of the delegated agent (*delegee*).
This not only holds in *Strong-Delegation* (where Y is aware of it and X counts on his awareness,
adhesion, and commitment, and there is some explicit or implicit communication of X's trust
and reliance; Section 2.6.1). It even holds in *Weak-Delegation* (where Y and his autonomous
behavior – in X's intention – is simply exploited by X), but in peculiar conditions.

6.3.1   The Case of Weak Delegation

As also shown in Section 2.6, we call the reliance simply based on exploitation for the
achievement of the task *weak delegation* (and express this with *W-Delegates(X Y τ)*). In it
there is no agreement, no request or even (intended) influence: *X is just exploiting in its plan*

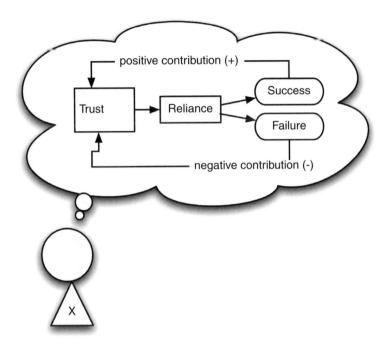

Figure 6.2 Simplified scheme of the performances' influences on Trust Evaluation

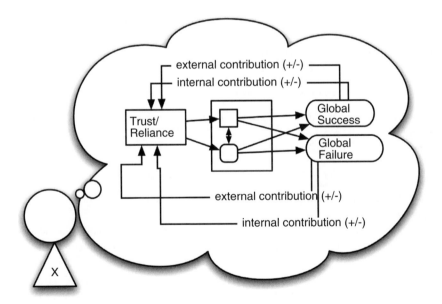

Figure 6.3 Realistic (with causal attribution) scheme of the performances' influences on Trust Evaluation

a fully autonomous action of Y. For a more complete discussion on the mental ingredients of the weak delegation see (Castelfranchi and Falcone, 1998).

The expression *W-Delegates(X Y τ)* represents the following *necessary* mental ingredients:

a) The achievement of τ (the execution of α and its result g) is a *goal* of X.

b) X believes that there exists another agent Y that has the *power of* (Castelfranchi, 1991) achieving τ.

c) X believes that Y will achieve τ in time and by itself (without any X's intervention).
   c-bis) X believes that Y *intends* (in the case that Y is a cognitive agent) to achieve τ in time and by itself, and that will do this in time and without any intervention of X.

d) X *prefers*[3] to achieve τ through Y.

e) The achievement of τ through Y is the choice (goal) of X.

f) X has the goal (*relativized* (Cohen and Levesque, 1987) to (e)) of not achieving τ by itself.

We consider (*a, b, c,* and *d*) what the agent X views as a '*Potential for relying on*' the agent Y, its *trust* in Y; and (*e* and *f*) what X views as the '*Decision to rely on*' Y. We consider 'Potential for relying on' and 'Decision to rely on' as two constructs temporally and logically related to each other.

We hypothesize that in *weak delegation* (as in any delegation) there needs to be a decision made based on trust and in particular there are two specific beliefs of X:

belief1$_X$: if Y makes the action then Y has a successful performance;

belief2$_X$: Y intends to do the action.

For example, X sees Y waiting at the bus stop, and – while running to catch the bus – she counts on Y to stop it.

As shown in Section 6.2, the trustworthiness of Y is evaluated by X using the formula (6.4). For the sake of simplicity we assume that:

$$DoT_{X,Y,\tau,\Omega} \equiv trustworthiness\ (Y\ \tau\ \Omega) \tag{6.9}$$

In other words: X has a perfect perception of Y's trustworthiness: X believes/knows Y's real trustworthiness.

The interesting stage in weak delegation is when:

$$Bel(X\neg Bel(Y\ W\text{-}Delegates\ (X,\ Y,\ \tau))) \cap Bel(Y\ W\text{-}Delegates(X, Y, \tau))^4 \tag{6.10}$$

in other words: there is a weak delegation by X on Y, but actually Y is aware of it (while X believes that Y is not).

The first belief (*belief1$_X$*) is very often true in weak delegation, while the second one (*belief2$_X$*) is necessary in the case we are going to consider. If *Bel(Y W-Delegates(X Y τ))*, this

---

[3] This means that, either relative to the achievement of τ or relative to a broader goal g' that includes the achievement of τ, X believes herself to be dependent on Y (see (Jennings, 1993), (Sichman *et al.*, 1994) and Section 2.9.1 on 'weak dependence').

[4] Other possible alternative hypoteses are:
$\neg Bel(X\ Bel(Y\ W\text{-}Delegates(X,\ Y,\ \tau))) \cap Bel(Y\ W\text{-}Delegates(X, Y, \tau))$ *or*
$Bel(X\ Bel(Y\neg W\text{-}Delegates(X, Y, \tau))) \cap Bel(Y\ W\text{-}Delegates(X, Y, \tau))$

belief could change Y's trustworthiness, either because Y will adopt X's goal as an additional motivation and accept such an exploitation, or because, on the contrary, Y will refuse such an exploitation, changing his behaviour and reacting to the delegation (there is in fact also a third case, in which this knowledge does not influence Y's behaviour and beliefs: we do not consider this case). After the action of delegation we have in fact a new situation Ω' (if delegation is the only event that influences the trustworthiness) and we can have two possible results:

i) the new trustworthiness of Y as for τ is greater than the previous one; at least one of the two possible elementary components is increased: OdA, OdW; so we can write:

$$\Delta trustworthiness(Y\ \tau) = F(OdA_{Y,\tau,\Omega'},\ OdW_{Y,\tau,\Omega'})\text{-}F(OdA_{Y,\tau,\Omega}, OdW_{Y,\tau,\Omega}) > 0$$
(6.11)

ii) Y's new reliability as for τ has reduced

$$\Delta trustworthiness\ (Y\ \tau) < 0 \tag{6.12}$$

In case (6.11) Y has adopted X's goal, i.e. he is doing τ also in order to let/make X achieve its goal g. Such adoption of X's goal can be for several possible motives, from instrumental and selfish, to pro-social.

The components' degree can change in different ways: the degree of ability (OdA) can increase because, for example, Y could invest more attention in the performance, use additional tools, new consulting agents, and so on; the degree of willingness (OdW) can increase because Y could have additional motives and a firm intention, and so on (the specific goal changes its level of priority).

In case (6.12) Y on the contrary reacts in a negative way (for X) to the discovery of X's reliance and exploitation; for some reason Y is now less willing or less capable of doing τ. In fact in case (ii) too, the reliability components can be independently affected: first, the degree of ability (OdA) can decrease because Y could be upset about the X's exploitation and Y's ability could be compromised; again, the willingness degree (OdW) can decrease (Y will have less intention, attention, etc.).

Notice that in this case the change of Y's reliability is not known by X. So, even if X has a perfect perception of previous Y's trustworthiness (that is our hypothesis), in this new situation – with weak delegation – X can have an under or over estimation of Y's trustworthiness. In other terms, after the weak delegation (and if there is a change of Y's trustworthiness following it) we have:

$$DoT_{X,Y,\tau,\Omega} \neq trustworthiness(Y\ \tau\ \Omega') \tag{6.13}$$

Let us show you the flow chart for the weak delegation (Figure 6.4): in it we can see how, on the basis of the mental ingredients of the two agents, the more or less collaborative behaviours of the trustee could be differently interpreted by the trustor. In the case of the mutual knowledge about the awareness of the weak-delegation (but not interpreted as a tacit request and agreement), the trustor could evaluate and learn if Y is a spontaneous collaborative agent (with respect that task in that situation) and how much Y is so collaborative (the value of Δ). In the case in which X ignores Y's awareness about the weak delegation, the trustor could evaluate the credibility of its own beliefs (both about Y's trustworthiness and about Y's awareness of the weak delegation) and, if the case, revises them.

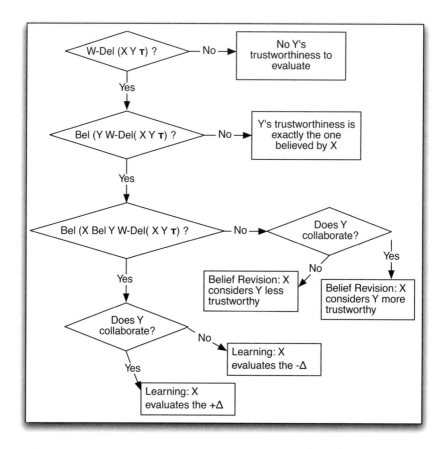

Figure 6.4 Flow-Chart resuming the different mental situations in weak-delegation

In Figure 6.5, it reiterated how weak delegation can influence the delegee's trustworthiness. Agent *X* has both a belief about *Y*'s trustworthiness and a hypothetical scenario of the utilities (in the case of success or failure) of all the possible choices it can do (to delegate to *Y* or to *W*, etc., or to not delegate and do it on its own or do nothing). On this basis it makes a weak delegation and maybe it changes *Y*'s trustworthiness. In the last case (changed trustworthiness of the trustee) maybe that *X*'s choice (done before *Y*'s action and of its spontaneous collaboration or of its negative reactions) results better or worst with respect to the previous possibilities. In other words, in the case in which the weak delegation changes *Y*'s trustworthiness (without *X* being able to foresee this change), the new trustworthiness of *Y* will be different from the expected one by *X* (and planned in a different decision scenario).

6.3.2 The Case of Strong Delegation

We call *strong delegation (S-Delegates(X Y τ))*, that *based on (explicit) agreement between X and Y.*

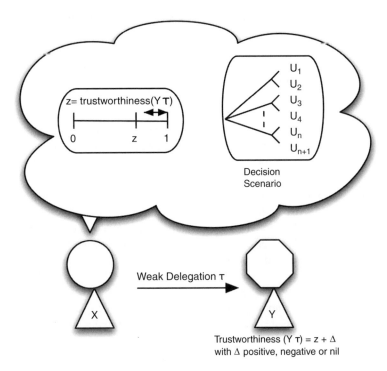

Figure 6.5 Potential Influence of the Weak-Delegation on the Decision Scenario. (Reproduced with kind permission of Springer Science+Business Media © 2001)

The expression *S-Delegates(X Y τ)* represents the following *necessary* mental ingredients:

a′) The achievement of τ is a *goal* of X.

b′) X believes that there exists another agent Y that has the *power of* achieving τ.

c′) X does not believe that Y will achieve τ by itself (without any intervention of X).

d′) X believes that if X realizes an action α′ there will be this result: Y will intend τ as the consequence of the fact that Y adopts X's goal that Y would intend τ (in other words, Y will be socially committed with X).

e′) X *prefers* to achieve τ through Y.

f′) X intends to do α′ relativized to (d′).

g′) The achievement of τ through Y is the goal of X.

h′) X has the goal (*relativized* to (g′)) of not achieving τ by itself.

We consider (a′, b′, c′, d′ and e′) what the agent X views as a '*Potential for relying on*' the agent Y; and (f′, g′ and h′) what X views as the '*Decision to rely on*' Y.
In this case we have: *MBel(X Y S-Delegates(X Y τ))*

i.e. there is a mutual belief of X and Y about the strong delegation and about the reciprocal awareness of it.

Like in the weak delegation, this belief could change the trustworthiness of Y, and also in this case we can have two possible results (we exclude also in this case the fact that this action does not have any influence on Y):

i) the new trustworthiness of Y as for τ is greater than the previous; so in this case we have the situation given from the formula (6.11): $\Delta trustworthiness(Y\,\tau) > 0$
ii) the new trustworthiness of Y as for τ is less than the previous one; so in this case we have the situation given from the formula (6.12): $\Delta trustworthiness(Y\,\tau) < 0$.

Why does Y's trustworthiness increase or decrease? In general, a strong delegation – if accepted and complete – increases the trustworthiness of the delegee because of its *commitment*.

This is in fact one of the motives why agents use strong delegation and count on Y's 'adhesion' (Section 2.8). However, it is also possible that the delegee loses motivation when he has to do something not spontaneously but by a contract, or by a role or duty, or for somebody else.

The important difference with the previous case is that now X knows that Y will have some possible reactions to the delegation and consequently X is expecting a new trustworthiness of Y (Figure 6.6): in some measure even if there is an increase in Y's trustworthiness it is not completely unexpected by X.

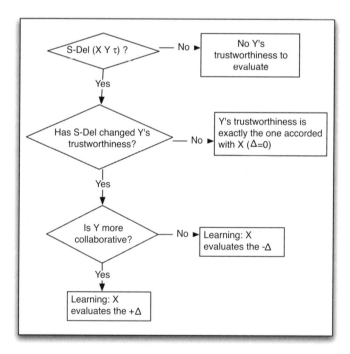

Figure 6.6   Flow-Chart resuming the different mental situations in strong-delegation

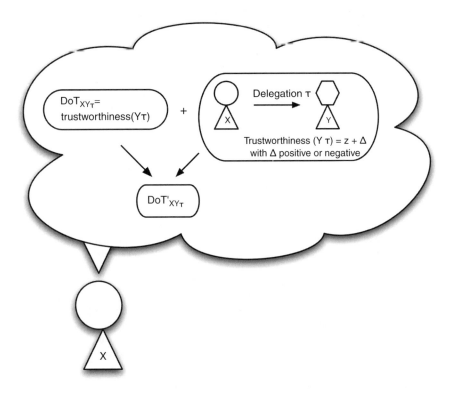

Figure 6.7   Changed Degree of Trust anticipated in strong-delegation. (Reproduced with kind permission of Springer Science+Business Media © 2001)

6.3.3   Anticipated Effects: A Planned Dynamics

This is the case in which the delegating agent X takes into account the possible effects of its 'strong delegation' and of the act of trusting (which becomes per se a 'signal'; Section 2.2.7) on Y's trustworthiness, in her decision, before she performs the delegation action. In this way, X changes her degree of trust in Y $(DoT_{XY\tau})$ before the delegation and on the basis of this her own action (Figure 6.7).

We analyze two cases:

i) the new degree of trust $(DoT'_{XY\tau})$ is greater than the old one $(DoT_{XY\tau})$: $DoT'_{XY\tau} > DoT_{XY\tau}$;
ii) the new degree of trust is lesser than the old one: $DoT'_{XY\tau} < DoT_{XY\tau}$.

Introducing a minimum threshold to delegate (σ) we can analyze the various cases by crossing the change of the DoT with the values of the threshold.

In other words, before performing a delegation action, an agent X could evaluate the (positive or negative) influence that its delegation will have on Y's trustworthiness and if this influence is relevant for either overcoming or undergoing the minimum level for delegating. In Table 6.2, all the possible situations and the consequent decisions of X are considered.

Table 6.2 Crossing different *DoTs* with minimum thresholds to delegate

	$DoT > \sigma$ $DoT - \sigma = \varepsilon' > 0$ (X would delegate Y before evaluating the effects of delegation itself)	$DoT < \sigma$ $DoT - \sigma = \varepsilon' < 0$ (X would not delegate Y before evaluating the effects of delegation itself)
$DoT' - DoT = \varepsilon > 0$ (X thinks that delegating Y would increase Y's trustworthiness)	$DoT' - \sigma = \varepsilon + \varepsilon' > 0$ $(\varepsilon > 0; \varepsilon > 0)$ **Decision to delegate**	$DoT' - \sigma = \varepsilon + \varepsilon' > 0$ $(\varepsilon > 0; \varepsilon' < 0; \varepsilon > \varepsilon')$ **Decision to delegate** $DoT' - \sigma = \varepsilon + \varepsilon' < 0$ $(\varepsilon > 0; \varepsilon' < 0; \varepsilon < \varepsilon')$ **Decision not to delegate**
$DoT' - DoT = \varepsilon < 0$ (X thinks that delegating Y would decrease Y's trustworthiness)	$DoT' - \sigma = \varepsilon + \varepsilon' > 0$ $(\varepsilon < 0; \varepsilon' > 0; \varepsilon < \varepsilon')$ **Decision to delegate** $DoT' - \sigma = \varepsilon + \varepsilon' < 0$ $(\varepsilon < 0; \varepsilon' > 0; \varepsilon > \varepsilon')$ **Decision not to delegate**	$DoT - \sigma = \varepsilon + \varepsilon' < 0$ $(\varepsilon < 0; \varepsilon' < 0)$ **Decision not to delegate**

In Table 6.3, we analyze the three cases deriving from X's decision to delegate as shown in Table 6.2. (In Table 6.3 we call TTE = Trustworthy than expected). Even in the case in which the trustee collaborates with the trustor, it may happen that the delegated task is not achieved; for example, because the expected additional motivation and/or abilities resulting from the delegation act are less effective than the trustor believed.

Another interesting way for increasing trustworthiness is through the self-confidence dimension, that we did not explicitly mention since it is part of the ability dimension. In fact, at least in human agents the ability to do α is not only based on skills (an action repertoire) or on knowing how (library of recipes, etc.), it also requires self-confidence that means the subjective awareness of having those skills and expertise, plus a general good evaluation (and feeling) of our own capability of success. Now the problem is that *self-confidence is socially influenced*, i.e. my confidence and trust in you can increase your self-confidence. So, I could strategically rely on you (letting you know that I'm relying on you) in order to increase your self-confidence and then my trust in you as for your ability and trustworthiness.

Finally, another interesting case in strong delegation is when there is the decision to rely upon Y but with diffidence and without any certainty that Y will be able to achieve τ. This is the case in which there is not enough trust but there is delegation (there are a set of possible reasons for this which we do not consider here). We are interested in the case in which Y realizes that anomalous situation (let us call this: diffidence). We have:

Table 6.3 Cases in which there has been delegation to Y

Trustee's Performance / Trustier's Mind	Good Collaboration	Bad Collaboration
$\epsilon + \epsilon' > 0$ $(\epsilon > 0; \epsilon' > 0)$	$\Delta > 0$ $\Delta > \epsilon$; (Y more TTE) $\Delta = \epsilon$; (Y equal TTE) $\Delta < \epsilon$; (Y less TTE)	$\Delta < 0 \cup \Delta = 0$ $\Delta = 0$; (Y less TTE) $\Delta < 0 \cap \|\Delta\| < \epsilon'$; (Y less TTE) $\Delta < 0 \cap \|\Delta\| > \epsilon'$; (**Y no trustworthy**)
$\epsilon + \epsilon' > 0$ $(\epsilon < 0; \epsilon' > 0)$	$\Delta > 0 \cup \Delta = 0$ (Y more TTE)	$\Delta < 0$ $\|\Delta\| > \epsilon \cap \|\Delta\| > \epsilon'$; (**Y no trustworthy**) $\|\Delta\| > \epsilon \cap \|\Delta\| = \epsilon'$; (Y less TTE) $\|\Delta\| > \epsilon \cap \|\Delta\| < \epsilon'$; (Y less TTE) $\|\Delta\| = \epsilon$; (Y equal TTE) $\|\Delta\| < \epsilon$; (Y more TTE)
$\epsilon + \epsilon' > 0$ $(\epsilon > 0; \epsilon' < 0)$	$\Delta < 0$ $\|\Delta\| > \epsilon$; (Y more TTE) $\|\Delta\| = \epsilon$; (Y equal TTE) $\|\Delta\| < \epsilon \cap \|\Delta\| < \epsilon + \epsilon'$; (**Y no trustworthy**) $\|\Delta\| < \epsilon \cap \|\Delta\| > \epsilon + \epsilon'$; (Y less TTE)	$\Delta < 0 \cup \Delta = 0$ (**Y no trustworthy**)

S-Delegates(X Y τ)

with $DoT_{XY\tau} \leq \sigma$; where σ is a minimal 'reasonable threshold' to delegate the task τ and

Bel (Y (DoT$_{XY\tau} \leq \sigma$))

Such a diffidence could be implicit or explicit in the delegation: it is not important.

Neither is it important, in this specific analysis, if

DoT$_{XY\tau} \neq$ trustworthiness(Y τ) or DoT$_{XY\tau}$ = trustworthiness(Y τ)
(in other words, if X's diffidence in Y is objectively justified or not).

This distrust could, in fact, produce a change (either positive or negative) in Y's trustworthiness:

- Y could be disturbed by such a bad evaluation and have a worst performance (we will not consider here, but in general in these cases of diffidence there is always some additional actions by the delegating agent: more control, some parallel additional delegation, and so on);

- Y, even if disturbed by the diffidence, could have a pridefully reaction and produce a better performance.

In sum, Y's trustworthiness could be different (with respect to X's expected one) and the cause of this difference could be both X's trust and/or X's distrust.

Let us in particular stress the fact that the predicted effect of the act of trusting Y on Y's possible performance can feedback on X's decision, and modify the level of trust and thus the decision itself:

- X's trust might be unsufficient for delegating to Y, but the predicted effects of trusting him might make it sufficient!
- Vice versa, X's static trust in Y might be sufficient for delegating to him, but the predicted effect of the delegation act on Y feeds back on the level of trust, decreases it, and makes it unsufficient for delegating.

6.4   The Dynamics of Reciprocal Trust and Distrust

The act of trusting somebody (i.e. the reliance) can also be an implicitly communicative act. This is especially true when the delegation is *strong* (when it implies and relies on the understanding and agreement of the delegee), and when it is part of a bilateral and possibly reciprocal relation of delegation-help, like in social exchange. In fact, in *social exchange X's* adoption of Y's goal is *conditional* to Y's adoption of X's goal. X's adoption is based on X's trust in Y, and vice versa. Thus, X's trusting Y for delegating to Y a task τ is in some sense conditional on Y's trusting X for delegating to X a task τ'. X has also to trust (believe) that Y will trust her and vice versa: *there is a recursive embedding of trust attitudes*. Not only this, but the measure of X's trusting Y depends on, varies with the decision and the act of Y's trusting X (and vice versa).

The act of trusting can have among its effects that of determining or increasing Y's trusting X. Thus, X may be aware of this effect and may plan to achieve it through her act of trusting. In this case, X must plan for Y to understand her decision/act of trusting Y. But, why does X wants to communicate to Y about her decision and action of relying on Y? In order to induce some (more) trust in Y. Thus the higher goal of that communication goal in X's plan is to induce Y to believe that 'Y can trust X since X trusts Y'. And this is eventually in order (to higher goal) to induce Y to trust X. As claimed in sociology (Gambetta, 1990) there is in social relations the necessity of actively promoting trust. 'The concession of trust' – which generates precisely that behaviour that seems to be its logical presupposition – is part of a strategy for structuring social exchange.

In sum, usually there is a circular relation, and more precisely a positive feedback, between trust in reciprocal delegation-adoption relations (from commerce to friendship). That – in cognitive terms – means that the (communicative) act of trusting and eventually delegating impacts on the beliefs of the other ('trust' in the strict sense) that are the bases of the 'reliance' attitude and decision producing the external act of delegating (Figure 6.8).

Analogously there is a positive feedback relation between distrust dispositions: as usually trust induces trust in the other, so usually distrust increments distrust. What precisely is the

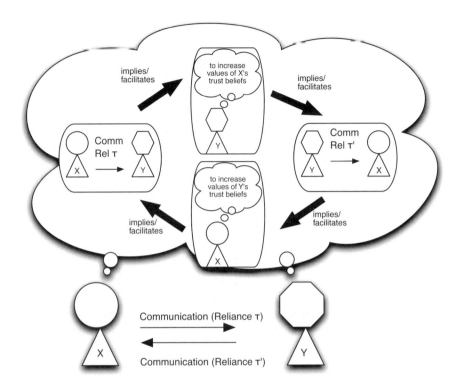

Figure 6.8   Reciprocal Trust and involved mental attitudes

mechanism of trust producing trust in exchange?[5] In my trust about your willingness to exchange (help and delegation) is included my trust that you trust me (and then delegate and then accept to help): if you do not trust me, you will not delegate me and then you will not adopt my goal. And vice versa. Now my trusting you (your believing that I trust you) may increase your trust in me as for delegating for exchange: since I delegate to you, (conditionally, by exchange) you can believe that I am willing to adopt your goal, so you can delegate to me. There are also other means to change Y's willingness to trust us and delegate to us (and then for example exchange with us). Suppose for example that X is not sure about Y trusting her because of some bad reputation or some prejudice about her. She can change Y's opinion about herself (ex. through some recommendation letter, or showing her good behavior in other interactions) in order to improve Y's trust in X and then Y's willingness, and then her own trust in delegating to him.

An interesting concept is the so-called *reciprocal trust* that is not simply *bilateral trust*. It is not sufficient that X trusts Y and Y trusts X at the same time. For example, X relies on Y for

---

[5] In general, in reciprocal relationship, trust elicits trust. This is also typical in friendship. If he trusts me (in relation to keeping secrets, not cheating, and not making fun of), he cannot have aggressive motives against me (he would expose himself to retaliation; he should feel antipathy, but antipathy does not support confidence). So, he must be benevolent, able to keep secrets, not cheat on me, and not make fun of me: this is my trust in reciprocating confidences.

stopping the bus and Y relies on X for stopping the bus, there is bilateral (unaware) trust and delegation; nobody stops the bus and both fail!

To have *reciprocal trust*, mutual understanding and communication (at least implicit) are needed: X has the goal that Y knows that she (will) trust Y, in order that Y trusts her, and that she trusts Y if and only if Y trusts her.

Exchange is in fact characterized by *reciprocal conditional trust*: how the act of trusting can increase Y's trust and then my own trust which should be presupposed by my act of trusting. However, no paradox or irrationality is there, since my prediction of the effect anticipates my act and justifies it. For example, X can be more sure (trust) about Y's motivation, because she is proposing to Y (or accepting from Y) a specific and reliable motive of Y for doing as expected: an *instrumental* goal-adoption for *selfish reasons* (see Section 2.8 A. Smith citation). Y has a specific interest and advantage for doing something 'for' X (provided that he believes that X will do as expected).

Given our agents X and Y and two possible tasks: τ and τ', we suppose that:

$$DoT_{XY\tau} < \sigma_1\;^6$$

The value of the $DoT_{XY\tau}$ changes on the basis of its components' variation. One of the ways to change the elementary components of $DoT_{XY\tau}$ is when the trustee (Y in our case) communicates (explicitly or implicitly) his own trust in the trustor (X) as for another possible task (τ') for example delegating the task τ' to X (relying upon X as for the task τ'). In our terms will be true the formula:

S-$Delegation(Y\ X\ \tau')$ (that always implies $MutualBel\ (X\ Y\ S$-$Delegation(Y\ X\ \tau')))$.

In fact, this belief has various possible consequences in X's mind:

i) there exists a dependence relationship between X and Y and in particular the Y's achievement of the task τ' depends on X. Even if it is important to analyze the nature of the Y's delegation as for τ'^7, in general X has the awareness to have any power on Y (and that Y believes this).

ii) if this delegation is spontaneous and in particular it is a special kind of delegation (for example it is a display of esteem) and X has awareness of this, i.e. $Bel\ (X\ (DoT_{YX\tau'} > \sigma_2))^8$, in general an abstract benevolence could arise in X as for Y.

iii) X could imply from point (i) the Y's unharmfullness (if the delegation nature permits it).

iv) trusting as a sign of goodwill and trustwortiness: Agents with bad intentions are frequently diffident towards the others; for example, in contracts they specify and check everything. Since they have non-benevolent intentions they ascribe similar attitudes to the others. This is why we believe that malicious agents are usually diffident and that (a risky abduction) suspicious agents are malicious. On such a basis, we also feel more trusting towards a non-diffident, trusting agent: this is a sign for us that it is goodwilling, non-malicious.

---

^{6} where σ_1 is the X's reasonable threshold for delegating to Y the task τ.

^{7} for example, if there is already an agreement between X and Y about τ' with reciprocal commitments and possible sanctions in the case in which there could be some uncorrect behaviour.

^{8} where σ_2 is the Y's reasonable threshold for delegating to X the task τ'.

Table 6.4 How delegation can influence reciprocal trust

	$DoT_{XY\tau} > \sigma_1$	$DoT_{XY\tau} < \sigma_1$
$Bel(X\ S\text{-}Del(YX\tau'))$ and $Bel(X\ DoT_{YX\tau'} > \sigma_2)$	Increases the X's degree of trust in Y as for τ	Increases the $DoT_{XY\tau}$ but $\neg$ Delegate (X Y τ)
		Increases the $DoT_{XY\tau}$ and Delegate (X Y τ)
$Bel(X\ \neg S\text{-}Del(Y\ X\ \tau'))$ and $Bel(X\ DoT_{YX\tau'} < \sigma_2)$	Decreases the $DoT_{XY\tau}$ but Delegate (X Y τ)	Decreases the X's degree of trust in Y as for τ
	Decreases the $DoT_{XY\tau}$ and $\neg$ Delegate (X Y τ)	

Each of the previous points allows the possibility that X delegates to Y τ: Going to analyze the specific components of the degree of trust we can say that:

- the point (i) could increase both $DoW_{XY\tau}$ and $DoA_{XY\tau}$;
- the point (ii) and (iii) decrease the value of σ_1 and may be increase both $DoW_{XY\tau}$ and $DoA_{XY\tau}$.

In other words, after Y's delegation we can have two new parameters:

$DoT'_{XY\tau}$ and σ_1' instaed of $DoT_{XY\tau}$ and σ_1 and it is possible that: $DoT'_{XY\tau} > \sigma_1$'. In Table 6.4 we display the different cases.

Another interesting case is when Y's diffidence in X is believed by X itself:

Bel (X DoT$_{Y X \tau'}$ < σ_2) and for this

Bel (X $\neg$ S-Delegates (Y X τ'))

Also, in this case, various possible consequences in X's mind are possible:

i) the fact that Y has decided not to depend on X could imply Y's willingness to avoid possible retaliation by X itself; so that X could imply Y's possible harmfulness.
ii) Y's expression of lack of estimation of X and the fact that *Bel (X DoT$_{Y X \tau'}$ ≤ σ_2)*, in general has the consequence that an abstract malevolence could arise in X as for Y.

Our message is trivial: the well-known interpersonal dynamics of trust (that trust creates a mirror trust, or diffidence induces diffidence), so important in conflict resolutions, in exchanges, in organizations, etc., must be grounded on their mental 'proximate mechanisms'; that is, on the dynamics of beliefs, evaluations, predictions, goals, and affects in X and Y' minds, and in X's theory of Y's mind and Y's theory of X's mind.

6.5   The Diffusion of Trust: Authority, Example, Contagion, Web of Trust

An interesting study about trust, partially different from the aspects so far analyzed, considers trust diffusion in groups and networks of agents as *meta-individual phenomenon*, and it is essentially based on problematics of contagion, for example, schemata, conventions, authority, and so on. In fact, trust is sometimes a *property of an environment*, rather than of a single agent or even a group: under certain conditions, the tendency to trust each other becomes diffused in a given context, more *like a sort of acquired habit or social convention than like a real decision*. These processes of 'trust spreading' are very powerful in achieving a high level of cooperation among a large population, and should be studied in their own right.

In particular, it is crucial to understand the subtle interaction between social pressures and individual factors in creating these 'trusting environments', and to analyze both advantages and dangers of such diffusive forms of trust.

In our analysis, these phenomena have also to be analyzed in terms of cognitive models, but in fact they follow slightly different rules from the ones considered in the individual interactions, and they concern a special kind and nature of goals and tasks.

In particular, we examine the point:

> *how widespread trust diffuses trust* (*trust atmosphere*), that is, how X's trusting in Y can influence Z trusting W or W', and so on. Usually this is a macro-level phenomenon and the individual agent does not calculate it.

Let us consider two prototypical cases, the two micro-constituents of the macro process:

i) Since X trusts Y, also Z trusts Y
ii) Since X trusts Y, (by analogy) Z trusts W.

We would like to underline the potential multiplicative effects of those mechanisms/rules: the process described in (i) would lead to a trust network like in Figure 6.9, while the process described in (ii) would lead to a structure like Figure 6.10.

6.5.1   Since Z Trusts Y, Also X Trusts Y

There are at least two mechanisms for this form of spreading of trust.

Agent's Authority (Pseudo-Transitivity)

This is the case in which, starting from the trustfulness that another agent (considered as an authority in a specific field) expresses about a trustee, other agents decide to trust that trustee.

Consider the situation:

Bel $(X \ DoT_{ZY\tau} > \sigma_2)$ (*where $DoT_{ZY\tau}$ is the degree of trust of Z on Y about the task τ*)

that is:

agent X believes that Z's degree of trust in Y on the task τ is greater than a reasonable threshold (following Z).

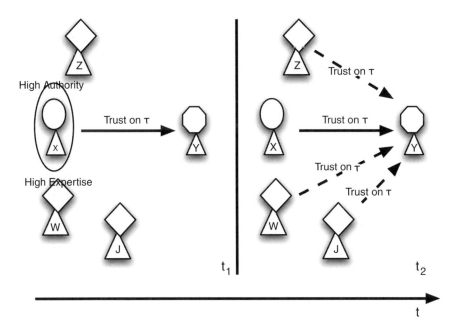

Figure 6.9   Emerging Trust in a specific trustee

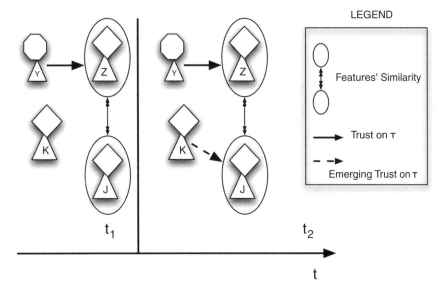

Figure 6.10   Emerging Trust by Analogy

Given this is X's belief, the question is: what about the $DoT_{XY\tau}$?

Is there a sort of *trust transitivity*? If so, what is its actual nature? (A frequent simplification about the trust transitivity is to consider it as a trivial question: if A trusts B and B trusts C, then A trusts C; in fact the problem is more complex and relates to the identification of the *real object to transfer* from an agent to another one).

Let us consider the case in which the only knowledge that X has about Y and about Y's possible performance on τ is given by Z. We hypothesize that:

$$DoT_{XY\tau} = Inf_X(Z\ Y\ ev(\tau)) * DoT_{XZev(t)}$$

where:

$DoT_{XZev(\tau)}$ is the X's degree of trust in Z about a new task $ev(\tau)$ that is the task of evaluating competences, opportunities, etc. about τ (see Figure 6.11). And $Inf_X(Z\ Y\ ev(\tau))$ represents

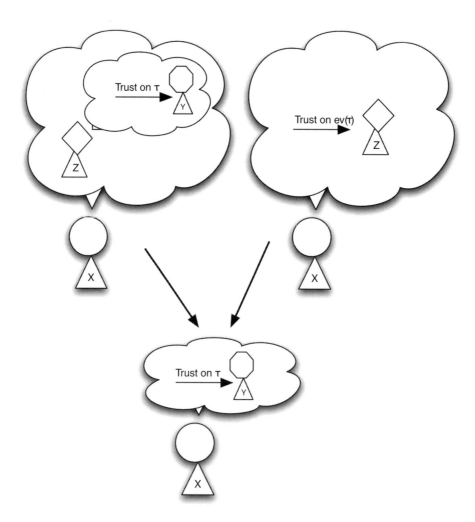

Figure 6.11   Pseudo-Transitivity: trusting through the others' trust

Z's potential influence in the task $ev(\tau)$ when τ is performed by Y (following X's opinion). This factor gives account of the different potential relationships among the (evaluating and evaluated) agents that can have an influence on the objectivity of the judgment of evaluation.

Notice that this pseudo-transitivity depends on subtle cognitive conditions. In fact, it is not enough that X trusts Z for adopting its trusting attitude towards Y; for us there is no real *transitivity* in trust: that X trusts Z and Z trusts Y does not imply that X trusts Y. Don't forget that trust is 'about' something; X's trust for Y is to do with some power, action, service. It has four arguments: not only X and Y but the task/goal and the context. About what does X trust Z? And about what does Z trust Y?

Suppose for example that X trusts Z as a medical doctor but considers Z a very impractical person as for business, and X knows that Z trusts Y as a stock-market broker agent; X has no reason to trust Y. On the contrary, if following X Z is a good doctor and he trusts Y as a good nurse, X can be learning to trust Y as a good nurse. What does this mean? This mean that X trusts Z as for a given *competence* in a given *domain* (Castelfranchi and Falcone, 1998b): Z is considered by X an 'authority' (or at least a good evaluator, expert) in this domain, this is why X considers Z's evaluation as highly reliable (Z is a trustworthy 'source' of evaluation).

So, since X trusts Y on the task τ we should have:

$$DoT_{XY\tau} = Inf_X(Z\ Y\ ev(\tau)) * DoT_{XZev(\tau)} > \sigma_1,$$

where σ_1 is the X's reasonable trust threshold for delegating τ.

We have a *pseudo-transitivity* (we consider it as cognitively-mediated transitivity) when:

1) $Inf_X(Z\ Y\ ev(\tau)) = 1$, in other words, there is no influence on Z by Y performing τ (following X) that could produce a non-objective judgment; and
2) $DoT_{XZev(\tau)} = DoT_{XZ\tau}$, in words, X has the same degree of trust in Z both on the task τ and on the task $ev(\tau)$; in other terms, X attributes the same DoT to Z both performing τ and trusting another agent about τ. This is a very common case in human activities, very often due to a superficial analysis of the trust phenomenon.

If the two above conditions are satisfied we can say that each time

$$DoT_{ZY\tau} > \sigma_2 \quad \text{and} \quad DoT_{XZ\tau} > \sigma_1 \text{ we will also have } DoT_{XY\tau} > \sigma_1.$$

In Table 6.5 we show the various possibilities of X's trusting or not of Y and how the two *DoTs* are combined.

Conformism

This mechanism is not based on a special expertise or authority of the 'models': they are not particularly expert, they must be numerous and just have experience and trust: since they do, I do; since they trust, I trust.

The greater the number of people that trust, the greater my feeling of safety and trust; (less perceived risk) the greater the perceived risk, the greater the necessary number of 'models' (Figure 6.12).

A good example of this is the use of credit cards for electronic payment or similar use of electronic money. It is a rather unsafe procedure, but it is not perceived as such, and this is

Table 6.5 Pseudo-Transitivity: Combining the Degrees of Trust

	$DoT_{ZY\tau} > \sigma_2$	$DoT_{ZY\tau} < \sigma_2$
$DoT_{X\,Z\,(ev)\tau} > \sigma_1$	X trusts Y as for τ	X does not trust Y as for τ
$DoT_{X\,Z\,(ev)\tau} < \sigma_1$	X trusts Y OR X does not trust Y as for τ	X trusts Y OR X does not trust Y as for τ

mainly due both to the utility and diffusion of the practice itself: it can be a simple conformity feeling and imitation; or it can be an explicit cognitive evaluation such as: 'since everybody does it, it should be quite safe (apparently they do not incur systematic damages)'. If everybody violates traffic rules, I feel encouraged to violate them too.

More formally, we have:

given $S=\{s_1,.\,.,\,s_n\}$ we are considering the set of agents belonging to the community as reference for trust diffusion.

if Bel $(X\ (DoT_{ZY\tau} > \sigma_2)\cap (DoT_{Z'Y\tau} > \sigma_3)\cap\ldots.\cap (DoT_{Z''Y\tau} > \sigma_n))$ then

$DoT_{XY\tau} > \sigma_1$.

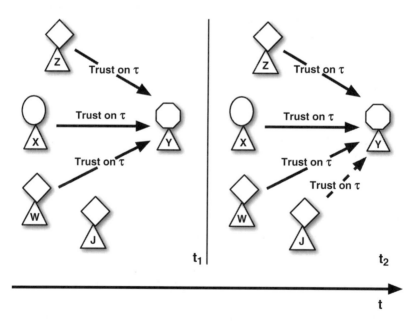

Figure 6.12 Trust Conformism

With $X, Z, \ldots, Z^n\ Y \in S$ and $\sigma_1, \ldots, \sigma_n$ respectively their thresholds for delegating.

A very crucial example of this trust-spreading mechanism is about circulating information. *The greater the number of (independent) sources of a piece of information, a reputation, a belief, the more credible it is.* (Castelfranchi, 1996)

In fact, belief acceptance is not an automatic process, it is subject to some tests and checks: one check is for its plausibility, coherence with previous knowledge, for the source reliability, etc. Thus, if several cognitive agents believe bel_1, probably bel_1 has been checked by each of them against its own direct experience, previous knowledge, and source evaluation, thus it is reasonable that it is more credible. Even in science, convergence is a criterion of validity. However, this is also a rather dangerous mechanism (of social bias and prejudice) since in fact sources in a social context are not independent and we cannot ascertain their independence, thus the number of sources can be just an illusion: there could be just a unique original source and a number of 'reporters'.

All these forms of trust spreading are converging in a given *target* of trust (being it an information source, a technology, a social practice, a company, a guy).

6.5.2   Since X Trusts Y, (by Analogy) Z Trusts W

This is a generalized form of trust contagion. It can be based on cognitive analogy or co-categorization:

- since X trusts Y (and X is expert) and W is like Y, Z will trust W
  or
- since everybody trusts some Y, and W is like/a Y, Z will trust W.

Where 'like' either means 'W is similar to Y as for the relevant qualities/requirements', or means 'W is of the same category of Y'; and 'some Y' means 'someone of the same category of Y or similar to Y'. Also in this case either a specific model and example counts, or a more generalized practice. For a more diffused and analytic treatment of this argument see the following Section 6.6.

6.5.3   Calculated Influence

As we said, usually, this trust contagion is a macro-level phenomenon and the individual agents do not calculate it, it is not intended. However, sometimes it is a strategic decision. Suppose for example that X believes herself to be a 'model' for Z, and that she wants Z to trust Y (or W). In this case X can deliberately make Z believe that he trusts Y, in order to influence Z and induce him to trust Y (or W).

The smart businessman might be aware of the fact that when he buys certain shares, a lot of other people will follow him, and he can exploit precisely this imitative behavior of his followers and speculate at their expense.

Trust Atmosphere

All the previous mechanisms (Section 6.5.1 through to Section 6.5.3) are responsible for that celebrated *'trust atmosphere'* that is claimed to be the basis of a growing market economy or of a working political regime. They are also very fundamental in computer mediated organizations, interactions (like electronic commerce), cooperation (CSCW), etc. and even in multi-agent systems with autonomous agents.

The emergence of a *'trust atmosphere'* just requires some generalization (see later Section 6.6): the idea (or feeling) not simply that 'X trusts Y' or/and 'Z trusts X/Y', but that 'everybody (in this context) trusts Y' or 'Z trusts everybody', or even 'here everybody trusts everybody'; or at least that 'a lot of people trust Z', 'a lot of people trust a lot of people here'.

6.6   Trust Through Transfer and Generalization

Would you buy a 'Volkswagen Tomato sauce'? And what about a FIAT tomato sauce? Why is it a bit more credible? And what about a Volkswagen water pump? Why is it appealing, perceived as reliable? How marketing managers and advertising people decide to sell a new product (good, service) under a well known brand? This is a serious problem in marketing, but, not based on a real model.

Would you fully trust a surgeon to recommend you a medication? Or a simple medical doctor to perform a serious surgical intervention? In general: *If X trusts Y for a given τ, will she Trust Y also for τ'?*

In this section, we analyze how it is possible to predict how/when an agent who trusts something/someone will therefore trust something/someone else, before and without direct experience. This is different from the models of trust just based on (or reduced to) a probability index or a simple measure of experience and frequency, we are interested in analyzing the trust concept so that we are able to cope with problems like: given X's evaluation about Y's trustworthiness on a specific task τ, what can we say about X's evaluation of Y's trustworthiness on a different but analogous task τ'? Given X's evaluation of Y's trustworthiness on a specific task τ, what can we say about X's evaluation of the trustworthiness of a different agent Z on the same task τ?

In fact, in our view only a cognitive model of trust, with its analytical power (as showed in Chapter 2), seems able to account for the inferential generalization of trustworthiness from task to task and from agent to agent not just based on specific experience and/or learning.

In general, trust derives, directly or indirectly, from the experience.[9] There are computational models of trust in which trust is conceived as an expectation sustained by the repeated direct interactions with other agents under the assumption that iterated experiences of success strengthen the trustor's confidence (Witkowski *et al.*, 2001), (Jonker and Treur, 1999). In the case of indirect experience, the more diffused case of study is the trust building on the basis of the others' valuations (reputation) ((Sabater and Sierra, 2001), (Jurca and Faltings, 2003), (Conte and Paolucci, 2002). A different and also interesting case of indirect experience for trust building (in some cases we can speak of attempts to rebuild, by other tools than observability, the direct experience), not particularly studied in these years, is based on the

---

[9] We have to say that there is also a part of trust that derives from some personality-based or cultural 'disposition' not based on previous experience.

inferential reasoning over the categories on which the world is organized (or could be thought to be organized): real and mental categories.

6.6.1 Classes of Tasks and Classes of Agents

In our model of trust we consider the trustor (X) and the trustee (Y) as single agents,[10] and the task (τ) as a specific task. For reasons of generality, optimization, economy, and scalability it would be useful to apply the trust concept not only to specific tasks and to single agents. In fact, it would be really useful and realistic to have a trust model that permits trust to be transferred among similar agents or among similar tasks. In this sense having as reference classes of tasks and classes of agents (as humans generally have) would be extremely important and effective. A good theory of trust should be able to understand and possibly to predict how/when an agent who trusts something/someone will therefore trust something/someone else, and before and without a direct experience. And, vice versa, from a negative experience of trustworthiness it could be possible to extract elements for generalizing about tasks and/or agents.

In this perspective we have to cope with a set of problems (grouped in two main categories):

1) Given X's evaluation about Y's trustworthiness on a specific task τ, what can we say about X's evaluation of Y's trustworthiness on a different but analogous task τ'? What should we intend for an 'analogous task'? When does the analogy work and when does it not work between τ and τ'? How is it possible to modify X's evaluation about Y's trustworthiness on the basis of the characteristics of the new task? How can we group tasks in a class? And so on.
2) Given X's evaluation about Y's trustworthiness on a specific task (or class of tasks) τ, what can we say about X's evaluation of the trustworthiness of a different agent Z on the same task (or class of tasks) τ? Which are the agent's characteristics that transfer (or not) the evaluation to different trustees?

In fact, these two sets of problems are strictly interwined with each other and their solutions require a more careful analysis of the nature of tasks and agents.

6.6.2 Matching Agents' Features and Tasks' Properties

In general, we can say that if an agent is trustworthy with respect to a specific task (or class of tasks) it means that, from the trustor's point of view, the agent has a set of *specific features* (resources, abilities and willingness) that are useful for that task (or class of tasks). But, what does it mean: useful for that task? We can say that, again depending on the trustor's point of view, a task has a set of *characterizing properties* requiring specific resources and abilities of various natures, which can be matched in some way with the agents' features. The attribution of the features to the agents, the right individuation of the tasks' properties and the match between the first and the second ones represent different steps for the trust building and are the bases for the most general inferential reasoning process for the trust generalization phenomenon.

---

[10] Either an 'individual' or a 'group' or an 'organization'.

The above described three attributions (features, properties and match) are essential for the success of trust building. For example, imagine the task of 'taking care of a baby during evening' (trustor: *baby's mother*; trustee: *baby-sitter*).

The *main properties* of the task might be considered:

a) to avoid dangers to the children;
b) to satisfy childrens' main physical needs;
c) to maintain a peaceful and reassuring climate by playing.

At the same time, we could appreciate several *main features* of the trustee:

1) careful and scrupulous;
2) lover of children;
3) able to maintain concentration for long time;
4) proactive;
5) impulsive, agitated and nervous.

The operation for evaluating the adequacy of the trustee to the task is mainly based on the match between the trustee features (that become '*qualities*' or '*defects*'; see Chapter 2) and the properties of the task. In the example, we can say that the feature number *(1)* is *good* for satisfying the properties *(a)* and *(b)*; the feature number *(2)* is *good* for satisfying the properties *(b)* and *(c)*; the feature numbers *(3 and 4)* are *good* for satisfying the properties *(a)* and *(b)*; the feature number *(5)* is *bad* for satisfying the properties *(a)* and *(c)*.

Both the properties of the task and the features of the trustee could be perceived from different trustors in different ways (think about the possible discussions in real life between a mother and a father about this). Not only this: the match could also be considered in a different way from different personalities and point of views. In addition, both the features of an agent and the properties of a task can be considered unchanged or not during time: it depends on the tasks, on the trustees and the trustors' perception/representation.

It is superfluous to be reminded that this kind of trust building is just one of the many ways in which to definine the agents' trustworthiness. Sometimes, the trustors do not know, except at a superficial level, the tasks' properties and/or the trustees' features (like when trust building is based on reputation or many cases of direct experiences).

The trust building based on the main inferential reasoning process, is then depending on several different factors and on their composition. When inferring the task's properties a trustor has to select the minimal acceptable values for the included indispensable ones (if there are any). At the same time, the trustor has to evaluate the potential trustee's features and verify their compatibility and satisfaction for the given task. These are complex attributions depending on the trustor and her trust model.

Starting from this kind of attribution we will analyze the phenomenon of generalization between similar tasks and similar agents. The essential informal 'reasoning' one should model can be simplified as follows:

- To what features/qualities of Y (the trustee) is its validity ascribable for the requirements of τ?

- Has Z the same relevant qualities? How much; how many? Does Z belong to the same class/set of Y, based on the relevant features?
- Does τ' share the same relevant requirements as τ? Does τ' belong to the same kin/class/set of services/goods as τ ?

6.6.3 Formal Analysis

In more systematic and formal terms we have tasks (τ) and agents (Ag): with $\tau \in T \equiv \{\tau_1, \ldots, \tau_n\}$, and $Ag \in AG \equiv \{Ag_1, \ldots, Ag_m\}$. We can say that each task τ can be considered composed of both a set of actions and the modalities of their running, this we call *properties*:

$$\tau \equiv \{p_1, \ldots, p_n\} \tag{6.14}$$

we consider this composition from the point of view of an agent (Ag_X):

$$Bel_{Ag_X} (\tau \equiv \{p_1, \ldots, p_n\}) \tag{6.15}$$

In general each of these properties could be evaluated with a value ranging between a minimum and maximum (i.e.: $0,1$): representing the complete failure or the full satisfaction of that action. So in general: $0 \le p_i \le 1$ with $i \in \{1, .., n\}$.

Of course, not all the properties of a task have the same relevance: some of them could be considered indispensable for the realization of the task, others could be considered useful in achieving greater success.

If we insert an apex c to all the properties that the agent (Ag_X) considers indispensable (*core properties*) for that task, we can write:

$$Bel_{Ag_X} (\tau \equiv \{p_1^c, \ldots, p_k^c\} \cup \{p_1, \ldots, p_m\}) \tag{6.16}$$

We call $\tau^C \equiv \{p^c{}_1, \ldots, p^c{}_k\}$ and $\tau^{NC} \equiv \{p_1, \ldots, p_m\}$, so

$$\tau = \tau^C \cup \tau^{NC} \tag{6.17}$$

The set of the core properties is particularly relevant for grouping tasks into classes and for extending the reasoning behind generalization or specification.

Analogously, we can define the set of the *features* f_{Ag_Y} for an agent (Ag_Y):

$$f_{Ag_Y} \equiv \{f_{Y1}, \ldots, f_{Yn}\} \tag{6.18}$$

we consider this composition from the point of view of an agent (Ag_X):

$$Bel_{Ag_X}(f_{Ag_Y} \equiv \{f_{Y1}, \ldots, f_{Yn}\}) \tag{6.19}$$

Also in this case, each of these features could be evaluated with a value ranging between a minimum and a maximum (i.e.: $0,1$): representing the complete absence or the full presence of that feature in the agent Ag_Y, from the point of view of Ag_X. So in general: $0 \le f_i \le 1$ *with* $i \in \{Y_1, \ldots, Y_n\}$.

Given the previous definitions, we can say that Ag_X could trust Ag_Y on the task τ if Ag_X has the following beliefs: (6.19), (6.16) and

$$Bel_{Ag_X} (\forall p_i \in \tau^C,\ \exists \{f_i\} \in \{f_{Y1}, \ldots, f_{Yn}\} | \{f_i\} \ne \emptyset \cap p_i \text{ is satisfied from } \{f_i\}) \tag{6.20}$$

where 'p_i is satisfied from $\{f_i\}$' means that the trustee (from the point of view of the trustor) has the right features for satisfying all the core properties of the task. In particular, the needed features are over the minimal threshold (σ_j) so that the main properties of the task will also be over the minimal threshold (ρ_i):

$$(\forall f_j \in \{f_i\}, \ (f_j > \sigma_j)) \ and \ (apply \ (\{f_i\}, \ \tau) \Rightarrow (\forall p_i \in \tau^C, \ p_i > \rho_i)) \qquad (6.21)$$

where the *apply* function defines the match between the agent's features and the task's properties. Also the different thresholds (σ_j and ρ_i) are depending on the trustor.

We have to say that even if it is possible to establish an objective and general point of view about both the actual composition of the tasks (the set of their properties, including the actual set of core properties for each task) and the actual features of the agents, *what is really important are the specific beliefs of the trustors about these elements*. In fact, on the basis of these *beliefs* the trustor would determine its trust. For this reason alone we have introduced as main functions those regarding the trustors' beliefs.

6.6.4   Generalizing to Different Tasks and Agents

Let us now introduce the reasoning-based trust generalization. Consider three agents: Ag_X, Ag_Y and Ag_Z (all included in AG) and two tasks τ and τ' (both included in T). Ag_X is a trustor and Ag_Y and Ag_Z are potential trustees.

Where:

$$\tau' \equiv \{p_1'^{c}, \ldots, \ p_k'^{c}\} \cup \{p_1', \ldots, \ p_m'\} = \tau'^{C} \cup \tau'^{NC}) \qquad (6.17bis)$$

and in general

$$(p_j' \neq p_j) \ with \ p_j' \in (\tau'^{C} \cup \tau'^{NC}) \ and \ p_j \in (\tau^{C} \cup \tau^{NC});$$

$$Ag_Z \equiv f_{Ag_Z} \equiv \{f_{Z1}, \ldots, f_{Zn}\} \qquad (6.18bis)$$

The first case (*caseA*) we consider is when Ag_X does not know either the τ's properties, or the Ag_Y features, but they trust Ag_Y on τ (this can happen for different reasons: for example, he was informed by others about this Ag_Y's trustworthiness, or simply he knows the successful result without assisting in the whole execution of the task, and so on). In more formal terms:

a1) $Trust_{Ag_X}(Ag_Y, \tau)$
a2) $\neg Bel_{Ag_X}(f_{Ag_Y} \equiv \{f_{Y1}, \ldots, f_{Yn}\})$
a3) $\neg Bel_{Ag_X}(\tau \equiv \{p_1^{c}, \ldots, p_k^{c}\} \cup \{p_1, \ldots, p_m\})$

In this case which kind of trust generalization is possible? Can Ag_X believe that Ag_Y is trustworthy on a different (but in some way analogous) task τ' (generalization of the task) starting from the previous cognitive elements (*a1, a2, a3*)? Or, can Ag_X believe that, another different (but in some way analogous) agent Ag_Z is trustworthy on the same task τ (generalization of the agent), again starting from the previous cognitive elements (*a1, a2, a3*)?

The problem is that the *analogies* (between τ and τ', and between Ag_Y and Ag_Z) are not available to Ag_X because they do not know either the properties of τ or the features

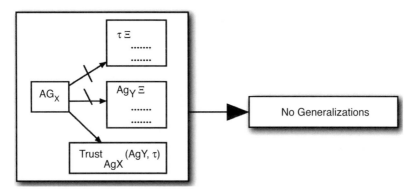

Figure 6.13   Generalization in case of Ag_X's ignorance about task's properties and trustee's features. (Reproduced with kind permission of Springer Science+Business Media © 2008)

of Ag_Y. So we can conclude (see Figure 6.13) that in *caseA* there is no possible rationale, grounded generalization to other tasks or agents: in fact the set of *(a1, a2, a3) do not permit generalizations.*

In fact, the only possibility for a rational generalization in this case is given from an indirect generalization by Ag_X . For example someone else can suggest to Ag_X that on the basis of his first trustworthy attitude he can trust another agent or another task because there is an analogy between them: in this case Ag_X, trusting this suggestion, acquires the belief for an indirect generalization. The second case (*caseB*) we consider is when Ag_X does not know Ag_Y's features, but he knows τ's properties, and he trust Ag_Y on τ (also in this case for several possible reasons different from inferential reasoning on the match between properties and the features). In more formal terms:

b1) $Trust_{AgX}(Ag_Y, \tau)$
b2) $\neg Bel_{AgX}(f_{AgY} \equiv \{f_1, \ldots, f_n\})$
b3) $Bel_{AgX}(\tau \equiv \{p_1^c, \ldots, p_k^c\} \cup \{p_1, \ldots, p_m\})$

Despite the ignorance about Ag_Y's features (*b2*), Ag_X can believe that Ag_Y is trustworthy on a different (but in some way analogous) task τ' (generalization of the task) just starting from the previous cognitive elements (*b1* and *b3*) and from the knowledge of τ "'s properties. He can evaluate the overlap among the core properties of τ and τ' and decide if and when to trust Ag_Y on a different task. It is not possible to generalize with respect to the agents because there is no way of evaluating any analogies with other agents.

So we can conclude (see Figure 6.14) that in the case (*B*) task generalization is possible: in fact the set *(b1, b2, b3) permits task generalizations but does not permit agent generalizations.*

Also, in this case we can imagine an *indirect* agent generalization. If Ag_X trusts a set of agents $AG1 \equiv \{Ag_Y, Ag_W \ldots, Ag_Z\}$ on a set of different but similar tasks $T1 \equiv \{\tau, \tau', \ldots, \tau^n\}$ (he can evaluate this similarity given his knowledge of their properties) he can trust each of the agents included in $AG1$ on each of the tasks included in $T1$.

The third case (*caseC*) we consider is when Ag_X does not know τ's properties, but he knows Ag_Y's features, and he trusts Ag_Y on τ (again for several possible reasons different from

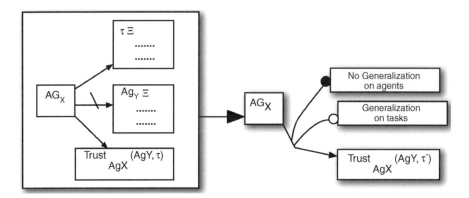

Figure 6.14 Generalization in case Ag_X knows only the trustee's features. (Reproduced with kind permission of Springer Science+Business Media © 2008)

inferential reasoning on the match between properties and the features). In more formal terms:

c1) $Trust_{AgX}(Ag_Y, \tau)$
c2) $Bel_{AgX}(f_{AgY} \equiv \{f_1, \ldots, f_n\})$
c3) $\neg Bel_{AgX}(\tau \equiv \{p_1^c, \ldots, p_k^c\} \cup \{p_1, \ldots, p_m\})$

Despite the ignorance about τ's features ($c3$), Ag_X can believe that a different (but in some way analogous) agent Ag_Z is trustworthy on the task τ (generalization of the agent) just starting from the previous cognitive elements ($c1$ and $c2$) and from the knowledge of Ag_Z's features. He can evaluate the overlap among the features of Ag_Y and Ag_Z and decide if and when to trust Ag_Z on τ. While, in this case, it is not possible to generalize a task because there is no way of evaluating any analogies with other tasks.

So we can conclude (Figure 6.15) that in the case (C) agent generalization is possible: in fact the set *(c1, c2, c3) permits agent generalizations but does not permit task generalizations.*

Exactly as in case B, also in this case we could imagine an *indirect* task generalization. If Ag_X trusts a set of different but similar agents $AG1 \equiv \{Ag_Y, Ag_W \ldots, Ag_Z\}$ (he can evaluate this

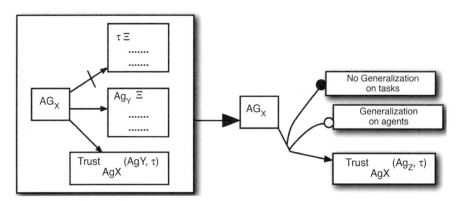

Figure 6.15 Generalization in case Ag_X knows only the task's properties. (Reproduced with kind permission of Springer Science+Business Media © 2008)

similarity given his knowledge of their features) on a set of tasks $T1 \equiv \{\tau, \tau', \ldots, \tau^n\}$ he can trust each of the agents included in $AG1$ on each of the tasks included in $T1$.

The fourth case ($caseD$) we consider is when Ag_X knows both τ's properties and Ag_Y's features, and they trust Ag_Y on τ (in this case for inferential reasoning on the match between properties and the features). In more formal terms:

d1) $Trust_{Ag X}(Ag_Y, \tau)$
d2) $Bel_{Ag X}(f_{Ag Y} \equiv \{f_1, \ldots, f_n\})$
d3) $Bel_{Ag X}(\tau \equiv \{p_1^c, \ldots, p_k^c\} \cup \{p_1, \ldots, p_m\})$

Ag_X can both believe:

- that a different (but in some way analogous) agent Ag_Z is trustworthy on the task τ (generalization of the agent) starting from the cognitive elements (d1 and d2) and from the knowledge of Ag_Z's features; or
- that Ag_Y is trustworthy on a different (but in some way analogous) task τ' (generalization of the task) starting from the cognitive elements (d1 and d3) and from the knowledge of τ' properties; or again
- that a different (but in some way analogous) agent Ag_Z is trustworthy on a different (but in some way analogous) task τ' (generalization of both the agent and the task).

So we can conclude (see Figure 6.16) that in the case (D) both task and agent generalization is possible: in fact the set *(d1, d2, d3) permits both agent and task generalizations.*

Note that we have considered in all the studied cases (*a1, b1, c1,* and *d1*) the fact that Ag_X *trusts* Ag_Y on the initial task τ. In fact, in case of *distrust* (or mistrust) we could receive analogous utility (in cases *B, C* and *D*) for the distrust (or mistrust) generalization to the other agents or tasks. In other words, Ag_X on the basis of his experience with Ag_Y on task τ could distrust agents similar to Ag_Y or/and tasks similar to τ.

The case in which, from a specific knowledge about agents, tasks and their matches, a trustor can infer complementary features or properties *contradicting* their previous positive or

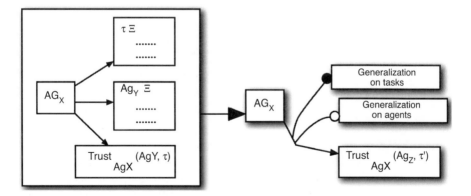

Figure 6.16 Generalization in case Ag_X knows both task's properties and trustee's features. (Reproduced with kind permission of Springer Science+Business Media © 2008)

negative beliefs is more complex. For this it is necessary a more subtle (fine-grained) analysis of features and properties.

Let us now try to refine the previous analysis. Starting from our characterization of the tasks, we can say that the similarity between different tasks is given by the overlap among the properties, in particular among their core properties (the indispensable relevant ones). So, the interesting thing is that, if the trustor believes that the trustee has the right features for a specific task, and another task has a significant overlap of its properties with that task (and specifically they share the core properties), then the trustor could also believe that the trustee has the right features for realizing the second task.

Also, in general, we can say that, given two tasks τ and τ' where both the formulas (6.17) and (6.17bis) are true, if the trustee (on the point of view of the trustor) has the right features for realizing τ and:

$$\tau'^C \subseteq \tau^C \tag{6.22}$$

then, on the trustor's point of view, that trustee also has the right features for realizing the task τ'.

So, even if the trustor had never seen the trustee operate on the task τ', it can infer its trustworthiness on that task. Note that the trustor could trust the trustee on a new task τ' also in the case in which they ignoring their features, but has experienced their trustworthiness about τ (in fact deducing them from the previous task in an indirect way).

In the case in which:

$$\tau'^C \not\subseteq \tau^C \tag{6.23}$$

there is at least one $p'_j \in (\tau'^C)$ that is different from all the $p_j \notin (\tau^C)$. It means that in this case the trustor can trust the trustee if: for each of the $p'_j \in (\tau'^C)$ either it is equal to one of the $p_j \in (\tau^C)$, or the trustor believes that it is satisfied from at least one of the trustee features ($f_{Ag_Y} \equiv \{f_1, ..., f_n\}$). In other words, the trustor can trust the Ag_Y on a different task τ' if they believe that:

- either all the core properties of τ' are included in the core properties of τ (where Ag_Y is believed trustworthy on the task τ);
- or for each property of τ' not included in τ Ag_X believes there is at least one feature of Ag_Y that satisfies that property (and, of course, Ag_Y is believed to be trustworthy on the task τ).

Of course, all the considerations made in this paragraph can also be applied to self-confidence and self-efficacy; Ag_X's trust about himself: 'Given Ag_X's success on task τ will Ag_X be able on task τ' ?'; 'Given Ag_Y's success on τ will Ag_X be able on τ' too?'.

6.6.5   Classes of Agents and Tasks

It is quite simple to extend the above reasoning to a class of tasks. We can define a class of tasks as the set of tasks sharing a precise set of (quite abstract) core properties. The different specifications of these properties define the different tasks belonging to the class. On the basis of the core properties of a class, we can select all the agents endowed with the appropriate features who will be trustworthy for the whole class of tasks.

If we call *KT* a class of tasks, we can say that it is characterized from a set of (quite abstract) properties. The tasks belonging to class *KT* are all those whose core properties are specifications of the abstract properties of *KT*. In more formal terms, in addition to (6.17) we have:

$$KT \equiv \{ap_1, \dots, ap_n\} \tag{6.24}$$

τ is a task belonging to the class *KT* if and only if:

$$(\forall ap_i \ (with \ n \geq i \geq 1) \ \exists p_j \ (with \ k \geq j \geq 1) | (p_j = ap_i) \vee (p_j \ is \ a \ \text{specification} \ of \ ap_i)) \tag{6.25}$$

where ap_i *(with $n \geq i \geq 1$)* is an abstract property.

For example, we can say that the class of tasks of 'taking care of people' is constituted from the following abstract properties:

a) to avoid dangers to the people;
b) to satisfy peoples' main physical needs;
c) to satisfy peoples' main psychological needs.

In the case of the task 'taking care of children', we can say that it is included in the class of 'taking care of people' because all of its core properties are specializations of the abstract properties of the main task class:

a) to avoid dangers to children;
b) to satisfy childrens' main physical needs;
c) to maintain a peaceful and reassuring climate by playing.

At the same time we could consider 'to take care of elderly people' another task included in this class, because the core properties in this case are also:

a) to avoid dangers to the elderly;
b) to satisfy the elderlies' main physical needs;
c) to maintain a peaceful and reassuring climate;

are specifications of the same abstract properties of the class.

The situation is more complex when a task is just partially included in a class. Consider for example the case of 'to take care of a house'. The core properties are:

a) to clean and order the things in the house;
b) to avoid damage to the things.

In these cases the analysis has to consider the potential trustee's features and the classes of attitudes they have. The match among classes of agents and classes of tasks can inform us about positive or negative attitudes of the agents belonging to that agent class with respect to the tasks belonging to that task class. Of course, going towards specifications of both agent classes and task classes permits us to establish better matches and confrontations about which agent is more adequate for which task.

6.7   The Relativity of Trust: Reasons for Trust Crisis

As we saw so far in this chapter, trust dynamics is a complex and quite interesting phenomenon. Its diverse nature depends on the many and interacting elements on which trust is essentially based (as shown in this book). One of the main consequences of a complex trust dynamics is the fact that it produces what we generally call *crisis of trust*. And they could be different and articulated on the basis of the causes from which they derive.

To analyze the situations in which trust relationship can enter into a crisis (and in case to collapse) in depth we have to take into consideration the different elements we have introduced into our trust model as showed in Chapters 2 and 3. In fact, there are interesting and complex dynamical interactions among the different basic elements for trusting that have to be considered when evaluating the trust crisis phenomena.

We show the different trust crises starting from the basic trust model and increase it with the complexities needed to describe the complete phenomenon of trust.

As we saw, we call the trustor (X) with a goal (g_X) that she wants achieve by a delegation to another agent (Y) assigning to him the task (τ). This delegation is based on two of X's main beliefs:

i)   The fact that Y has the features for achieving the task (in our model they are represented by competences, skills, resources, tools, but also by willingness, persistence, honesty, morality, and so on); these features must be sufficient to achieve the involved task (we made use of thresholds for measuring this sufficiency (see Chapter 3); and these thresholds were dependant on both the goal's relevance, the potential damages in case of failure, the personality of the agent, etc.).

ii) The fact that the task τ is believed appropriate for achieving a world state favouring (or directly achieving) the goal g_X (see Figure 6.17).

In this first simplified trust model, X can 'revise' her own trust in Y on τ (and then go in a trust crisis) on the basis of different reasons:

i)   she can change her own beliefs on Y about his features (for example, X does not evaluate Y sufficiently able and/or motivated for the task τ);

ii) X can change her own beliefs about the appropriateness of the task τ for the achievement of the considered goal: maybe the action α is not useful (in X's view) for achieving the world state p; or may be p no longer satisfies the goal g_X;

iii) in X's mind the value (and then the relevance) of the goal g_X can change or it is suddenly achieved by other means.

In fact, a more developed and appropriate model of trust gives us additional elements for the analysis. We have to consider the constitutive components of the two main attitudes of Y (competence and willingness). As shown in Figure 2.15 in Chapter 2, there are many sources for these main beliefs. And they can change. At the same time, X could/should consider how the context/environment in which the task has to be realized changes in its turn: in this way introducing facilitant or impeding elements with respect to the original standard (or previously evaluated) situation (see Figure 6.18).

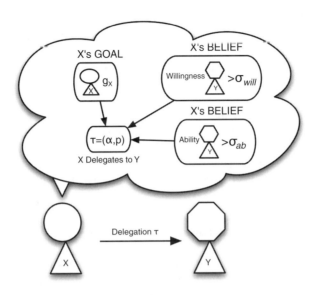

Figure 6.17 X's Mental Ingredients for Delegating to Y the task τ

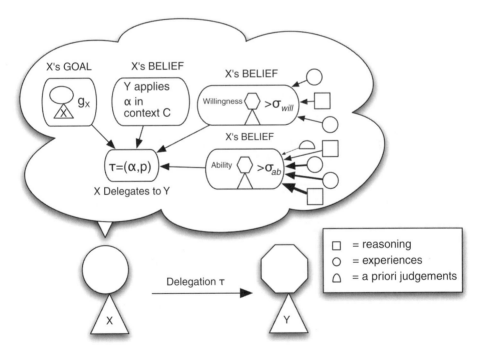

Figure 6.18 X's Mental Ingredients for Delegating to Y the task τ (introducing more beliefs and the Context)

In this new picture (see Figure 6.18) X's crisis of trust can be based on different causes:

i) due to a deeper analysis of *her own* beliefs about the reasons supporting Y's willingness and abilities, X can evaluate (analyzing better strengths and weakness of her beliefs) the appropiateness of her previous judgment;
ii) due to a deeper analysis of *her own* beliefs about the appropiateness of the delegated task to the achievement of the goal; in particular in the better analyzed and defined context/environment (X realizes that (given that context) action α does not achieve the state p; and/or p does not include/determine g_X).
iii) due to the decrease of the goal's value g_X. In particular, the cost of delegation (in its general terms, not only in economic sense) is not comparable with (is not paid from) the achievement of the goal.

Again increasing and sophisticating the trust model we have the situation shown in Figure 6.19.

In this new scenario it is not only the presence of Y (with his features deeply analyzed) that is considered in the specific contex/environment, but also the availabilty of other potential trustees (Z, W in Figure 6.19). Their presence represents an additional opportunity for X (depending from X's judgments about W and Z and about their potential performances in the specific environment) of achieving her own goal. In fact this opportunity can elicit X's crisis

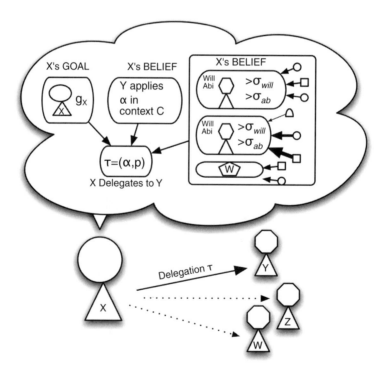

Figure 6.19 X's Mental Ingredients for Delegating to Y the task τ (considering also other potential trustees)

of trust in Y. This kind of crisis is a bit different from those previously shown: X could again evaluate Y as able to realize the task, but she considers whether other available agents would be more suitable delegates (W or Z).

We can resume the reasons of this crisis in the following way:

i) given new X's beliefs on Y's features, on the context, and on features of other agents, we can have a trust crisis in Y with delegation of the task to another agent (Z or W in Figure 6.19).

ii) due to a deeper analysis of *her own* beliefs about the appropiateness of the delegated task to the achievement of the goal; in particular in the better analyzed and defined context/environment (X realizes that (given that context) action α does not achieve the state p; and/or p does not include/determine g_X). These considerations are also true in the case in which there is more than one agent available for delegation: in fact, there is no added value (nor is there unsufficient added value) with the presence of more potential trustees.

iii) due to the decrease of the goal's value g_X. In particular, the cost of delegation (in its general terms, not only in economic sense) is not comparable with (is not paid from) the achievement of the goal. And this is true even in the case of a presence of more potential trustees.

In the final, more complete, version of the trust model we introduce the question of the contemporary presence of X's goals, evaluating the competition among them and the consequent dynamics for the evaluation of the priorities (Figure 6.20).

In this case the decrease of a goal's value (say g) and the increase of the value of another one (say g') could elicit X's trust crisis toward Y with respect to τ. This crisis is not based on Y's intervened inadequacy (depending on his own features, or on the new conflicting context, or on the presence of other more efficient and valued competitors). But the problem is that changing the priorities among X's goals, also changes the task X has to delegate and maybe Y has not got the right features for this new task (at least in X's beliefs).

This last example shows how our trust model is able to reconcile the two main cognitive ingredients: *beliefs* and *goals*. On this basis it can produce relevant forecasts: trust can change or collapse on both the brows (and they are very different phenomena). Current models (expecially in social, political and economical fields) neglect the relevant role of the *goals*, superficially disregarding the implications of a deep analysis between beliefs and goals differences.

Resuming and concluding this paragraph on the trust crisis we would like underline this difference:

a) On the one hand, there could be *beliefs crisis* (*revision*): change of opinion, reception of new information and evidences about Y's features, abilities, virtues, willingness, persistence, honesty, loyalty, and so on. As a consequence X's evaluation and trust can collapse.

b) Very different, on the other hand, is the *crisis of goals*: if X no longer has that goal, she does not want achieve it, the crisis of trust between X and Y is very different. It has in fact concluded the presupposition, the assumption and the willingness of the cooperation, of the delegation.

X does not think about being dependent on Y: she is not more interested about what Y does or is able to do: it is irrelevant for her. The detachment is more basic and radical: referring to

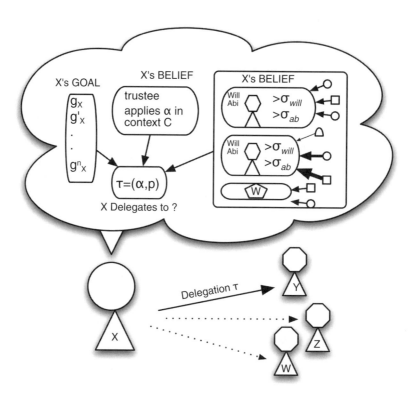

Figure 6.20 *X*'s Mental Ingredients for Delegating a task (considering other potential trustees, and different goals)

our model is the main network (the *dependence network* in our terms), that is falling, not only the (also very but less important) trust network (see also section 6.7.1).

6.8 Concluding Remarks

Strategies and devices for trust building should take into account the fact that social trust is a very dynamic phenomenon both in the mind of the agents and in society; not only because it evolves in time and has a history, that is *A*'s trust in *B* depends on *A*'s previous experience and learning with *B* itself or with other (similar) entities. We have in fact explained how *trust is influenced by trust* in several rather complex ways. In particular we have discussed three crucial aspects of such a dynamics and we have characterized some mechanism responsible for it and some preliminary formalization of it. We have modelled:

a) *How A's trusting B and relying on it in situation* Ω *can actually (objectively) influence B's trustworthiness in* Ω. Either trust is a self-fulfilling prophecy that modifies the probability of the predicted event; or it is a self-defeating strategy by negatively influencing the events. Both *B*'s ability (for example through *B*'s increased self-confidence) and *B*'s willingness and disposition can be affected by *A*'s trust or distrust and delegation. *B* can for example accept A's tacit exploitation and adopt A's goal, or negatively react to that. A can be aware of and take into account the effect of its own decision in the very moment of that decision. This make the decision of social trusting more complicated and 'circular' than a trivial decision of relying or not on a chair.

b) *How trust creates a reciprocal trust, and distrust elicits distrust*; for example because B knows that A's is now dependent on it and it can sanction A in case of violation; or because B believes that bad agents are suspicious and diffident (attributing to the other similar bad intentions) and it interprets A's trust as a sign of lack of malevolence. We also argued that the opposite is true: A's trust in B could induce lack of trust or distrust in B towards A, while A's diffidence can make B more trustful in A.

c) *How diffuse trust diffuses trust* (*trust atmosphere*), that is how A's trusting B can influence C trusting B or D, and so on. Usually, this is a macro-level phenomenon and the individual agent does not calculate it. We focused on *pseudo-transitivity* arguing how indirected or mediated trust always depends on trust in the mediating agent: my accepting X's evaluation about Z or X's reporting of Z's information depends on my evaluation of Z's reliability as evaluator and reporter. In other words this is not a logical or automatic process, but it is cognitively mediated.

We also discussed a more basic form of trust contagion simply due to diffusion of behaviours and to imitation because of a feeling of safety. Usually, these are macro-level phenomena and the individual agents do not calculate it.

d) *How trust can be transferred among agents on the basis of generalization of both tasks and agent's features*, that is how it is possible to predict how/when an agent who trusts something/someone will therefore trust something/someone else, before and without a direct experience. It should be clear that any theory of trust just based on or reduced to a probability index or a simple measure of experience and frequency (of success and failure) cannot account for this crucial phenomenon of a principled trust transfer from one agent to another or from one task to another. Only an explicit attributional model of the 'qualities' of the trustee that make her 'able' and 'willing' to (in the trustor's opinion), and of the 'requirements' of τ as related to the trustee's qualities, can provide such a theory in a principled way.

Our cognitive model of trust, with its analytical power, seems able to account for the inferential generalization of trustworthiness from task to task and from agent to agent not just based on specific experience and/or learning. It should also be clear how important and how productive this way of generating and propagating trust beyond direct experience and reputation should be.

References

Baier, A. (1994) Trust and its vulnerabilities in *Moral Prejudices*, Cambridge, MA: Harvard University Press, pp. 130–151.

Barber, S. and Kim, J. (2000) Belief Revision Process based on trust: agents evaluating reputation of information sources, *Autonomous Agents 2000 Workshop on 'Deception, Fraud and Trust in Agent Societies'*, Barcelona, Spain, June 4, pp. 15–26.

Birk, A. (2000) *Learning to trust, Autonomous Agents 2000 Workshop on 'Deception, Fraud and Trust in Agent Societies'*, Barcelona, Spain, June 4, pp. 27–38.

Bratman, M. E. (1987) *Intentions, Plans, and Practical Reason*. Cambridge, MA: Harvard University Press.

Castelfranchi, C. (1991) Social Power: a missed point in DAI, MA and HCI. In *Decentralized AI*. Y. Demazeau & J.P. Mueller (eds.) (Elsevier, Amsterdam) pp. 49–62.

Castelfranchi, C. Reasons: belief support and goal dynamics. *Mathware & Soft Computing*, 3: 233–247, 1996.

Castelfranchi, C. and Falcone, R. (1998) Towards a theory of delegation for agent-based systems, *Robotics and Autonomous Systems*, SI on Multi-Agent Rationality, Elsevier Editor, 24(3-4): 141–157.

Castelfranchi, C. and Falcone, R. (1998) Principles of trust for MAS: cognitive anatomy, social importance, and quantification. Proceedings of the International Conference on Multi-Agent Systems (ICMAS'98).

Cohen, P. and Levesque, H. (1987) Rational interaction as the basis for communication. Technical Report, N°89, CSLI, Stanford.

Conte, R. and Paolucci, M. (2002) *Reputation in Artificial Societies. Social Beliefs for Social Order*. Boston: Kluwer Academic Publishers.

Falcone, R. and Castelfranchi, C. (2001) The socio-cognitive dynamics of trust: does trust create trust? In *Trust in Cyber-societies* R. Falcone, M. Singh, and Y. Tan (eds.), LNAI 2246 Springer. pp. 55–72.

Festinger, L. (1957) *A Theory of Cognitive Dissonance*. Stanford, CA: Stanford University Press.

Gambetta D. (1998) Can we trust trust? In *Trust: Making and Breaking Cooperative Relations*, edited by D. Gambetta. Oxford: Blackwell.

Gambetta, D. (ed.) (1990) *Trust*. Basil Blackwell, Oxford.

Jennings. N. R. Commitments and conventions: The foundation of coordination in multi-agent systems. *The Knowledge Engineering Review*, 3: 223–250, 1993.

Jonker, C. and Treur, J. (1999) Formal analysis of models for the dynamics of trust based on experiences, *Autonomous Agents '99 Workshop on 'Deception, Fraud and Trust in Agent Societies'*, Seattle, USA, May 1, pp. 81–94.

Jurca, R and Faltings, B. (2003) Towards incentive-compatible reputation management, In R. Falcone, S. Barber, L. Korba and M. Singh (eds.) *Trust, Reputation, and Security: Theories and Practice*, LNAI 2631 Springer. pp. 138–147.

Sabater, J. and Sierra, C. (2001) Regret: a reputation model for gregarious societies. In Proceedings of the First International Joint Conference on Autonomous Agents and Multi-Agent Systems, pp. 475–482. ACM Press, NewYork.

Sichman, J. R. Conte, C. Castelfranchi, Y., Demazeau (1994) A social reasoning mechanism based on dependence networks. In *Proceedings of the 11th ECAI*.

Weiner, B. (1992) *Human Motivation: Metaphors, Theories and Research*, Newbury Park, CA: Sage Publications.

M. Witkowski, A. Artikis, and J. Pitt (2001) Experiments in building experiental trust in a society of objective-trust based agents. In R. Falcone, M. Singh, and Y. Tan (eds.) *Trust in Cyber-societies: Integrating the Human and Artificial Perspectives*, LNAI 2246 Springer. pp. 111–132.

7

Trust, Control and Autonomy: A Dialectic Relationship

In this chapter we are going to analyze the relationships between trust, control and autonomy: in particular, we are interested in showing how *trust and control are strictly interwined*, how their relationships are dynamic and influence the autonomy of the involved agents. We will also analyze the concept of 'adjustability' of both autonomy and delegation, and how it is dependent, elicited and guided from the previous notions of control and trust and from their interactions.

7.1 Trust and Control: A Complex Relationship

The relationship between trust and control is quite relevant both for the very notion of trust and for modelling and implementing trust-control relationships among autonomous systems; but it is not trivial at all.

On the one hand, it is true that where/when there is monitoring and control there is no trust (or at least there is less trust than without control), and vice versa: when/where there is a deliberate absence of control, there is trust (or at least there is more trust than in the case in which it has been necessary to insert control). However, this refers to a restricted notion of trust: i.e., what we call 'trust in Y', which is just a part, a component of the global trust needed because of relying on the action of another agent. We claim that *control is antagonistic of this strict form of trust* (internal trust, see Chapter 2); but also that it can complete and complement strict trust (in Y) for arriving at a *global* trust. In other words, putting control and guarantees is an important step towards trust-building; it produces a *sufficient* trust, when trust in Y's autonomous willingness and competence would not be enough. We also argue that *control requires new forms of trust*: trust in the control itself or in the controller, trust in Y as for being monitored and controlled; trust in possible authorities (or third parties; Section 7.1.5); and so on.

Finally, we show that paradoxically control might not be antagonistic of strict trust in Y, but it could even create trust, increase it by making Y more willing or more effective.

Trust Theory   Cristiano Castelfranchi and Rino Falcone
© 2010 John Wiley & Sons, Ltd

We will show how, depending on the circumstances, control makes Y more reliable or less reliable; control can either decrease or increase Y's trustworthiness and the internal trust. We will also analyze two kinds of control, characterized by two different functions: *pushing or influencing control* aimed at preventing violations or mistakes, versus *safety, correction or adjustment control* aimed at preventing failure or damages after a violation or a mistake.

A good theory of trust cannot be complete without a good theory of control and of their reciprocal interactions.

7.1.1   To Trust or to Control? Two Opposite Notions

The relation between trust and control is very important and perhaps even definitory; however it is everything but obvious and linear. On the one hand, some definitions delimit trust precisely thanks to being its opposite. But it is also true that monitoring and guarantees make me more confident when I do not have enough trust in my partner. And what is *confidence* if not a broader form of trust?[1]

On the other hand, it appears that the 'alternative' between control and trust is one of the main *tradeoffs* in several domains of information technology and computer science, from Human Computer Interaction to Multi-Agent Systems, Electronic Commerce, Virtual Organisations, and so on, precisely as in human social interaction.

Consider, for example, the problem of mediating between two such diverging concepts as control and autonomy (and the trust on which the autonomy is based) in the design of human-computer interfaces (Hendler, 1999):

'One of the more contentious issues in the design of human-computer interfaces arises from the contrast between *direct manipulation interfaces* and *autonomous agent-based systems*. The proponents of direct manipulation argue that a human should always be in control – steering an agent should be like steering a car – you're there and you're active the whole time. However, if the software simply provides the interface to, for example, an airlines booking facility, the user must keep all needs, constraints and preferences in his or her own head. (. . .) A truly effective internet agent needs to be able to work for the user when the user isn't directly in control.'

Consider also the naive approach to security and reliability in computer mediated interaction, just based on strict rules, authorization, cryptography, inspection, control, etc. (Castelfranchi, 2000) which can be in fact self-defeating for improving Electonic Commerce, Virtual Organisation, Cyber-Communities (Nissenbaum, 1999).

The problem is that the trust-control relationship is both conceptually and practically quite complex and dialectic. We will try to explain it both at the conceptual and modelling level, and in terms of their reciprocal dynamics.

7.1.2   What Control Is

'Control' is a (meta) action[2]:

---

[1] *'Do you trust this system/company/aircraft/drug . . . !?' 'Yes I do! There are so many controls and safety measures . . !'*

[2] We will call *control activity* the combination of two more specific activities: monitoring and intervention.

(a) aimed at ascertaining whether another action has been successfully executed or if a given state of the world has been realized or maintained (*monitoring, feedback*);
(b) aimed at dealing with the possible deviations and unforeseen events in order to positively cope with them and adjusting the process (*intervention*).

When the trustor is delegating (see Section 2.6) a given object-action, what about its control activity? Considering, for the sake of simplicity, that the control action is executed by a single agent, if *Delegates(Ag₁ Ag₂ τ)* there are at least four possibilities:

i) Ag_1 delegates the control to Ag_2: the trustor delegates both the task and the control on the task realization to the trustee;
ii) Ag_1 delegates the control to a third agent;
iii) Ag_1 gives up the control: nobody is delegated to control the success of α;
iv) Ag_1 maintains the control for itself.

Each one of these possibilities could be either about (a) (monitoring, feedback) or (b) (intervention), and could be either explicit, or implicit (in the delegation of the action, in the roles of the agents – if they are part of a social structure – in the previous interactions between the trustor and trustee, etc.).

To understand the origin and the functionality of control it is necessary to consider that Ag_1 can adjust the run-time of its delegation to Ag_2 if it is in the position of:

a) receiving in time the necessary information about Ag_2's performance (*feedback*);
b) intervening on Ag_2's performance to change it before its completion (*intervention*).

In other words, Ag_1 must have some form of control on and during Ag_2's task realization. *Control* requires feedback plus intervention (see Figure 7.1).
Otherwise no adjustment is possible. Obviously, the feedback useful for a run-time adjustment must be provided in time for the intervention. In general, the feedback activity is the

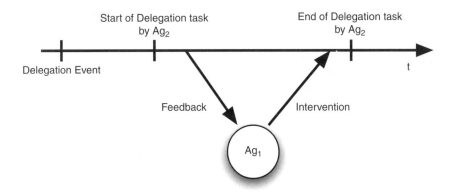

Figure 7.1 Control channels for the client's adjustment. (Reproduced by Permission of © 2001 IEEE)

precondition for an intervention; however it is also possible that either only the feedback or only the intervention will hold.[3]

Feedback can be provided by observation of Ag_2's activity (inspection, surveillance, monitoring), or by regularly sent messages by Ag_2 to Ag_1, or by the fact that Ag_1 receives or observes the results/products of Ag_2's activity or their consequences.

As for *Intervention* we consider five possible kinds:

i) *stopping the task* (the delegation or the adoption process is suddenly interrupted by the trustor);

ii) *substitution* (an intervention by the trustor allocates part of the (or the whole) task either to the trustor themselves or to a third agent);

iii) *correction of delegation* (after the intervention by the trustor, the task is partially or totally changed: the intervention transforms/changes the delegated task without any change of task allocation to other agents);

iv) *specification or abstraction of delegation* (after the intervention by the trustor, the task is more or less constrained; this is a specific case of the previous kind (*correction of delegation*));

v) *repairing of delegation* (the intervention by the trustor leaves the task activity unchanged but it introduces new actions (that have to be realized by either the trustee, or the trustor or some other agent) necessary to achieve the goal(s) of the task).

Imagine that Ag_1 and Ag_2 have decided to prepare a dinner at home, and Ag_1 delegated the task of cooking 'pasta with pesto' to Ag_2 while Ag_1 is preparing two tomato eggs; we have:

- case (i) when for example suddenly Ag_1 stops this delegation to Ag_2 (maybe Ag_1 is no longer hungry, she feels unwell, someone else has brought pizza to their house, and so on);
- case (ii) when for example Ag_1 decides to prepare the pesto herself (maybe Ag_2 is not able to find the ingredients, he is too slow, he is not able to mix the different parts correctly);
- case (iii) when for example Ag_1 sees that the basilico is finished and suggests to Ag_2 that they (or she prepares herself) the 'aglio e olio' as a sauce for spaghetti;
- case (iv) when for example Ag_1 seeing that Ag_2 is not completely happy about the spaghetti with pesto says to him to prepare spaghetti with the sauce he prefers;
- case (v) when for example Ag_1 seeing that the pesto sauce prepared by Ag_2 is not enough for two people, prepares an additional quantity of pesto.

Each of these interventions could be realized through either a *communication act* or a *direct contribution* to the task by the trustor.

The *frequency of the feedback on the task* could be:

- *purely temporal* (when the monitoring or the reporting is independent of the structure of the activities in the task, they only depend on a temporal choice);
- *linked with the working phases* (when the activities of the task are divided into phases and the monitoring or the reporting is connected with them).

---

[3] Sometimes we want to monitor the delegated action or its result not in time and in order for intervention. But just for the future; for confirming or correcting out trust in Y (see later).

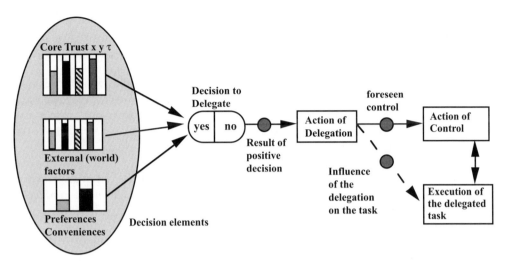

Figure 7.2   Decision, delegation and control

Also the *frequency of intervention* is relevant. As explained above, the intervention is strictly connected with the presence of the monitoring/reporting on the task, even if, in principle, both the intervention and the monitoring/reporting could be independently realized. In addition, also the frequencies of intervention and of monitoring/reporting are correlated. More precisely, the frequency of intervention could be:

1) *never;*
2) *just sometimes* (phase or time, a special case of this is at the end of the task);
3) *at any phase or at any time.*

Figure 7.2 shows how the control action impacts on the execution of the task after the trustor's delegation to the trustee. Plans typically contain control actions of some of their actions (Castelfranchi and Falcone, 1994).

7.1.3   Control Replaces Trust and Trust Makes Control Superflous?

As we said before, a perspective of duality between trust and control is very frequent and at least partially valid (Tan and Thoen, 1999). Consider for example the definition of (Mayer *et al.*, 1995):[4]

> The willingness of a party to be vulnerable to the actions of another party based on the expectation that the other party will perform a particular action important to the trustor, irrespective of the ability to monitor or control that other party

---

[4] About our more analytic considerations on this definition see Chapter 1 in this book.

This captures a very intuitive and common sense use of the term trust (in social interaction). In fact, it is true – in this limited sense – that if you control me 'you don't trust me!'; and it is true that if you do not trust me enough (to count on me) you would like to monitor, control and enforce me in some way.

In this view, control and normative '*remedies*' 'have been described as weak, impersonal *substitutes* for trust' (Sitkin and Roth, 1993), or as '*functional equivalent* ... mechanisms' (Tan and Thoen, 1999): 'to reach a minimum level of confidence in cooperation, partners can use trust and control to complement each other' (Beamish, 1988).[5]

We have some problems with respect to this view:

- on the one hand, it is correct: it captures something important. However, in such a complementariety, how the control precisely succeeds in augmenting confidence, is not really modelled and explained.
- on the other hand, there is something reductive and misleading in such a position:
  - it reduces trust to a strict notion and loses some important uses and relations;
  - it ignores different and additional aspects of trust also *in* the trustee;
  - it misses the point of considering control as a way of increasing the strict trust in the trustee and his trustworthiness.

We will argue that:

- firstly, control is *antagonistic* to strict trust;
- secondly, it requires new forms of trust including *broad trust* to be built;
- thirdly, it *completes and complements* it;
- finally, it can even create, *increase* the strict/internal trust.

As the reader can see it is quite a complex relationship.

7.1.4   Trust Notions: Strict (Antagonist of Control) and Broad (Including Control)

As said we agree on the idea that (at some level) trust and control are antagonistic (one eliminates the other) but complementary. We just consider this notion of trust – as defined by Mayer – too restricted. It represents for us the notion of trust in a strict sense, i.e. applied to the agent (and in particular to a social agent and to a process or action), and strictly relative to the 'internal attribution', to the internal factor. In other words, it represents the 'trust *in Y*' (as for action α and goal g) (see Section 2.6.1). But this trust – when enough for delegation – implies the 'trust *that*' (g will be achieved or maintained); and anyway it is part of a broader trust (or non-trust) that g.[6] We consider both forms of trust. Also the trust (or confidence) *in Y*, is, in fact, just the trust (expectation) *that Y* is able and will do the action α appropriately

---

[5] Of course, as (Tan and Thoen, 1999) noticed, control can be put in place by default, not because of a specific evaluation of a specific partner, but because of a generalized rule of prudence or for lack of information. (See later, on the level of trust as insufficient either for uncertainty or for low evaluation).

[6] Somebody, call this broader trust 'confidence'. But in fact they seem quite synonymous: there is confidence *in Y* and confidence *that g*.

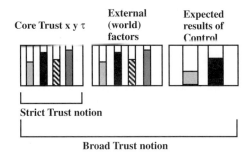

Figure 7.3   Control complements strict trust

(that I expect for its result *g*). But the problem is: are such an ability and willingness (the 'internal' factors) enough for realizing *g*? What about conditions for successfully executing *α* (i.e. the opportunities)? What about other concurrent causes (forces, actions, causal process consequent to *Y*'s action)? If my trust is enough for delegating to *Y*, this means that I expect (trust) that *g* will probably be realized.

We propose a broader notion of trust including all my expectations (about *Y* and the world; including actions of other agents, and including possible control activity on *Y*) such that *g* will be eventually true thanks (also) to *Y*'s action; and a strict notion of trust as 'trust in' *Y*, relative only to the internal factors (see Figure 7.3).

This strict notion is similar to that defined by Mayer (apart from the lack of the competence ingredient), and it is in contrast, in conflict with the notion of control. If there is control then there is no trust. But on the other hand they are also two complementary parts, as for the broad/global trust: control supplements trust.[7]

In this model, trust in *Y* and control of *Y* are *antagonistic*: where there is trust there is no control, and vice versa; the larger the trust the less room for control, and vice versa; but they are also *supplementary*: one remedies to the lack of the other; they are parts of one and the same entity. What is this attitude that can either be built out of trust or out of control? It is confidence, i.e. trust again, but in a broader sense, as we formalized it.[8]

In our view we need these two levels and notions of trust. With this in mind, notice that control is both *antagonist* to (one form of trust: the *strict* one) and *consituent* to (another form of trust: the *broader* one). Obviously, this schema is very simplistic and just intuitive. We will make this idea more precise. However, let us note immediately that this is not the only relation between strict-trust and control. Control is not only aimed at supplementing and 'completing' trust (when trust in *Y* would not be enough); it can also be aimed precisely at augmenting the internal trust in *Y*, *Y*'s trustworthiness.

---

[7] Control – especially in collaboration – cannot be completely eliminated and lost, and delegation and autonomy cannot be complete. This, not only for reasons of confidence and trust, but for reasons of distribution of goals, of knowledge, of competence, and for an effective collaboration. The trustor usually has at least to know whether and when the goal has been realized or not (Castelfranchi and Falcone, 1994).

[8] This holds for a fully delegated task. It is clear that for coordination between X and Y in a multi-agent plan, X *has to* monitor Y (and vice versa) even if she trusts him a lot.

7.1.5   Relying on Control and Bonds Requires Additional Trust: Three Party Trust

To our account of trust one might object that we overstate the importance of trust in social actions such as contracting, and organizations; since everything is based on delegation and delegation presupposes enough trust. In fact, it might be argued – within the duality framework – that people put contracts in place precisely because they do *not* trust the agents they delegate tasks to. *Since there is no trust people want to be protected by the contract.* The key in these cases would not be trust but the ability of some authority to assess contract violations and to punish the violators. Analogously, in organizations people would not rely on trust but on authorization, permission, obligations and so forth.

In our view (Castelfranchi and Falcone, 1998) this opposition is fallacious: it seems that trust is only relative to the character or friendliness, etc. of the trustee. In fact, in these cases (control, contracts, organizations) we are just dealing with *a more complex and specific kind of trust.* But trust is always crucial.

As Emile Durkheim claims 'A contract is not sufficient by itself, but is only possible because of the regulation of contracts, which is of social origin' ((Durkheim, 1893) p. 162), and this social background includes trust, social conventions and trust in them, and in people respecting them, the authorities, the laws, the contracts (see Chapter 9).

We put control in place only because we believe that the trustee will not avoid or trick monitoring, will accept possible interventions, will be positively influenced by control. We put a contract in place only because we believe that the trustee will not violate the contract, etc. These beliefs are nothing but *trust.*

Moreover, when true contracts and norms are there, this control-based confidence requires also that *X trusts* some authority or its own ability to monitor and to sanction *Y*, see (Castelfranchi and Falcone, 1998). *X* must also trust procedures and means for control (or the agent delegated to this task).

To be absolutely clear, we consider this level of trust as a *three party relationship*: it is a relation between the client *X*, the contractor *Y* and the authority *A*. And there are three trust

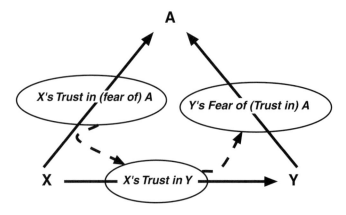

Figure 7.4   Three party relationships among Trustor, Trustee and Authority

sub-relations in it see Figure 7.4):

- *X trusts Y* by believing that *Y* will do what is promised because of his honesty or because of his respect/fear toward *A*;
- *X trusts A* and its ability to control, to punish etc. and relies on *A* for this;
- *Y trusts A* (both when *Y* is the client and when he is the contractor) the same beliefs being the bases of his respect/fear toward *A* (that is: trusting a threatening agent!).

In other words, *X* relies on a form of *paradoxical* trust of *Y* in *A*: *X* believes that *Y* believes that *A* is able to control, to punish, etc. Notice that *Y*'s beliefs about *A* are precisely *Y*'s trust in the authority when he is the client. When *Y* is the contractor the same beliefs are the bases of his respect/fear toward *A*.

We can also say that, in addition to the *X*'s trust in *Y* based on the internal (*Y*'s competence and willingness) and contextual-environmental reasons believed by *X*), there is a part of *X*'s trust in *Y* based on the fact that the other two relationships are true (and believed by the agents) and are the relationships of *X* and *Y* with the authority *A*.

In sum, in contracts and organizations it is true that *personal trust* in *Y* may not be enough, but what we put in place is a higher level of trust which is our trust in the authority but also our trust in *Y* as for acknowledging, worrying about and respecting the authority. Without this trust in *Y* the contract would be useless. This is even more obvious if we think of possible alternative partners in contracts: how to choose among different contractors with the same conditions? Precisely on the basis of our degree of trust in each of them (both, trust about their competence, but also trust about their reliability, their respecting the contract).

As we have already said, these more complex kinds of trust are just richer specifications of the reasons for '*Y*'s doing what we expect: reasons for *Y* 's predictability which is based on his willingness; and reasons for his willingness (he will do α, either because of his selfish interest, or because of his friendliness, or because of his honesty, or because of his fear of punishment, or because of his institutional and normative respect: several different bases of trust).

More formally (simplifying with respect to the external conditions and concentrating on the core trust, as shown in Figures 2.11, 2.12, and 2.13 in Chapter 2):

X's mental state in *Trust(X, Y, τ)* is essentially constituted by:

$$Bel_X Can_Y(\alpha, p) \wedge Bel_X WillDo_Y(\alpha, p) \tag{7.1}$$

X's mental state in *Trust(X, A, τ')* is essentially constituted by:

$$Bel_X Can_A(\alpha', p') \wedge Bel_X WillDo_A(\alpha', p') \tag{7.2}$$

Y's mental state in *Trust(Y, A, τ')* is essentially constituted by:

$$Bel_Y Can_A(\alpha', p') \wedge Bel_Y WillDo_A(\alpha', p') \tag{7.3}$$

Where τ is the task that *Y* must perform for *X;* τ' the task that *A* must perform for *X* towards *Y*, i.e. check, supervision, guarantee, punishment, etc.

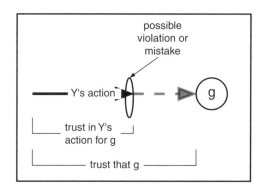

Figure 7.5   Trust in Y's action versus Trust in the final Result (achieving the goal)

More precisely and importantly in X's mind there is a belief and a goal (thus an expectation) about this trust of Y in A:

$$Bel_X(Trust(Y, A, \tau')) \wedge Goal_X(Trust(Y, A, \tau'))^9 \qquad (7.4)$$

And this expectation gives an important contribution to X's trust in the contractor.

X trusts Y by believing that Y will do what is promised because of his honesty or because of his respect/fear toward A. In other words, X relies on a form of paradoxical trust of Y in A: X believes that Y believes that A is able to control, to punish, etc. Of course, normally a contract is bilateral and symmetric, thus the point of view of Y's should be added, and his trust in X and in A as for monitoring X.

7.1.6   How Control Increases and Complements Trust

As we saw, in some sense control complements and surrogates trust and makes broad trust notions (see Figure 7.3) sufficient for delegation and betting. How does this work? How does control precisely succeed in augmenting confidence?

Our basic idea, is that strict-trust (trust *in* Y) is not the complete scenario; to arrive from the belief that 'Brings Y about that action α' (it is able and willing, etc.) to the belief that 'eventually g', something is lacking: the other component of the global trust: more precisely, the trust in the 'environment' (external conditions), including the intervention of the trustor or of somebody else. *Control can be aimed at filling this gap* between Y's intention and action and the desired result 'that g' (Figure 7.5).

However, does control only augment the broad trust? Not necessarily: the relationship is more dialectic. It depends on the kind and aim of control. In fact, it is important to understand that trust (also trust *in* Y) is not an ante-hoc and static datum (either sufficient or insufficient for delegation before the decision to delegate). It is a dynamic entity (see Chapter 6 in this

---

[9] As we said, here the use of the predicate 'Trust' is a bit inappropriate and misleading in a way. Actually, this is not trust but 'fear' of A. We want here just to stress that the basic cognitive constitutents are the same: an *evaluation* and an *expectation* towards A. It just depends on the Goal implied by the expectation if this is fear or trust: it is trust when Y is the trustor relying on X (and A); it is fear when Y is the trustee.

volume). For example there are effects, feedbacks of the decision to delegate on its own pre-condition of trusting Y. Analogously the decision to put control can affect Y's trustwortiness and thus the strict-trust whose level makes control necessary! Thus the schema: 'trust plus control' is rather simplistic, static, a-dialectic; since the presence of control can modify and affect the other parameters. As we wrote, there are indeed two kinds and functions of control: let us analyze these more deeply.

7.1.7   Two Kinds of Control[10]

(A) ***Pushing or influencing control:*** **preventing violations or mistakes**
 The first kind or function of control is aimed at operating on the 'trust in Y' and more precisely at increasing it by increasing Y's (perceived) trustworthiness. It is aimed in fact at reducing the probability of Y's defaillance, slips, mistakes, deviations or violation; i.e., at preventing and avoiding them. Behind this kind of surveillance there is at least one of the following beliefs:
 i) if Y is (knows to be) surveilled his performance will be better because either he will put more attention, or more effort, or more care, etc. in the execution of the delegated task; in other words, *he will do the task better* (there will be an influence on Y's ability); or
 ii) if Y is (knows to be) surveilled he will be more reliable, more faithful to his commitment, less prone to violation; in other words, *he probably will have a stronger intention to do the task* (there will be an influence on Y's willingness).

Since X believes this, by deciding to control Y (and letting Y know about this) she increases her own evaluation/expectation (i.e., her trust) of Y's willingness, persistence, and quality of work. As we can see in Figure 7.6, one of the control results is just to change the core trust of X on Y about τ.
 More formally we can write:

$$Bel_Y(Control(X\ Y\ \tau)) \supset \Delta Trustworthiness(Y\ \tau) \tag{7.5a}$$

where $Control(X\ Y\ \tau)$ measures the level of control by X on Y about the task τ. While $\Delta Trustworthiness(Y\ \tau)$ measures the corresponding Y's variation of trustworthiness. In other words, if Y believes that X controls him about τ a set of Y's attitudes will be introduced (consciously or unconsciously) by Y himself during his performance of τ.
 In addition, if:

$$Bel_X(Bel_Y(Control(X\ Y\ \tau)) \supset \Delta Trustworthiness(Y\ \tau)) \tag{7.5b}$$

then $DoT^*_{XY\tau}$ (the X's degree of trust in Y about τ including the knowledge of the control presence) might be different from the $DoT_{XY\tau}$ (the one without (X's believed) control).
 In other words, these additional attitudes can change Y's attention, effort, care, reliability, correctness, etc. and consequently produce a positive, negative, but also null contribution to X's degree of trust in Y about τ (depending from the expectation of X).

---

[10] As we said, there is a third form of control (or better of monitoring) merely aimed at Y's evaluation. If this mere monitoring (possibly hidden to Y) is for a future adjustment off-line (for changing or revocating the delegation next time) this form of control becomes of the second (B) class: control for adjustment, for correction.

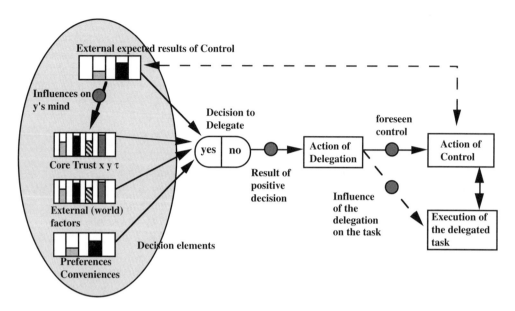

Figure 7.6 The expectation about control results influences the decision of trusting

This form of control is essentially *monitoring* (inspection, surveillance, reporting, etc.), and can work also without any possibility of *intervention*. Indeed, *it necessarily requires that Y knows about being surveilled.*[11] This can be just a form of 'implicit communication' (to let the other see/believe that we can see him, and that we know that he knows, etc.), but frequently the possibility of some explicit communication over this is useful (*'Don't forget that I see you!'*). Thus, some form of *intervention* can also be necessary: and as a consequence there would be a communication channel.

B) *Safety, correction or adjustment control*: **preventing failure or damages**
This control is aimed at preventing dangers due to *Y*'s violations or mistakes, and is aimed in general at the possibility of having adjustment of delegation and autonomy of any type ((Falcone and Castelfranchi, 2001), (Castelfranchi and Falcone, 2000b)). In other words, it is not only for repairing but for correction, through advice, new instructions and specifications, changing or revoking tasks, direct reparation, recovery, or help, etc.

For this reason this kind of control is possible only if some intervention is allowed, and requires monitoring (feedback) run-time.
In general *X* believes that the probability of achieving *g* when it is possible to intervene – *Pr*(achieve(g))* – is greater than without this possibility: *Pr(achieve(g))*:

$$Bel_X(Pr^*(achieve(g)) > Pr(achieve(g)))$$ (7.6)

---

[11] It is also necessary that *Y* cares about *X*'s evaluation. Otherwise this control has no efficacy. A bad evaluation is some sort of 'sanction', however it is not an 'intervention' – except if *X* can communicate it to *Y* during its work – since it does not interrupt or affect *Y*'s activity.

This distinction is close to the distinction between 'control for prevention' and 'control for detection' used by (Bons *et al.*, 1998). However, they mainly refer to legal aspects of contracts, and in general to violations. Our distinction is related to the general theory of action (the function of control actions) and delegation, and it is more general.

The first form/finality of control (*kindA*) prevents not only violations (in case of norms, commitments, or contracts) but also missed execution or mistakes (also in weak delegation where there are no obligations at all).

The second form/finality (*kindB*) is not only for sanctions or claims, but for timely intervening and preventing additional damages, or remedying and correcting. 'Detection' is just a means; the real aim is intervention for safety, enforcement or compensation.[12] Moreover, an effect (and a function/aim) of the second form of control can also be to prevent violation; this happens when the controlled agent knows or believes – before or during his performance – that there will be 'control for detection' and he worries about this (sanctions, reputation, lack of autonomy, etc.).

7.1.8   Filling the Gap between Doing/Action and Achieving/Results

Let's put the problem in another perspective. As we said, trust is the background for delegation and reliance i.e., to 'trust' as a decision and an action; and it is instrumental to the satisfaction of some goal. Thus the trust in Y (sufficient for delegation) implies the trust that g (the goal for which X counts on Y) will be achieved.

Given these two components or two logical steps scenario, we can say that the first kind of control is pointing to, is impinging on the first step (trust in Y) and is aimed at increasing it; while the second kind of control is mainly pointing to the second step and is aimed at increasing it, by confirming the achievement of g also in case of (partial) defaillance of Y.

In this way the control (monitoring plus intervention) complements the trust in Y which would be insufficient for achieving g, and for delegating; this additional assurance (the possibility to correct work in progress of Y's activity) makes X possible to delegate to Y the goal g. In fact, in this case X is not only counting on Y, but X counts on a potential multi-agent plan that includes her own possible actions.

As we can see from the formula (7.5a) in Section 7.1.7 the important thing is that Y *believes* that the control holds, and not if it really holds.[13] For example, X could not trust Y enough and communicate to him the control: this event modifies Y's mind and X's judgment about Y's trustworthiness. Thus, in trust-reliance, without the possibility of intervention for correction and adjustment, there is only one possible way to achieve g, and one activity (Y's activity) on which X bets (Figure 7.7).

Meanwhile, if there is control for correction/adjustment, the achievement of g is committed to Y's action plus X's possible action (intervention), X bets on this combination (Figure 7.8).

A very similar complementing or remedying role are guarantees, protections and assurance. I do not trust the action enough, and I put protections in place to be sure about the desired

---

[12] Different kinds of delegation (weak, mild, strong: see Section 2.9.1) allow for specific functions of this control. There will be neither compensation nor sanctions in weak delegation (no agreement at all), while there will be intervention for remedy.

[13] This is the actual power of the Gods.

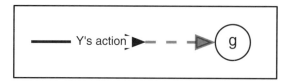

Figure 7.7   The gap between action and expected results

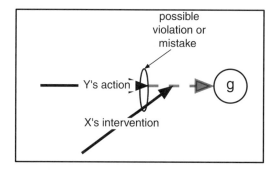

Figure 7.8   Intervention in the gap

results. For example, I do not trust driving a motorcycle without a crash-helmet, but I trust doing so with it.

7.1.9   The Dynamics

It is important to reinforce the idea that the first form/aim of control is oriented at increasing the *reliability* of Y (in terms of fidelity, willingness, keeping promises, or in terms of carefulness, concentration and attention) and then it is a way of increasing X's trust in Y which should be a presupposition not an effect of my decision:

- X believes that (if X watches over Y) Y will be more committed, willing and reliable; i.e. the strength of X's trust-beliefs in Y and thus X's degree of trust in Y are improved.

This is a very interesting social (moral and pedagogical) strategy. In fact it is in opposition to another well known strategy aimed at increasing Y's trustworthiness; i.e., 'trust creates trust' (see Chapter 6).[14]

In fact, the reduction/renouncing of control is a strategy of 'responsabilization' of Y, aimed at making it more reliable, more committed. Those stategies are in conflict with each other. When and why do we choose to make Y more reliable and trustworty through responsabilization

---

[14] Resuming and simplifying trust creates trust in several senses and ways. The decision to trust Y can increase X's trust in Y, via several mechasnisms: cognitive dissonance; because X believes that Y will be responsabilized; because X believes that Y will feel more self-confident; because X believes that Y will trust X and then be more willing to do good. The decision to trust Y can increase Y's trust in Y, via several mechanisms: Y has power over X that makes himself vulnerable and dependent; Y feels that if X is not diffident probably he is not malicious; Y perceives a positive social attitude in X and this elicits his goodwill; and so on.

(renounce to surveillance), and when through surveillance? A detailed model of how and why trust creates/increases trust is necessary in order to answer this question.

Should we make our autonomous agents (or our cyber-partners) more reliable and trustworthy through responsabilization or through surveillance?

We will not have this doubt with artificial agents, since their 'psychology' will be very simple and their effects will not be very dynamic. At least for the moment with artificial agents control will complement insufficient trust and perhaps (known control) will increase commitment. However, those subtle intertaction problems will certainly be relevant for computer-mediated human interaction and collaboration.

7.1.10   Control Kills Trust

Control can be bad and self-defeating, in several ways.

- There might be misunderstandings, mistakes, and incompetence and wrong intervention by the controller ('who does control controllers?') (in this case $Pr^*(achieve(g)) < Pr(achieve(g))$.
- Control might have the opposite effect than function (*kindA*), i.e. instead of improving performance, it might make performance worse. For example by producing anxiety in the trustee or by making him wast time and concentration for preparing or sending feedbacks (case in which $Trustworthiness^*(Y\,\tau) < Trustworthiness(Y\,\tau)$).
- It can produce a breakdown of willingness. Instead of reinforcing commitment and willingness, control can disturb it because of a bad reaction or rebellion, or because of delegation conflicts (Castelfranchi and Falcone, 1998) and need for autonomy; or because of the fact that distrust creates distrust (also in this case $Trustworthiness^*(Y\,\tau) < Trustworthiness (Y\,\tau)$).
- It can 'signal' (Section 6.3) to Y a lack of confidence and thus impact on (decrease) Y's self-confidence and self-esteem, negatively affecting his performance or commitment.

Here we mainly care about the bad effect of control on trust, which lets us see these dynamics. As trust virtuously creates trust, analogously the trust of Y in X, that can be very relevant for his motivation (for example in the case of exchange and collaboration), can decrease because X exibits not so much trust in Y (by controlling Y).

- X is too diffident, does this mean that X is malicious and Machiavellan? Since X suspects so much about the others would she herself be ready to deceive? Thus if X distrusts Y, Y can become diffident about X.
- Otherwise: X is too rigid, not the ideal person to work with.[15]
- Finally, if the agents rely on control, authority, norms they relax the moral, personal, or affective bonds, i.e. one of the strongest basis for interpersonal trust. Increasing control procedures in organizations and community can destroy trust among the agents, and then make cooperation, market, organization very bad or impossible, since a share of risk acceptance and of trust is unavoidable and vital.

---

[15] Control could also increase Y's trust in X, as a careful person, or a good master and boss, etc.

In sum, as for the dynamics of such a relation, we explained how:

- X's Control over Y denounces and derives from a lack of X's trust *in Y;*
- X's Control over Y can increase X's trust *in Y;*
- X's Control over Y increases X's trust in deciding to delegate to Y (her global trust);
- Control over Y by X can both increase and decrease Y's trust in X; in case that control decreases Y's trust in X, this should also affect X's trust in Y (thus this effect is the opposite of the second one);
- X's control over Y improves Y's performance, or makes it worse;
- X's control over Y improves Y's willingness, or makes him more demotivated.

7.1.11   Resuming the Relationships between Trust and Control

As we saw, relationships between trust and control are rather complicated. In this paragraph (see also Figure 7.9) we resume the different role that control can play with respect to trust.

In fact, as shown in Figure 7.9 the control can increase or decrease and in both the cases we can evaluate the potential influence on the two aspects of trust (strict and broad trust).

7.2   Adjusting Autonomy and Delegation on the Basis of Trust in Y

In this part we are going to analyze the complex scenario in which a cognitive agent (an agent with its own beliefs and goals) has the necessity to decide if and how to delegate/adopt a task to/for another agent in a given context. How much autonomy is necessary for a given task. How could this autonomy be changed (by both the trustor and the trustee) during the realization of the task. How *trust* and *control* play a relevant role in this decision and how important are their relationships and reciprocal influences.

Autonomy is very useful in cooperation (why someone should have an intelligent collaborator without exploiting its intelligence?) and even necessary in several cases (situatedness, different competence, local information and reactivity, decentralization, etc.), but it is also risky because of misunderstandings, disagreements and conflicts, mistakes, private utility, etc. A very good solution to this conflict is to maintain a high degree of interactivity *during* the collaboration, by providing *both* the man/user/client and the machine/delegee/contractor the possibility of taking the initiative in interaction and help (*mixed initiative* (Ferguson and Allen, 1998), (Hearst, 1999)) and of *adjusting* (Hexmoor, 2000) the kind/level of delegation and help, and the degree of autonomy run time.

We will analyze a specific view of autonomy which is strictly based on the notions of delegation and adoption (Castelfranchi and Falcone, 1998). In fact, in several situations the multi-agent plan, the cooperation between the delegating agent (*delegator*) and the delegated one (*delegee*), requires a strict collaboration and a control flow between the partners, in order to either maintain the delegator's trust or avoid breakdowns, failures, damages, and unsatisfactory solutions.

Software and autonomous agents will not only be useful for relieving human agents from boring and repetitive tasks; they will be mainly useful for situations where delegation and autonomy are necessary ('*strong dependence*', Section 2.9) because the user/client/delegator does not have the local, decentralized and updated knowledge, or the expertise, or the

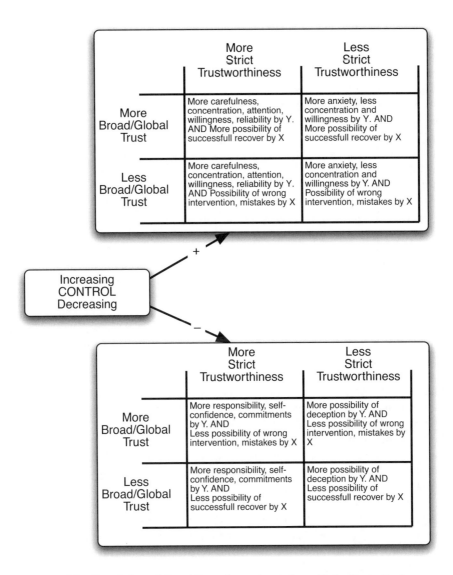

Figure 7.9 How Strict and Broad Trust change in relation with the Control's change

just-in-time reactivity, or some physical skill that requires some local control-loop. Thus autonomy and initiative are not simply optional features for *agents*, they are necessary requirements, and obligatory directions of study.

However, *control cannot be completely lost and delegation cannot be complete*, not only for reasons of confidence and trust, but for reasons of distribution of goals, of knowledge, of competence, and for an effective coordination. In this sense the possibility of controlling and adjusting the autonomy of the agents is becoming a growing and interesting field of research.

It has to be clear that this problem is central in any collaboration – between individuals, organizations, etc. – and that this theory is aimed at the general, not just for AI. Our claim in fact is that: *in designing how to adjust the level of autonomy and how to arrive at a dynamic level of control, it is necessary to have an explicit and general theory of (trust-based) delegation, which specifies different dimensions and levels of delegation, and relates the latter to the notion and the levels of autonomy.*

Thus, we propose our plan-based analysis (Pollack, 1990) of levels of delegation and levels of help, and discuss a related notion of autonomy. In several cases of collaboration among agents an *open delegation* is required, that is the delegation 'to bring it about that . . .'. The agent is supposed to use its knowledge, intelligence, ability, to exert some degree of discretion.

Given that the knowledge of the delegating agent/user (client) concerning the domain and the helping agents is limited (both incomplete and incorrect) the 'delegated task' (the request or the elicited behavior) might not to be so useful for the delegator itself. Either the expected behavior is useful but cannot be executed, or it is useless or self-defeating, or dangerous for the delegator's other goals, or else there is a better way of satisfying the delegator's needs; and perhaps the helping agent is able to provide greater help with its knowledge and ability, going beyond the 'literally' delegated task. We will call *extension of help* or *critical help* this kind of help. To be really helpful this kind of agent must take the initiative of opposing (not for personal reasons/goals) the other's expectations or prescriptions, either proposing or directly executing a different action/plan. To do this it must be able to recognize and reason with the goals, plans and interests of the delegator, and to have/generate different solutions.

Open delegation and over/critical help distinguish a *collaborator* from a simple tool, and presupposes intelligence and autonomy (discretion) in the agent. However, of course, there is a trade-off between pros and cons both in *open delegation* and in *extended(critical)-help*: the more intelligent and autonomous the delegee (able to solve problems, to choose between alternatives, to think rationally and to plan) the less *passively obedient* it is.[16] So, possible conflicts arise between a client and its contractor; conflicts which are due either to the intelligence and the initiative of the contractor or to an inappropriate delegation by the client. We are interested here only in the conflicts originating from the agent's willingness to collaborate and to help the other in a better and more efficient way: a kind of *collaborative conflict*. We do not consider the contractor's *selfish reasons* for modifying delegation (because the nature of the conflict, negotiation, etc. would be different).[17]

It is worth specifying that this work is aimed at providing a theoretical framework, i.e. the conceptual instruments necessary for analyzing and understanding interaction with autonomous entities. As has just been said, we assume that this framework is useful not only for organization theory or management, but also for a principled engineering, i.e. for getting systems designed not only on the basis of empirical data and practical experience, but also on the basis of a more complete view and typology, and of some prediction and explanation. The role of the dynamics of trust (Chapter 6) and control in this conflict and in the adjustment of the level of Y's autonomy is clear.

We also suggest some criteria about when and why to adjust the autonomy of an agent, and preliminary hints about necessary protocols for adjusting the interaction with agents.

---

[16] Obviously a very autonomous but stupid agent is even worse.

[17] In this chapter there is at least one of these reasons that should be taken into account: when the contractor/trustee adjusts the delegation for having more autonomy.

7.2.1 The Notion of Autonomy in Collaboration

For the purpose of this book we use a practical and not very general notion of autonomy.[18] In particular, we refer to the *social* autonomy in a *collaborative* relationship among agents. We distinguish between:

- a *meta-level autonomy* that denotes how much the agent is able and in condition to negotiate over the delegation or indeed to change it (in this regard, a slave, for example, is not autonomous: he cannot negotiate or refuse);
- a *realization autonomy*, that means that the agent has some discretion in finding a solution to an assigned problem, or a plan for an assigned goal.

Both are forms of goal-autonomy, the former at the higher level, the latter at the sub-goals (instrumental) level. For definition of different kinds of autonomy, including some of the dimensions we consider, see also (Huhns and Singh, 1997).

The lower the control of the client/trustor (monitoring or intervention) on the execution, the more autonomous is the contractor. In this context then, *autonomy means the possibility of displaying or providing an unexpected behavior (including refusal) that departs from the requested (agreed upon or not) behavior. The autonomous agent can be either entitled or not to perform such an unexpected behavior.*[19]

7.2.2 Delegation/Adoption Theory

We introduced the delegation notion in Section 2.9. Here we use that concept, integrating it with the notion of adoption and developing the theory of adjustable autonomy.

---

[18] We do not consider here some important aspects of autonomy (that could be adjusted) like the agent's independence or self-sufficiency. For an analytical discussion on the notion of *autonomy* in agents and for a more principled definition, see (Martin and Barber, 1996), (Castelfranchi, 2000b), and (Castelfranchi, 1995).

[19] In this book we do not discuss in detail another very important distinction between:

- being *practically in condition of* doing something (refusing, negotiating, changing and doing something else), i.e. what we would like to call ≪practical possibility≫; and
- being *deontically in condition of* doing something, i.e. to be entitled, permitted in the strong sense, i.e. the ≪deontic possibility≫.

An agent can have the former without the latter, or vice versa (see (Castelfranchi, 2000b)). In fact, there are two kinds of lack of power (hence, of dependence and autonomy): one based on practical conditions, the other based on deontic conditions. In deontic autonomy, an agent is permitted to do/decide/ interpret/ infer/ etc. Not only is it practically able and in condition to, but it can do this without violating a social or legal norm, or the user/designer prescriptions. As there are two kinds of autonomy there are two kinds of 'empowerment' (giving autonomy): deontic empowerment versus practical, material empowerment (Jones and Sergot, 1996). Therefore, an additional dimension of *adjustment* should be taken into account that is, the deontic one. The delegator (or the delegee) can attempt to modify (restrict or enlarge) either what the delegee is practically and actually able to do independently of the other, or what it is entitled to do. For example when a delegee re-starts negotiation, instead of directly modifying the task, it is implicitly asking some sort of permission, or agreement. Obviously enough, in strong delegation (contract relation, see later) the assignment of a task τ to the delegee implicitly entails giving it the permission to do τ. Adjusting the entitled space of freedom, or adjusting the practical space of freedom, is an interesting difference, but we cannot examine it in this book. Notice that this theory would imply the same plan-based dimensions of delegation and help.

Formal Constructs

Several formal constructs are needed in the following. Let $Act = \{\alpha_1, \ldots, \alpha_n\}$ be a set of *actions*, and $Agt = \{Ag_1, \ldots, Ag_m\}$ a set of *agents*. The *general plan library* is $\Pi = \Pi^a \cup \Pi^d$, where Π^a is the abstraction hierarchy rule set and Π^d is the decomposition hierarchy rule set. An action $\alpha' \in Act$ is called *elementary action* in Π if there is no rule r in Π such that α' is the left part of r. We will call *BAct* (*Basic Actions*) the set of elementary actions in Π and *CAct* (*Complex Actions*) the remaining actions in *Act*.

Given α_1, α_2 and Π^d, we introduce the *Dom-c(α_1 α_2)* operator to say that α_1 *dominates* α_2 (or α_2 *is dominated by* α_1) in Π^d: *Dom-c(α_1 α_2)=True* if there is a set of rules $(r_1, \ldots, r_m)$ in Π^d, such that: $(\alpha_1=Lr_1)\wedge(\alpha_2\in Rr_m)\wedge(Lr_i\in Rr_{i-1})$, where: Lr_j and Rr_j are, respectively, the left part and the right part of the rule r_j and $2\leq i\leq m$ (in the same way it is possible to define the *Dom-a(α_1 α_2)* operator considering the abstraction hierarchy rule set Π^a). We denote Π_{Agx} as the Ag_x's plan library, and Act_{Agx}, the set of actions known by Ag_x. The set of irreducible actions (through decomposition or specification rules) included in Π_{Agx} is composed of two subsets: the set of actions that Ag_x believes to be elementary actions ($BAct_{Agx}$) and the set of actions that Ag_x believes to be complex but for which it has no reduction rules ($NRAct_{Agx}$: *Non Reduced actions*). Then $BAct_{Agx}$ is included in *Act* and possibly $BAct_{Agx}$ is included or coincides with *BAct*. In fact, given an elementary action, an agent may (or may not) know the body of that action. We define S_{Agx} as the *skill set* of Ag_x, the actions in $BAct_{Agx}$ whose body is known by Ag_x (action repertoire of Ag_x).[20] We call R the operator that, when applied to an action α, returns the set of the *results* produced by α.

Definition of Delegation and Adoption

The domain of MAS, collaboration (Haddadi, 1996), and teamwork are already familiar with the notion of delegation. However, our analysis is grounded on more basic notions (Hexmoor, 2000). In addition, our delegation theory is not limited to explaining and modeling interpersonal relationships; the basic concepts of our definition also apply to (and are necessary even if not sufficient for) other important concepts such as:

- *institutional delegation*, in which the delegator transfers to the delegee not just some task but also some right, obligation, responsibility, power and so on (Jones and Sergot, 1996). Of course, this notion is richer than our basic concept (see below).
- *roles and prescriptions in organizations*, roles can be analyzed also as sets of delegated tasks (Castelfranchi and Falcone, 1997).

In our model, *delegation and goal adoption are characterized in terms of the particular set of mental states (cognitive ingredients) of the agents involved in the interaction*. Informally, *in*

---

[20] In sum, an agent Ag_x has its own plan library, Π_{Agx}, in which some actions ($CAct_{Agx}$ and $NRAct_{Agx}$) are complex actions (and it knows the reduction rules of $CAct_{Agx}$) while some other actions ($BAct_{Agx}$) are elementary actions (and it knows the body of a subset - S_{Agx} - of them).

delegation (reliance) an agent Ag_1 needs or likes an action of another agent Ag_2 and includes it in its own plan (see Section 2.9).

In *adoption (help) an agent Ag_2 acquires and has a goal as (long as) it is the goal of another agent Ag_1, that is, Ag_2 has the goal of performing an action because this action is included in the plan of Ag_1.* So, also in this case Ag_2 plays a part in Ag_1's plan (sometimes Ag_1 has no plan at all but just a need, or a goal) since Ag_2 is doing something for Ag_1.

We consider the action/goal pair $\tau=(\alpha,g)$ as the real object of delegation,[21] and we called it a 'task'. Then by τ, we will refer to the action (α), to its resulting world state (g), or to both. We introduce an operator of delegation with three parameters:

$$Delegates(Ag_1\ Ag_2\ \tau) \tag{7.7}$$

where Ag_1, Ag_2 are agents and $\tau=(\alpha,g)$ is the task. This means that Ag_1 delegates the task τ to Ag_2. In analogy with delegation we introduce the corresponding operator for adoption:

$$Adopts(Ag_2\ Ag_1\ \tau) \tag{7.8}$$

This means that Ag_2 adopts the task τ for Ag_1: Ag_2 helps Ag_1 by caring about τ.

Dimensions of Delegation and Adoption

We consider three main dimensions of delegation/adoption: *interaction-based, specification-based*, and *control-based* types of delegation/adoption (Castelfranchi and Falcone, 1998). Let us analyze these cases more in detail.

- *Interaction-based types of delegation.* Three general cases may be given: *weak, mild* and *strong delegation.* They represent different degrees of strength of the established delegation. In the following we synthesize (more formal details can be find in Section 2.9) the mental ingredients of trust in the different delegation actions.

W-Delegates is the operator for representing *weak delegation.* So the expression:
W-Delegates($Ag_1\ Ag_2\ \tau$) represents the *necessary* mental ingredients for Ag_1 trusting Ag_2 on the task τ, shown in Figure 2.11 and resumed in a less formal way in Table 7.1.

We consider in Table 7.1 (a, b, c, and d) what the agent Ag_1 views as a '*Potential for relying on*' agent Ag_2, its *trust* in Ag_2; and (e and f) what Ag_1 views as the '*Decision to rely on*' Ag_2.

---

[21] We assume that *delegating an action necessarily implies delegating some result of that action* (i.e. expecting some results from Ag_2's action and relying on it for obtaining those results). Conversely, *to delegate a goal state always implies the delegation of at least one action (possibly unknown to Ag_1) that produces such a goal state as a result* (even when Ag_1 asks Ag_2 to solve a problem, to bring it about that g without knowing or specifying the action, Ag_1 necessarily presupposes that Ag_2 should and will do some action and relies on this).

Table 7.1 Mental Ingredients in Weak-Delegation (pseudo-formal description)

a) The achievement of τ is a *goal* of Ag_1.

b) Ag_1 believes that there exists another agent Ag_2 that has the *power of* achieving τ.

c) Ag_1 believes that Ag_2 will achieve τ in time and by itself (without Ag_1's intervention).
   (if Ag_2 is a cognitive agent, Ag_1 believes that Ag_2 *intends* to achieve τ.).

d) Ag_1 *prefers*[22] to achieve τ through Ag_2.

e) The achievement of τ through Ag_2 is the choice (goal) of Ag_1.

f) Ag_1 has the goal (*relativized* (Cohen and Levesque, 1987) to (e)) of not achieving τ by itself.

Table 7.2 Mental Ingredients in Mild-Delegation (pseudo-formal description)

a'$\equiv$a; b'$\equiv$b; d'$\equiv$d; e'$\equiv$e; f'$\equiv$f; (referring to a, b, d, e, and f as described in Table 7.1)

c') Ag_1 does not believe that Ag_2 will achieve τ by itself (without Ag_1's intervention).

g') Ag_1 believes that if Ag_1 realizes an action α' then it is be more probable that Ag_2 intends τ.
   But Ag_2 does not adopt Ag_1's goal that Ag_2 intends τ.

h') Ag_1 intends to do α' relativized to (e').

We consider 'Potential for relying on' and 'Decision to rely on' as two constructs temporally and logically related to each other .[23]

M-Delegates is the operator for representing *mild delegation*.

M-Delegates(Ag_1 Ag_2 τ) represents the necessary mental ingredients of trust shown in Figure 2.12 and resumed in less formal way in Table 7.2.

We consider in Table 7.2 (a', b', c', d' and e') what agent Ag_1 views as a *'Potential for relying on'* agent Ag_2; and (f', g' and h') what Ag_1 views as the *'Decision to rely on'* Ag_2.[24]

S-Delegates is the operator for representing *strong delegation*. So the expression *S-Delegates(Ag_1 Ag_2 τ)* represents the *necessary* mental ingredients of trust as shown in Figure 2.13 and resumed in less formal way in Table 7.3:

We consider in Table 7.3 (a'', b', c'', d'' and e'') what agent Ag_1 views as a *'Potential for relying on'* agent Ag_2; and (f'', g'' and h'') what Ag_1 views as the *'Decision to rely on'* Ag_2.

For a corresponding analysis of adoption, and for how the kind of interaction between client and contractor influences the adoption itself see (Castelfranchi and Falcone, 1998).

---

[22] This means that Ag_1 believes that either the achievement of τ or a broader goal g' that includes the achievement of τ, implies Ag_1 to be dependent on Ag_2. Moreover (d) implies Ag_1's goal that Ag_2 achieves τ.

[23] As for weak delegation it is interesting to analyze the possibilities of Ag_2's mind. We should distinguish between two main cases: Ag_2 knows $W - Delegates(Ag_1 Ag_2$ $\tau)$ and Ag_2 does not know $W - Delegates(Ag_1 Ag_2\tau;)$. In other words, a weak delegation is possible even if the delegee knows it. Either this knowledge has no effect (the achievement of Ag_1's goal is just a side-effect known by Ag_2) or this knowledge changes Ag_2's goal: Ag_2 can either arrive at spontaneous and unilateral help or to a reactive, hostile attitude.

[24] In analogy with what we have said in weak delegation, also in mild delegation we should distinguish between two main cases about the possible mental states of Ag_2: Ag_2 knows $M-Delegates(Ag_1$ Ag_2 $\tau)$ and Ag_2 does not know $M-Delegates(Ag_1$ Ag_2 $\tau)$. So, it is possible to have a mild delegation even if the delegee knows it and if consequently it changes its own behavior to favor or to hamper the success of it.

Table 7.3   Mental Ingredients in Strong-Delegation (pseudo-formal description)

$a'' \equiv a$; $b'' \equiv b$; $c'' \equiv c$ $d'' \equiv d$; $e'' \equiv e$; $f'' \equiv f$; (referring to a, b, c, d, e, and f as described in Table 7.1)

g'') Ag_1 believes that if Ag_1 realizes an action α' there will be this result: Ag_2 will intend τ as the consequence of the fact that Ag_2 adopts $Ag_1's$ goal that Ag_2 intends τ (in other words, Ag_2 will be socially committed to Ag_1).

h'') Ag_1 has the goal (*relativized* to (e')) of not achieving τ by itself.

- *Specification-based types of delegation/adoption.* How is the task specified in delegation and how does this specification influence the contractor's autonomy? The object of delegation/adoption (τ) can be minimally specified (*open delegation*), completely specified (*closed delegation*) or specified at any intermediate level. Let us consider two cases:
  i) *Merely Executive (Closed) Delegation:* here the client (or the contractor) believes it is delegating (adopting) a completely specified task; what Ag_1 expects from Ag_2 is just the execution of a sequence of elementary actions (or what Ag_2 believes Ag_1 delegated to it is simply the execution of a sequence of elementary actions).[25]
  ii) *Open Delegation:* when the client (contractor) believes it is delegating (adopting) a non completely specified task: either Ag_1 (Ag_2) is delegating (adopting) an abstract action, or it is delegating (adopting) just a result (i.e. a state of the world).[26] Ag_2 can realize the delegated (adopted) task by exerting its autonomy. We can have several possible levels of openness of the delegated (adopted) task.
- *Control-based types of Delegation.* In this case we distinguish the delegation on the basis of the level of control it implies. At one extreme we have 'full control' (in fact the delegee is always under control during the realization of the delegated task) while at the other extreme we have 'no control' (in fact the delegee is never under control during the realization of the delegated task). As we have seen there are two main kinds in the control dimension: monitoring and intervention. Both have to be considered as influencing the delegation (we do not consider here a more detailed analysis of their different influences on delegation).

In Figure 7.10 we summarize the three main dimensions of delegation: each characterizes the variability of delegation action. The delegee's autonomy decreases towards the origin of the Cartesian space within the solid. Each of these dimensions implies, in fact, a specific aspect of the delegee's autonomy about the task.

7.2.3   The Adjustment of Delegation/Adoption

Run-Time Adjustment

We can consider the adjustment of autonomy (the revision of delegation) in three different time periods:

---

[25] More formally, either $\alpha \in S_{Ag1}$, or $\alpha \in BAct_{Ag1}$ ($\alpha \in S_{Ag2}$, or $\alpha \in BAct_{Ag2}$,), or g is the relevant result of α and $\alpha \in S_{Ag1}$ or $\alpha \in BAct_{Ag1}$ ($\alpha \in S_{Ag2}$, or $\alpha \in BAct_{Ag2}$,).

[26] More formally, either $\alpha \in CAct_{Ag1}$, or $\alpha \in NRAct_{Ag1}$ (either $\alpha \in CAct_{Ag2}$, or $\alpha \in NRAct_{Ag2}$); and also when g is the relevant result of α and $\alpha \in CAct_{Ag1}$ or $\alpha \in NRAct_{Ag1}$ ($\alpha \in CAct_{Ag2}$, or $\alpha \in NRAct_{Ag2}$).

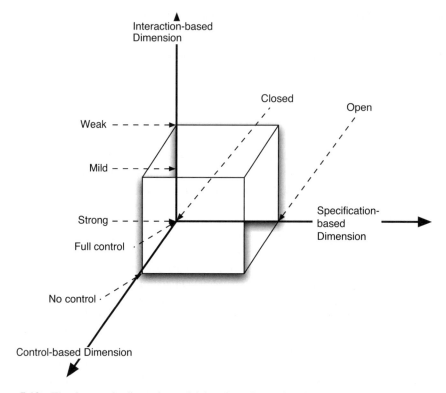

Figure 7.10   The three main dimensions of delegation. (Reproduced by Permission of © 2001 IEEE)

i) *After the delegation event,* but *before the execution of the task;*
ii) *Run-time,* with *work-in-progress;*
iii) *At the end of the performance* and the realization of the task; in this case the adjustment of autonomy will have an effect only on a future delegation (we can consider this case as a case of *learning*).

We will focus here on *run-time* adjustment, this is particularly important in human-machine interaction and in multi-agent cooperation, and call it simply *adjustment*. We will first examine the problem of adjustable autonomy *in a broad sense*, i.e. as adjusting the level and the kind of delegation/adoption (in our terminology *delegation conflicts* (Castelfranchi and Falcone, 2000c)). We claim that this is the right theoretical frame for understanding also the adjustment of autonomy in a strict sense, since *any autonomy adjustment requires a delegation adjustment*, but not vice versa (see next section).

In the following, we will analyze the general reasons for delegation/adoption adjustment. Let us here consider the taxonomy of the adjustments (some of which will be neglected because meaningless), their nature and their importance. Each of the possible adjustments is *bilateral*, i.e. either the client or the contractor can try to modify the previous delegation.

Table 7.4   Adjustments with respect to the interaction dimension

Line Number	Agent that has the initiative of the adjustment	Starting State	Final State
1	Delegator	Weak delegation	Mild delegation
2	Delegator	Weak delegation	Strong delegation
3	Delegator	Mild delegation	Strong delegation
4	Delegator	Mild delegation	Weak delegation
5	Delegator	Strong delegation	Weak delegation
6	Delegator	Strong delegation	Mild delegation
7	Delegee	Weak delegation	Mild delegation
8	Delegee	Weak delegation	Strong delegation
9	Delegee	Mild delegation	Strong delegation
10	Delegee	Strong delegation	Mild delegation
11	Delegee	Strong delegation	Weak delegation
12	Delegee	Mild delegation	Weak delegation
13	Delegator	Weak adoption	Strong adoption
14	Delegator	Strong adoption	Weak adoption
15	Delegee	Weak adoption	Strong adoption
16	Delegee	Strong adoption	Weak adoption

Source: Reproduced by Permission of © 2001 IEEE.

Delegation/Adoption Adjustments with Respect to the Interaction Dimension

As described in Table 7.4 there are (with respect to the interaction dimension) several possibilities of adjustment; they are determined by:

- the *agent* who has the initiative of the adjustment;
- the *starting state* (the kind of delegation or adoption acting in that given instant and that the *agent* intends to modify);
- the *final state* (the kind of delegation or adoption to which *agent* intends to arrive).

A few cases shown in Table 7.4 deserve some comments.

- *Line 1*: can be inferred from the difference between the mental ingredients of weak (see Table 7.1) and mild (see Table 7.2) delegation: in fact, c is replaced by c', g' and h'. In other words, Ag_1 does not believe that Ag_2 will achieve τ without any influence and so decides to realize an action α' that could produce this influence. In this case there is still no social commitment (Castelfranchi, 1996) by Ag_2: Ag_2 does not adopt Ag_1's goal that Ag_2 intends τ. In this case there is no sufficient trust and the trustor decides, for achieving the task, to introduce additional influences on the trustee.
- *Line 2*: g'' and h'' are added beliefs. In other words, Ag_1 tries to achieve τ through a social commitment of Ag_2: for this it realizes α'.
- *Lines 8 and 9*: could represent the willingness of Ag_2 to convert Ag_1's exploitation into a clear social commitment between them.
- *Lines 13-14*: are linked with the initiative of Ag_1 in the case in which Ag_1 is aware of Ag_2's adoption.

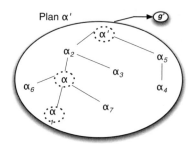

Figure 7.11   Composition Relationship

Delegation/Adoption Adjustments with Respect to the Specification Dimension

Also in these cases we must consider the intervention of both the client/trustor and the contractor/trustee with regard to delegation and adoption, respectively. Before analyzing the different cases included in this dimension, we show (also graphically) the meaning of the relationship about action composition (we called it *Dom-c*: *Dom-c(α' α)* defines this relationship between α' and α, where α is a component action of α').

Given Figure 7.11 we can say that the plan (complex action) α' gives the results for achieving the goal g'. This complex action is constituted from a set of different actions related with each other by the composition relationship (for example: *Make-Dinner* is constituted at a first level from *Buy-food*, *Prepare-food* and *Cook-food*; each of this action is, in its turn, constituted from other elementary or complex actions, and so on).

In fact, each of the actions shown in Figure 7.11 produces (temporary or final) results: see Figure 7.12. Temporary results will not be present in the results produced by the plan at the end of its execution (while final results will be present).

Delegee's Adjustments

The reasons for the delegee's adjustments can be of different nature and related to selfish or cooperative goals. In any case, these reasons cannot be irrespective of the evaluation of his own

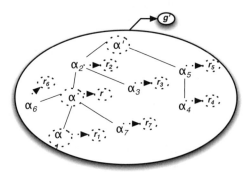

Figure 7.12   Composition Relationship with the evidence of the component action results

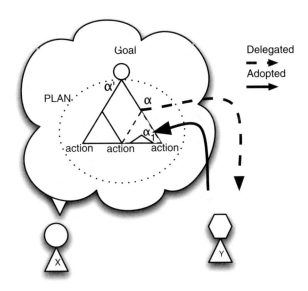

Figure 7.13　Sub-help

attitudes in that specific task (a sort of selftrust). Suppose that *Delegates(Ag₁ Ag₂ τ)* and τ is included in a more general *Ag₁*'s plan aimed at achieving goal g' through a complex action α'. Moreover, *Dom-c(α' α)*, $\tau=(\alpha,g)$, and $\tau'=(\alpha',g')$. We have three main delegee's adjustments:

- Reduction of help

  Here the *delegee provides less help on τ than delegated*. If

  $$Adopts(Ag_2\ Ag_1\ \tau_1) \wedge Dom\text{-}c(\alpha\ \alpha_1) \qquad (7.9)$$

with $\tau_1=(\alpha_1,g_1)$, the delegee reduces the task to a subpart of the requested one (see Figure 7.13). For example *Ag₁* delegates *Ag₂* to prepare a main course for a dinner and bring it to her house. *Ag₂* only buys the ingredients but does not cook them.

A sub-help is not necessary a help (although lower than expected). Maybe the realized action and the achieved subgoal are completely useless (in a plan the coordination among the actions and their results are also very important).

For example, in the case of the previous example, maybe the ingredients cannot be cooked at *Ag₁*'s house because of a problem with the kitchen.

- Extension of help

  Here the delegee provides more help on τ than delegated. If

  $$Adopts(Ag_2\ Ag_1\ \tau_1) \wedge Dom\text{-}c(\alpha_1\ \alpha) \wedge (Dom\text{-}c(\alpha'\ \alpha_1)\ OR\ (\alpha' \equiv \alpha_1)) \qquad (7.10)$$

with $\tau_1=(\alpha_1,g_1)$; the delegee goes beyond what has been delegated by the client without changing the delegator's plan (Figure 7.14). In fact, the delegee chooses a task that satisfies a

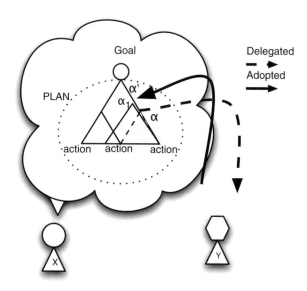

Figure 7.14 Over-help

higher level task (within the general delegator's intended plan) compared with the delegated task.

An example for this case is when Ag_1 delegates Ag_2 to cook 'spaghetti with pesto' but Ag_2 prepares the whole dinner.

- Critical help

It is the case in which the *delegee provides a qualitatively different action/help than expected* (what has been delegated). Let us analyze some subcases:

- *Simple critical help*

$$Adopts\,(Ag_2\ Ag_1\ \tau_x) \wedge g \in R(\alpha_x) \qquad (7.11)$$

with $\tau_x=(\alpha_x,g)$; the delegee achieves the goal(s) of the delegated plan/action, but it changes that plan/action (Figure 7.15). An example of *Simple Critical help* is when Ag_1 delegates Ag_2 to cook 'spaghetti with pesto' but Ag_2 has already bought cooked 'spaghetti with pesto' (he changes his own actions but the final result of them is supposed to be the same).

- *Critical overhelp*

$$Adopts\,(Ag_2\ Ag_1\ \tau_x) \wedge g_1 \in R(\alpha_x) \wedge Dom\text{-}c(\alpha_1\ \alpha) \wedge (Dom\ \text{-}c(\alpha'\ \alpha_1)\ OR\ (\alpha' \equiv \alpha_1))$$
$$(7.12)$$

with $\tau_1 = (\alpha_1,g_1)\ \tau_x = (\alpha_x,g_1)$, $R(\alpha_x)$ the set of results produced by α_x (see Figure 7.16); the delegee implements both a simple critical help and an extension of help (it chooses a task that

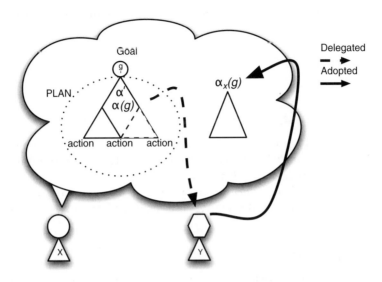

Figure 7.15   Simple Critical Help

satisfies a higher level task with respect to the task delegated, and achieves the goal(s) of this higher task, while changing the expected plan).

An example of *Critical Overhelp* is when Ag_1 delegates Ag_2 to cook 'spaghetti with pesto' (a subplan of preparing a dinner, thinking of cooking the other courses herself and in this way achieving the goal of offering a dinner to some old friends). But Ag_2 reserves a famous

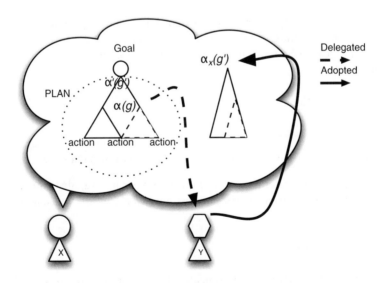

Figure 7.16   Critical Over-Help

restaurant in the town (in this way conserving Ag_1's goal of offering a dinner to some old friends).

- *Hyper-critical help*

$$Adopts\,(Ag_2\;Ag_1\;\tau_1) \wedge (g_1 \neq g') \wedge (g_1 \neq g) \wedge (g_1 \in I_{Ag1}) \qquad (7.13)$$

with $\tau_1 = (\alpha_1, g_1)$, and I_{Ag1} is the set of interests of Ag_1. Ag_2 adopts goals or interests of Ag_1 that Ag_1 themselves did not take into account: by doing so, Ag_2 neither performs the delegated action/plan nor achieves the results that were delegated.

A typical example of *Hyper-critical Help* is when Ag_1 asks Ag_2 for a cigarette and Ag_2 says to Ag_1 'you must not smoke'. In this way Ag_2 is dictating an interest of Ag_1 (to be healthy).

Delegator's Adjustment

The reasons for the delegator's adjustments can be different and related to selfish or cooperative goals. In any case, these reasons cannot be irrespective of the delegator's trust in the trustee about that specific task. Suppose that *Delegates(Ag₁ Ag₂ τ)*, and that Ag_1 intends to change that delegation. Suppose also that Ag_1 is achieving goal g' through plan τ', with $Dom\text{-}c(\alpha'\;\alpha)$. We can have five main delegator's adjustments:

- Reduction of delegation

    It is the case in which there is a new delegation:

$$Delegates(Ag_1\;Ag_2\;\tau_1) \wedge Dom\text{-}c(\alpha\;\alpha_1) \qquad (7.14)$$

with $\tau_1 = (\alpha_1, g_1)$, the delegator adjusts the original delegation, by *reducing the task that the contractor must realize* (the client reduces the task to a subpart of the previous requested task).

    For example, the delegator no longer trusts the trustee to complete the more complex action (see Figure 7.17).

- Extension of delegation

$$Delegates\,(Ag_1\;Ag_2\;\tau_1) \wedge Dom\text{-}c(\alpha_1\;\alpha) \wedge (Dom\text{-}c(\alpha'\;\alpha_1)\;OR\;(\alpha' \equiv \alpha_1)) \qquad (7.15)$$

with $\tau_1 = (\alpha_1, g_1)$, the delegator adjusts its delegation in such a way that its *new request goes beyond what has been originally delegated without changing the previous plan* (see Figure 7.18).

- Modification of delegation

    In an analogy with the delegee's adjustments, which consists of four subcases (modification of the previous delegated task just changing the previous goal; modification of the previous

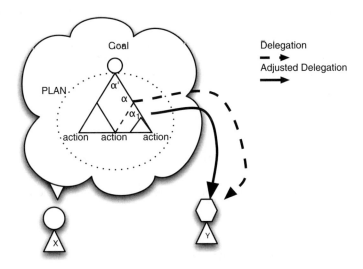

Figure 7.17   Riduction of Delegation

delegated task considering an over-goal and changing the plan to obtain that over-goal; modification of the previous delegated task considering a sub-goal and changing the plan to obtain that sub-goal; modification of the previous delegated task changing both the plan and the goal).

• Openness of delegation

$$Delegates\,(Ag_1\ Ag_2\ \tau_x) \wedge Dom\text{-}a(\alpha_x\ \alpha) \tag{7.16}$$

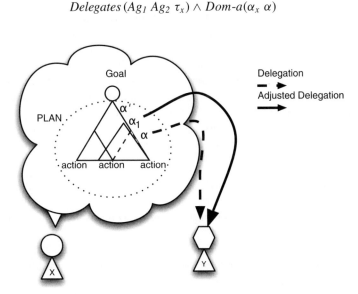

Figure 7.18   Extension of Delegation

In other words, the delegator adjusts their own delegation so that the new delegated plan is more abstract.

For example Ag_1 changes her delegation to Ag_2 from 'prepare a dinner composed of spaghetti with pesto' and 'chicken with french fries' to 'I need something to eat' (Ag_1 can cook a different dinner or can buy something to eat or can invite Ag_1 to a restaurant, and so on).

- Closing of delegation

$$Delegates\,(Ag_1\ Ag_2\ \tau_x) \wedge Dom\text{-}a(\alpha\ \alpha_x) \qquad (7.17)$$

In other words, the delegator adjusts its own delegation so that the new delegated plan is more specified.

7.2.4 Channels for the Bilateral Adjustments

For adjusting delegation and help, channels and protocols are necessary. As we have said, on the trustor/client's side, they are useful for monitoring (reporting, observing, inspecting), and intervention (instructions, guidance, helps, repair, brake); on the trustee/delegee's side it is useful to have some space for discretion and practical innovation. For both client and contractor, are useful channels and protocols for communication and re-negotiation during the role-playing and the task execution.

The Trustor's Side

As we have already written, Ag_1 must have some form of *control* on and during Ag_2's task realization, otherwise no adjustment is possible. Obviously, the feedback, i.e. monitoring, is useful for a run-time adjustment and must be provided in time for the intervention. In general, the feedback activity is the precondition for an intervention.

In multi-agent systems, in order to guarantee agents a dynamic adjustment of delegation and their mixed initiative we have to provide such an *infrastructure*, while in human-computer interaction we have to provide the user with those two channels.

When Ag_2 has the initiative (it is starting with an adoption action), if Ag_1 wants to change this adoption it needs a *communication channel* with Ag_2.

The Trustee's Side

Ag_2 can run-time adjust the delegation of Ag_1 (and its own autonomy) if it is in condition of either:

a) having a *communication channel* for *(re-)starting negotiation* by offering/proposing a different level of help to Ag_1; or
b) having enough practical freedom to *directly change* the action (this condition should be by definition a characteristic of autonomous agents).

Feedback from

	TRUSTOR	TRUSTEE
TRUSTOR	*INSPECTION, SURVELLANCE*	*REPORT*
TRUSTEE	*GUIDANCE*	*SELF-MONITORING*

Feedback to

Figure 7.19 How feedback determines different kinds of control. (Reproduced by Permission of © 2001 IEEE)

Trustees should not necessarily negotiate or give advice in order to change their delegated tasks; they might have full initiative. This entails meta-level autonomy. Of course, the trustee must also have feedback about its own execution, but this is true in general for goal directed actions.

To sum up, if an agent has the initiative of a delegation/adoption then, in order to adjust that initiative, it is not obliged to communicate with the other agent. As for the necessary *feedback for adjustment* we can distinguish between: *inspection, report, guidance,* and *self-monitoring* (Figure 7.19).

Considered the kinds of intervention action showed in Section 7.1.2, we can say that each of these interventions could be realized through either a *communication act* or a *direct action* on the task by the intervening agent (Table 7.5).

7.2.5   Protocols for Control Adjustments

Starting from our model it is also possible to identify some guidelines for designing interaction in human-machine interaction or in multi-agent systems. For example, our model makes it

Table 7.5   Different kinds of client intervention

	Client's message	Client's direct action
Stopping the task	Stop	Stopping intervention
Substitution	I do it	It realizes an action of the task
Correction of delegation	Change that action with this other	It introduces constraints such that an action is changed with another
Specification of delegation	Make that plan in this way	It introduces constraints such that a plan is specified
Repairing of delegation	Add this action of the task	It introduces constraints such that a new action must be realized to favor success of the task

Source: Reproduced by Permission of © 2001 IEEE.

clear that agent control requires either communication or a simple action-perception loop. On the one hand, Ag_1 can monitor Ag_2 without any communication (without sending any specialized message or signal), by simply observing it; at the same time Ag_1 can also influence Ag_2 by physical/practical actions and interventions on Ag_2 or on the world. For example Ag_1 can brake and stop Ag_2. On the other hand, Ag_1 can monitor Ag_2 thanks to messages sent by Ag_2 to Ag_1 (reports), and can influence Ag_2 by sending them messages (instructions, warnings, etc.).

Examples of Monitoring

Let us show you how the *monitoring* actions can be expressed in praxis and in communication.
PRAXIS:

inspection (visiting the environment in which Ag_2 is working to ascertain if everything is as expected);

internal inspection (inspecting some inside agent data to check its reasoning, agenda, plan library, plan, etc.);

surveillance (by sample) observing Ag_2's activity and partial results, and the environment for avoiding damages;

detecting analyzing some traces of Ag_2's activity in order to (abductively) check whether its behaviour has been correct and at what stage it is.

COMMUNICATION:

report requests ('let me know what is happening'; 'any news?');

inspective questions ('is everything as expected?' 'what are you doing?' 'is p true?').

Examples of the Intervention

Let us now show you how the *intervention* actions can be expressed in praxis and in communication.
PRAXIS:

substitution, Ag_1 performs (part of) an action previously allocated to Ag_2;

support, Ag_1 modifies the conditions of the world so that Ag_2 can successfully perform its action or damages can be prevented;

brake, Ag_1 stops Ag_2's activity (either by external obstacles or directly acting upon/in Ag_2's body or software);

tuning, Ag_1 modifies and corrects Ag_2's action (either by external obstacles or directly acting upon/in Ag_2's body or software);

repair, Ag_1 acts in order to repair damages as a result of Ag_2's action and to recover from failures.

COMMUNICATION:

alert/warning, Ag_1 alerts Ag_2 to unexpected events or possible danger;

advice, Ag_1 provides Ag_2 with some possible recipe, solution or better action just as a piece of advice (Ag_2 is free to accept or not);

instructions, Ag_1 gives instruction to Ag_2 about how to proceed (Ag_1 is specifying (partially closing) the previously 'open' delegation);

threats, Ag_1 threats Ag_2 to induce it to do what Ag_2 should do;

reminding, Ag_1 reminds Ag_2 about what it should do; *stop*, Ag_1 orders Ag_2 to stop its activity; *abort*, Ag_1 stops delegating to Ag_2.

An interesting development of this work would be to model when and why a given form of intervention (for example: to stop Ag_2) is useful or better than others; and what kind of feedback (for example: surveillance versus report) is appropriate for a given task level of trust and possible kind of intervention.

7.2.6   From Delegation Adjustment to Autonomy Adjustment

As we said, a delegation adjustment does not always produce a change in the trustee's autonomy (by limiting, restricting or, vice versa, enlarging, expanding it). The main causes of autonomy adjustment are the following:

- there is a change of Ag_2's entitlement at the meta-level (Ag_2 can refuse, negotiate, change the delegation); or it takes such an initiative even though it is not entitled (*meta-autonomy adjustment*);
- the new task is more or less *open* than the former (*realization-autonomy adjustment*);
- there is more or less control on Ag_2 (*control-dependent autonomy adjustment*);
- there is a change in the strength of delegation (*interaction-dependent autonomy adjustment*).

Each of these autonomy adjustments can be *bilateral* (realized by either the client or the contractor or both) and *bidirectional* (either augmenting or reducing the autonomy itself). These adjustments are, at least in a significant part, strictly linked with the Ag_1's trust in Ag_2 and/or with the selftrust of Ag_2 himself.

7.2.7   Adjusting Meta-Autonomy and Realization-Autonomy of the Trustee

By crossing the first two kinds of adjustment (meta-autonomy adjustment and realization-autonomy adjustment) with the delegation adjustments, we obtain the results shown in Table 7.6: rows 1–3 show the adjustments of delegation by the trustee (trustee's adjustments)

Table 7.6 How autonomy changes while adjusting delegation and help

	Meta autonomy	Autonomy of realization
Reduction of Help	*Increased*	*Equal*
Extension of Help	*Increased*	*Increased or Equal*
Critical Help	*Increased*	*Increased*
Reduction of Delegation	*Equal*	*Reduced of Equal*
Modification of Delegation	*Equal*	*Increased or Equal*
Critical Delegation	*Equal*	*Increased or Equal or Reduced*
Openess of Delegation	*Equal*	*Increased*
Closing of Delegation	*Equal*	*Reduced*

Source: Reproduced by Permission of © 2001 IEEE.

while rows 4–8 show the adjustments by the trustor (trustor's adjustments) on its own previous delegation. In particular, we can see that:

- *When there is a trustee's adjustment there is always a change of its meta-autonomy* (the trustee decides to change the trustor's delegation); while not always there is a change in its realization autonomy. For example, in the *reduction of help,* realization autonomy remains the same because the trustee realizes just a part of the delegated task (but this part was also included in the previously delegated task). In other words, the trustee does not change autonomy as for how to realize τ. Conversely in the *extension of help*, there are two possibilities: i) the trustee has more realization autonomy when the adopted plan includes some (not delegated) part which is not completely specified (thus the delegee has more discretion in its realization); ii) the trustee has the same realization autonomy if the adopted plan does not need more discretion than the delegated one. Finally, in *critical help*, given its possibility to choose new actions, there is always more realization autonomy.
- *When there is a trustor's adjustment the trustee's meta-autonomy never changes* (in fact, the trustor itself takes the initiative to modify the delegation). As for the trustee's realization autonomy we can say that: in the *reduction of delegation* case, Ag_2's autonomy of execution (if its discretionary power is reduced with the new delegation) is reduced or it remains unchanged (suppose that the old task was completely specified in all details). In the *extension of delegation* case, either the autonomy of realization increases (if the new task presupposes some action – not included in the old one – with a certain degree of openness) or it remains unchanged (if this new task was completely specified in all details). In the *critical delegation* case, the autonomy of realization of the trustee increases or not depending respectively on whether the new actions are more or less open than the old ones. In the *openness of delegation* case, the autonomy of realization of the trustee always increases (openness is in fact a factor that increases the discretion of the trustee). Vice versa, in the case of *closing of delegation*, the trustee's autonomy of realization is always reduced.

7.2.8   Adjusting Autonomy by Modyfing Control

As already observed a very important dimension of autonomy is the control activity of the adopted/delegated task. Given that control is composed of feedback plus intervention, adjusting it means having to adjust (at least one of) its components.

Adjusting the Frequency of the Feedback

We showed in Section 7.1.2 as the *frequency of the feedback on the task* can be:

- *purely temporal* (when the monitoring (*Mon*) or the reporting (*Rep*) is independent of the structure of the activities in the task);
- *linked with the task phases* (when the activities of the task are divided in phases and the *Mon* or the *Rep* is connected with them).

Trustor and trustee can adjust the frequency of their feedback activity in three main ways:

- by *changing the temporal intervals* fixed at the beginning of the task delegation or task adoption (when the *Mon/Rep* is purely temporal);
- by *changing the task phases* in which the *Mon/Rep* is realized with respect to those fixed at the beginning of the task delegation;
- by *moving from* the purely temporal *Mon/Rep* to the task phases *Mon/Rep* (or vice versa).

Adjusting the Frequency and Kind of Intervention

As explained in Section 7.1.2, the intervention is strictly connected with the presence of the *Mon/Rep* on the task, even if, in principle, both the intervention and the *Mon/Rep* could be independently realized. In addition, the occurrence of intervention and *Mon/Rep* are also correlated. More precisely, the intervention can occur:

1) *never*;
2) *just sometimes* (during some phase or at specified times, a special case of this is at the end of the task);
3) *at any phase or at any time (depending on the necessity)*.

The *adjustment of the frequency of intervention* by the trustor is an important case of adjustment of the trustee's autonomy. Suppose that at the beginning there is an agreement about the fact that the established frequency of intervention is *never*, and suppose that the trustor intervenes once or twice during the trustee's task realization: the trustee's autonomy has been reduced. In general, a trustee is more autonomous if the frequency of the trustor's intervention is low. So the adjustments by the trustor in this direction (low frequency of interventions) produce an increase of trustee's autonomy. If the trustor adjusts the possible kind of intervention established at the beginning of delegation this might increase or reduce the trustee's autonomy depending on this adjustment.

7.2.9 When to Adjust the Autonomy of the Agents

We will examine in this section the general principles (criteria) for adjusting (restricting or expanding) the trustee's autonomy by both the trustor/client/delegator, and the trustee/contractor/delegee.

Table 7.7   Reducing the trustee's autonomy

WHEN (classes of reasons):

- Ag_1 believes that Ag_2 is not doing (in time) what Ag_1 has delegated to it; and/or
- Ag_1 believes that Ag_2 is working badly and makes mistakes (because of lack of competence, knowledge, control, etc.); and/or
- Ag_1 believes that there are unforeseen events, external dangers and obstacles that perhaps Ag_2 is not able to deal with; and/or
- Ag_1 believes that Ag_2 is going beyond its role or task, and Ag_1 is not happy about this (because of lack of trust or of conflict of power)[27]

THEN (reduction of autonomy) Ag_1 will reconsider its delegation to Ag_2, and Ag_2's level of autonomy in order to reduce it by either specifying the plan (task) or by introducing additional control, or constraining the interaction (strong delegation), etc.

Table 7.8   When to expand the trustee's autonomy

WHEN (classes of reasons):

- Ag_1 believes that Ag_2 is doing or can do better than previously expected (predicted); and/or
- Ag_1 believes that the external conditions are more favorable than expected; and/or
- Ag_1 believes that Ag_2 is working badly and makes mistakes (because of lack of flexibility, or because of too much control, etc.) and/or
- Ag_1 believes that Ag_2 can do more than previously assigned, or can find its own situated way of solving the problem

THEN (expansion of autonomy) Ag_1 will change the delegation to Ag_2, and Ag_2's level of autonomy in order to expand it by either letting the plan (task) less specified or reducing control or making the interaction weaker, etc.

Adjusting the Autonomy of Trustee

In this preliminary identification of reasons for autonomy adjustment, we prefer a more qualitative and simplified view, not necessarily related with a probabilistic framework like the one we will use in the following. Of course, to be more precise, one should specify that what changes is the subjective probability assigned to those events (beliefs). For example, at the time of the delegation, Ag_1 believed that the probability of Ag_2's mistakes was *pm* (and this expectation was compatible with the decision of delegating a given degree of autonomy), while Ag_1 realizes that this probability has changed (higher or lower than expected).

Let us simplify the issue in Table 7.7, Table 7.8, Table 7.9, and Table 7.10.

Trust as the Cognitive Basis for Adjusting Autonomy

What we have just seen (principles and reasons for bilateral delegation and autonomy adjustment) can be considered from the main perspective of trust.

---

[27] Notice that in all those cases the trustee's expectations on which trust and reliance were based are disappointed.

Table 7.9 When to limit one's own autonomy

Let us now consider some (*collaborative*) reasons of adjustment on the delegated agent's side.
WHEN (classes of reasons):

- Ag_2 comes to believe that it is not able to do the complete task (level of self-confidence); and/or
- Ag_2 comes to believe that there are unforeseen events, external dangers and obstacles that are difficult to deal with

THEN (reduction of autonomy)
Ag_2 will reconsider the received delegation (for example providing sub-help and doing less than delegated) and its level of autonomy in order to reduce it by either asking for some specification of the plan (task) or for the introduction of additional control (example: 'give me instructions, orders; monitor, help, or substitute me').

Table 7.10 When to expand one's own autonomy

WHEN (classes of reasons):

- Ag_2 gets to a grounded belief that it is able or in condition of doing more or providing a better solution for the delegated goal (within Ag_1's plan, or also with regard to Ag_1's other desires and interests), and
- it is not forbidden or it is (explicitly or implicitly) permitted by Ag_1 that Ag_2 takes such a collaborative initiative, and/or
- Ag_2 believes that Ag_1 will accept and enjoy its initiative (because convenience largely exceeds surprise or distress)

THEN (expansion of autonomy)
Ag_2 will reconsider the received delegation and level of autonomy in order to go beyond those limits by directly providing, for example, over-help or critical-help (doing more and better).

(When the 2nd and 3rd conditions above are not realized, Ag_2 could take the initiative of communicating by offering the new solution or asking for a permission, and in fact for re-negotiating the delegation).

Trust, being the mental ground and counterpart of delegation, plays the main role in adjustment: limiting autonomy is usually due to a trust crisis (Section 6.7), while expanding autonomy is usually due to an increased trust.

In Section 3.4 and Section 3.6 we have shown how *changing the credibility degree of some beliefs should change the final choice* about the delegation (and the same holds for the utilities and for the control). Resuming and concluding: *The trustor's adjustment reflects a modification in her mental ingredients.* More precisely, the trustor either *updates or revises* their delegation beliefs and goals, i.e.:

a) either they revise their *core trust beliefs* about Ag_2 (the latter's goals, capabilities, opportunities, willingness);

b) or they revise their *reliance beliefs* about: i) their dependence on Ag_2, or ii) their preference to delegate to Ag_2 rather than to do the job themselves, or to delegate to Ag_3 (a third agent) or to renounce the goal;

c) or they changes their risk *policy* and is more or less likely to accept the estimated risk (this means that the trustor changes either their set of utilities ($U(Ag_1,t_0)$) or their set of thresholds. In other words, either Ag_1's trust on Ag_2 is the same but their preferences have changed (including their attitude towards risk), or Ag_1 has changed their evaluations and predictions about relying on Ag_2.

The modifications showed in the cases (a, b, c) might produce delegation adjustments but also they could suggest to the trustor the introduction of control actions (either as monitoring or as intervention). So the relationships among trust, control, autonomy, and delegation are very complex and not so simple to predict: also the trustor and trustee personalities can play a relevant role in these relationships.

7.3   Conclusions

As already shown, the relationships among *Trust*, *Control* and *Autonomy* are very complex and interesting. *Autonomy* is very useful in collaboration and even necessary in several cases but it is also risky – because of misunderstandings, disagreements and conflicts, mistakes, private utility, etc. The utility and the risk of having an autonomous collaborator can be the object of a trade-off by maintaining a high level of interactivity *during* the collaboration, by providing *both* the trustor/delegator/client and the trustee/delegee/contractor with the possibility of having initiative in interaction and help (*mixed initiative*) and of *adjusting* the kind/level of delegation and help, and the degree of autonomy run-time. This also means providing channels and protocols – on the delegator's side – for *monitoring* (reporting, observing, inspecting), and for *intervention* (instructions, guidance, helps, repair, brake); and – on the delegee's side – providing some room for discretion and practical innovation; for both client and contractor, channels and protocols are needed for communication and *re-negotiation* during the role-playing and the task execution.

Our model also provides a principled framework for adjusting autonomy on the basis of the degree of trust and of the control's level of the trustor. In particular we have shown that in order to adjust autonomy one should in fact adjust the delegation/help relationship. Thus a precise characterization of different dimensions of delegation and of goal-adoption is necessary. Moreover, we argued that adjustment is *bi-directional* (one can expand or reduce the delegee's autonomy) and is *bilateral*; not only the trustor or the delegator but also an adaptive/intelligent delegee, the trustee (the 'agent') can change or try to change its level of autonomy by modifying the received delegation or the previous level/kind of help. This initiative is an additional and important aspect of its autonomy. We showed how trust, being also the mental ground and counterpart of delegation, plays a major role in adjustment: limiting autonomy is usually due to a trust crisis, while expanding autonomy is usually due to an increased trust. Collaborative conflicts are mainly due to some disagreement about the agent's trustworthiness.

We assume that this theoretical framework can also be useful for developing principled systems.

We have outlined:

- the criteria about when and why to adjust the autonomy of an agent (for example, when one believes that the agent is not doing (in time) what it has been delegated to do and/or is working badly and makes; and/or one believes that there are unforeseen events, external dangers and obstacles that perhaps the agent is not able to deal with); and
- possible protocols of both monitoring and inspection, and of physical or communicative intervention, that are necessary for control and adjustment.

A very important dimension of such an interaction has been neglected: the normative dimension of empowerment and autonomy (entitlement, permission, prohibition, etc.) which is related to a richer and institutional relation of delegation. Also, this dimension is a matter of run-time adjustment and must be included as a necessary component when modeling several forms of interactions and organizations.

Another important issue for future works is the acceptable limits of the agent's initiative in helping. Would, for example, our personal assistant be too intrusive by taking care of our 'interests' and 'needs' beyond and even against our request (*Hyper-critical help*)? Will the user/client like such a level of autonomy or would they prefer an obedient slave without initiative? Let us leave this question unanswered as it is enough to have characterized and delimited the complex framework of such an issue.

Finally, we will leave for another book a rather important clarification for engineering: does the implementation of such a model necessarily require deliberative agents?

In fact our framework for collaboration and adjustable autonomy is presented in terms of cognitive agents, i.e. of agents who have propositional attitudes, reason about plans, solve problems, and even assume an 'intentional stance' by having a representation of the mind of the other agent. This can be exemplified via some kind of BDI agent, but in fact it is more general (it does not only apply to a specific kind of architecture). We present our framework in a cognitive perspective because we want to cover the higher levels of autonomy,[28] and also the interaction between a human user and a robot or a software agent, or between humans. However, the basic ontology and claims of the model could also be applied to non-cognitive, merely rule-based agents.

Obviously, a cognitive agent (say a human) can delegate in a weak or mild sense a merely rule-based entity. Strong delegation based on mutual understanding and agreement cannot be used, but it can be emulated. The delegated device could have interaction protocols and reactive rules such that if the user (or another agent) asks to do something – given certain conditions – it will do that action. This is the procedural emulation of a true 'goal adoption'.

Our notions could in fact be just embedded by the designer in the rules and protocols of those agents, making their behavior correspond functionally to delegation or adoption, without the 'mental' (internal and explicit) goal of delegating or of helping. One could, for example, have fixed rules of over-help like the following ones:

---

[28] In our view, to neglect or reduce the mental characterization of delegation (allocation of tasks) and adoption (to help another agent to achieve its own goals) means, on the one hand, to lose a set of possible interesting kinds and levels of reliance and help, and, on the other hand, not to completely satisfy the needs and the nature of human interaction that is strongly based on these categories of cooperation.

$<$if Ag_x asks for departure time, provide departure time & gate$>$ (over-answering);

$<$if Ag_x asks for action α that has result r & not able to perform α & able to perform α' with the same result r, then perform $\alpha'>$

The previous behaviors are in fact a kind of *over-help* although the performing agent does not conceive any help (the real adopter is the programmer writing such a rule).

The same remains true in the case of a rule-based delegated agent: the agent could have simple *rules* for abstaining from doing everything itself α ' while inducing – via some protocols – the needed action in another agent, or for abstaining from doing by itself α' when receiving information (communication protocol; observation) about another agent already doing the needed action.

In sum, several basic phenomena and issues (of delegating, adopting, monitoring, intervening, changing delegation, etc.) are held and recognized also by non-cognitive agents and can be incorporated in a procedural emulation of a really social interaction.

References

Beamish, P. (1988) *Multinational Joint Ventures in Developing Countries*. London: Routledge.

Bons, R., Dignum, F., Lee, R., Tan, Y.H. (1998) A formal specification of automated auditing of trustworthy trade procedures for open electronic commerce, *Autonomous Agents '99 Workshop on 'Deception, Fraud and Trust in Agent Societes'*, Minneapolis, USA, May 9, pp.21–34.

Castelfranchi, C. (1995) Guaranties for Autonomy in Cognitive Agent Architecture. In M. Wooldridge and N. Jennings (eds.) *Intelligent Agents*. Springer. LNAI 890, 56–70.

Castelfranchi, C. (1996) Commitment: from intentions to groups and organizations. In *Proceedings of ICMAS'96*, S. Francisco, June, AAAI-MIT Press.

Castelfranchi, C. (2000) Formalizing the informal? Invited talk DEON2000 Toulouse.

Castelfranchi, C. (2000) Founding Agent's Autonomy on Dependence Theory, in *Proceedings of ECAI'00*, Berlin, August 2000.

Castelfranchi, C. and Falcone, R. (1994) Towards a theory of single-agent into multi-agent plan transformation. The 3rd Pacific Rim International Conference on Artificial Intelligence (PRICAI94), Beijing, China, 16–18 August, pp. 31–37.

Castelfranchi, C. and Falcone, R. (1997) From task delegation to role delegation, in M. Lenzerini (Editor), *AI*IA97: Advances in Artificial Intelligence*, Lecture Notes in Artificial Intelligence, 1321. Springer-Verlag pp. 278–289.

Castelfranchi, C. and Falcone, R. (1998) Towards a theory of delegation for agent-based systems, *Robotics and Autonomous Systems*, Special issue on Multi-Agent Rationality, Elsevier Editor, 24 (3-4):.141–157.

Castelfranchi, C. and Falcone, R. (1998) Principles of trust for MAS: cognitive anatomy, social importance, and quantification, Proceedings of the International Conference on Multi-Agent Systems (ICMAS'98), Paris, July, pp. 72–79.

Castelfranchi, C. and Falcone, R. (2000) Social trust: a cognitive approach, in *Deception, Fraud and Trust in Virtual Societies* by Castelfranchi, C. and Yao-Hua, Tan (eds.), Kluwer Academic Publisher.

Castelfranchi, C. and Falcone, R. (2000) Trust and control: a dialectic link, *Applied Artificial Intelligence Journal*, Special Issue on 'Trust in Agents' Part 1, Castelfranchi, C., Falcone, R., Firozabadi, B., Tan, Y. (Editors), Taylor & Francis 14 (8).

Castelfranchi, C. and Falcone, R. (2000) Conflicts within and for collaboration, in C. Tessier, L. Chaudron and H. J. Muller (eds.) *Conflicting Agents: Conflict Management in Multi Agent Systems*, Kluwer Academic Publishers, pp. 33–61.

Cohen, Ph. and Levesque, H. (1987) 'Rational Interaction as the Basis for Communication'. Technical Report, N°89, CSLI, Stanford.

Durkheim, E. (1997) *The Division of Labor in Society*, New York: The Free Press.

Falcone, R, (2001) Autonomy: theory, dimensions and regulation, in C. Castelfranchi and Y. Lesperance (eds.) *Intelligent Agents VII, Agent Theories Architectures and Languages*, Springer, pp. 346–348.

Falcone, R. and Castelfranchi, C. (2001) Social trust: a cognitive approach, in *Trust and Deception in Virtual Societies* by Castelfranchi, C. and Yao-Hua, Tan (eds.), Kluwer Academic Publishers, pp. 55–90.

Falcone, R. and Castelfranchi, C. (2002) Issues of trust and control on agent autonomy, *Connection Science*, special issue on 'Agent Autonomy in Groups', 14 (4): 249–263.

Ferguson, G. and Allen, J. (1998) TRIPS: An Integrated Intelligent Problem-Solving Assistant, *Proc. National Conference AI (AAAI-98), AAAI Press*, Menlo Park, Calif.

Haddadi, A. (1996) *Communication and Cooperation in Agent Systems*, The Springer Press.

Hearst, M. (editor) (1999) Mixed-initiative interaction - Trends & Controversies, IEEE Intelligent Systems, September/October 1999.

Hendler, J. (1999) Is there an Intelligente Agent in your Future?, http://helix.nature.com/webmatters/ agents/agents.html.

Hexmoor, H. (editor) (2000) Special Issue on autonomy control software, *Journal of Experimental & Theoretical Artificial Intelligence*, 12 (2), April-June 2000.

Huhns, M. and Singh, M. (1997) Agents and multiagent systems: themes, approaches, and challenge, in *Reading in Agents* (Huhns, M. and Singh, M. Editors) Morgan Kaufmann Publishers, Inc., San Francisco, California.

Jones, A. J. I. and Sergot, M. (1996) A formal characterisation of institutionalised power. *Journal of the Interest Group in Pure and Applied Logics*, 4 (3).

Martin, C. E. and Barber, K. S. (1996) Multiple, Simultaneous Autonomy Levels for Agent-based Systems, in *Proc. Fourth International Conference on Control, Automation, Robotics, and Vision*, Westing Stamford, Singapore, pp. 1318–1322.

Mayer, R. C., Davis, J. H., Schoorman, F. D. (1995) An integrative model of organizational trust, *Academy of Management Review*, 20 (3): 709–734.

Nissenbaum, H. (1999) Can trust be secured online? A theoretical perspective; http://www.univ.trieste.it/ ~dipfilo/etica_e_politica/1999_2/nissenbaum.html

Pollack, M. (1990) Plans as complex mental attitudes in Cohen, P. R., Morgan, J. and Pollack, M. E. (eds.), *Intentions in Communication*, MIT Press, USA, pp. 77–103.

Sitkin, S. B., and Roth, N. L. (1993) Explaining the limited effectiveness of legalistic 'remedies' for trust/distrust. *Organization Science*. 4:367–392.

Smith, R. G. (1980) The contract net protocol: High-level communication and control in a distributed problem solver. *IEEE Transactions on Computers*.

Tan, Y. H. and Thoen, W. (1999) Towards a generic model of trust for electronic commerce, *Autonomous Agents '99 Workshop on 'Deception, Fraud and Trust in Agent Societes'*, Seattle, USA, May 1, pp. 117–126.

van der Vecht, B. (2008) Adjustable autonomy, controlling influences on decision making, *Doctoral Thesis*, Utrecht University, The Netherlands.

8

The Economic Reductionism and Trust (Ir)rationality

Trust is a traditional topic in economics, for obvious reasons: economic relationships, first of all exchange, presuppose that X relies on Y for receiving what she needs (an item or a service, work or salary); she has to trust Y on both competence and quality, and on his reliability (credibility, honesty, fidelity). Trust is the presupposition of banks, money, commerce, companies, agency, contracts, and so on.

So, trust has been the subject of several approaches by economists and social scientists ((Williamson, 1993), (Axelrod and Hamilton, 1981) (Yamagishi, 2003) (Pelligra, 2005; 2006]). Many of them (often out of a desire to find a simple measure, some quantification,[1]) are very reductive, both from a psychological and a social point of view; the notion/concept itself is usually restricted and sacrificed for the economic framework.

In addition, a lot of interesting considerations on trust have developed (a 'trust game' for example, (Joyce *et al.*, 1995], (Henrich *et al.*, 2004]) around game theory: a recently growing domain (Luce and Raiffa, 1957) (Axelrod and Hamilton, 1981), (Shoham and Leyton-Brown, 2009).

In this chapter we aim to discuss some of those ideas (which appear, quite diffused, in other disciplines), showing why they are too reductive and how they miss fundamental aspects of the trust phenomenon, crucial even in economics.

We will discuss the particular (and for us not acceptable) formulation of trust by Deutsch (Deutsch, 1985) and in general the relationship between trust and irrationality (of the trustee or of the trustor); the very diffused reduction of trust to subjective probability; Williamson's eliminativistic position (Williamson, 1993); trust defined in terms of risk due to Y's temptation and opportunism; and as an irrational move in a strategic game; the reductive notion of trust in the trust game, and in some socio-economic work (Yamagishi, 2003)); why (the *act of*) trust cannot be mixed up and identified with the *act of cooperating* (in strategic terms); the wrong foundational link between trust and reciprocity.[2]

---

[1] We call this attitude: *'quanificatio precox'*.

[2] See for example (Pelligra, 2005) and (Castelfranchi, 2009).

Trust Theory   Cristiano Castelfranchi and Rino Falcone
© 2010 John Wiley & Sons, Ltd

8.1   Irrational Basis for Trust?

8.1.1   Is Trust a Belief in the Other's Irrationality?

The predictability/reliability aspect of trust has been an object of study in social sciences, and they correctly stress the relationship between sincerity, honesty (reputation), friendliness, and trust. However, sometimes this has not been formulated in a very linear way; especially under the perspective of game theory and within the framework (the mental syndrome) of the *Prisoner Dilemma*.

Consider for example the notion of trust as used and paraphrased in Gambetta's interdisciplinary book on trust, on the basis of (Deutsch, 1958). It can be enounced like this: 'When I say that I trust *Y*, I mean that *I believe that, put on test, Y would act in a way favorable to me, even though this choice would not be the most convenient for him AT THAT MOMENT'*.[3]

So formulated, (considering subjective rationality) *trust is the belief that Y will choose and will behave in a non-rational way!* How might he otherwise choose against his interest? To choose what is perceived as less convenient? This is the usual dilemma in the *prisoner dilemma game*: the only rational move is to defeat.

Since trust is one of the pillars of society (no social exchange, alliance, cooperation, institution, group, is possible without trust), should we conclude *that the entire society is grounded on the irrationality of the agents*: either the irrationality of *Y*, or the irrationality of *X* in believing that *Y* will act irrationally, against his better interest!

As usual in arguments and models inspired by rational decision theory or game theory, together with 'rationality', 'selfishness' and 'economic motives' (utility, profit) are also smuggled. We disagree about this reading of rationality: when I trust *Y* in strong delegation (social commitment by *Y*) I'm not assuming that he – by not defeating me – will act irrationally, i.e. against his interests. Perhaps he acts 'economically irrationally' (i.e. sacrificing in the meanwhile his *economic* goals); perhaps he acts in an unselfish way, preferring some altruistic or pro-social or normative motive to his selfish goals; but he is not irrational because he is just following his subjective preferences and motives, and those include friendship, or love, or norms, or honesty, avoiding possible negative feelings (guilt, shame, regret, . . .), etc. It is not such a complicated view to include within the subjective motives (rewards and outcomes) the real spectrum of human motives, with their value, and thus determining our choices, beyond the strictly (sometimes only with a short life) 'economic' outcomes.

Thus when *X* trusts *Y* she is simply assuming that other motivations (his values) will in any case prevail over his economic interests or other selfish goals. So we would like to change

---

[3] See also recently (Deutsch, 1958): '. . . . *a more limited meaning of the term implies that the trustworthy person is aware of being trusted and that he is somehow bound by the trust which is invested in him. For this more specific meaning, we shall employ the term 'responsible.' Being responsible to the trust of another implies that the responsible person will produce 'X' (the behavior expected of him by the trusting individual), even if producing MY' (behavior which violates the trust) is more immediately advantageous to him.'*.   See   also (Bacharach and Gambetta, 2001) *'We say that a person 'trusts someone to do X' if she acts on the expectation that he will do X when both know that two conditions obtain:*

- *if he fails to do X she would have done better to act otherwise, and*
- *her acting in the way she does gives him a selfish reason not to do X.'*. Notice in this definition the important presence of: 'to do', 'acts', 'expectation'.

the previous definition to something like this: 'When I say that I trust Y, I mean that *I believe that, put on test, Y would act in a way favorable to me, even though this choice would not be the most convenient for his private, selfish motives at that moment, but the <u>adopted</u> interests of mine will – for whatever reason and motive – prevail'.*[4]

At this level, *trust is a theory and an expectation about the kind of motivations the agent is endowed with, and about which will be the prevailing motivations in case of conflict.* This preserves the former definition (and our definition in Chapter 1) by just adding some specification about the *motives* for Y's reliability: for example, the beliefs about Y's morality are supports for the beliefs about Y's intention and persistence. I not only believe that he will intend and persist (and then he will do), but I believe that he will persist *because of certain motives*, that are more important than other motives that would induce him to defection and betrayal. And these motives are already there: in his mind and in our agreement; I don't' have to find new incentives, to think of additional prizes or of possible punishments. If I am doing so (for example, promising or threatening) I don't really trust Y (yet) (see Section 9.5.2).

After an agreement we trust Y because of the advantages we promised (if it is the case), but also or mainly because we believe that he has other important motives (like his reputation, or to be honest, or to respect the laws, or to be nice, or to be helpful, etc.) for behaving as expected.

This is the crucial link between 'trusting' and the image of 'a good person'. 'Honest' is an agent who prefers his goal of not cheating and not violating norms to other goals such as pursuing his own private benefits.[5] Social trust is not only a mental 'model' of Y's cognitive and practical capacities, it is also a model of his motives and preferences, and tells us a lot about his morality.

In this framework, it is quite clear why we trust friends. First, we believe that as friends they want our good, they want to help us; thus they will both take on our request and will keep their promise. Moreover, they do not have reasons for damaging us or for hidden aggressing us. Even if there is some conflict, some selfish interest against us, friendship will be more important for them. We rely on the *motivational strength* of friendship.

As we explained in Chapter 2 these beliefs about Y's virtues and motives are sub-species of trust, not just beliefs and reasons supporting our trust in Y.

Rational Is Not Equal to Selfish

As we have just seen, it is incorrect to (implicitly) claim that the trustor relies on the trustee's irrationality (not acting out of self-interest, not pursuing their own goals and rewards); *the trustor is just ascribing to the trustee motives (perhaps unselfish ones) which will induce Y to behave conformingly to X's expectations.*

---

[4] Notice that this is a definition only of the most typical social notion of trust, where X relies on Y's *adoption* of her goals (Chapter 2). As we know, we admit weaker and broader forms of trust where Y is not even aware of X's delegation.

[5] It is important to remind us that this is not necessary and definitional for trust: I can trust Y for what I need, I can delegate him, *just relying on his selfish interests* in doing as expected. Like in commerce (see Adam Smith's citation in Chapter 2, Section 2.8).

However, is it irrational to count upon non 'official', non economic (monetary), non selfish, and sometimes merely hidden and internal rewards of Y? Is this equal to assuming and counting on the trustee's irrationality? Our answer is: Not at all!

The identification of 'rationality' with 'economic rationality' is definitely arbitrary and unacceptable. Rationality is a merely formal/procedural notion; has nothing to do with the contents (Castelfranchi, 2006]. There are no rational 'motives'. No motive can be per se irrational. It is up to the subject to have one motive or another. Economic theory cannot *prescribe* to people what are the right motives to have and to pursue; it can just prescribe *how* to choose among them, given their *subjective* value and probability. Unless Economics admits not being the science of (optimal) resource allocation and rational decisions (given our preferences or, better, motives) (Lionel Robins' view), but being the science of making money, where money is the only or dominating *valid* 'motive' of a 'rational' agent.

When X decides to trust Y (where Y is a psychological agent), she is *necessarily* ascribing to Y some *objectives*, and some *motive*, which predict Y's expected behavior. These objectives are very diverse, and some of them are non-visible and even *violating the official 'game' and the public rewards*. Consider the following common sense example.

Subjective versus Objective Apples (Rewards)

X and Y are at a restaurant and receive two apples: one ($apple_1$) is big, mature, nice; the other one ($apple_2$) is small and not so beautiful. Y chooses and takes the better one: $apple_1$. X is manifestly a bit disappointed by that. Y – realizing X's reaction – says: *'Excuse me, but which apple would have you chosen if you had chosen first?'* X: *'I would have taken apple$_2$, leaving you apple$_1$'*, Y: *'And I let you have apple$_2$! So why you are unhappy?'*

Why is X unhappy if he would have taken $apple_2$ has in fact got $apple_2$, when they are in fact one and the same apple? Actually (from the subjective and interactional perspective, not from the official, formal, and superficial one) they are two very different apples, *with different values* and providing very different rewards.

- $apple_2$, when spontaneously chosen by X, is $apple_2$ (that material apple with its eating value) plus:
  - a *sacrifice*, compared with the other possibility (a negative reward, but a *voluntary sacrifice* that implies the following items),
  - an *internal gratification* for being polite and kind (a positive reward), and
  - an *expectation for the recognition of this from Y*, and some gratitude or approval (a positive reward).
- $apple_2$, when left to X by Y, is – for X – $apple_2$ (that material apple with its eating value) plus:
  - a *sacrifice*, compared with the other possibility (a negative reward, but an *imposed sacrifice*, which is worse than a spontaneous one),
  - an *impolite act towards X*, a lack of respect from Y's side (a negative reward).

Thus, in the two cases, X *is not eating the same apple*; their flavor is very different. $Apple_2$ 'has a different value in the two cases', in the first case there are fewer negative rewards and more positive rewards.

As we said, to take into account those motives and rewards in Y, is not seeing Y as 'irrational'; Y can be perfectly *subjectively* rational (given his real, internal rewards).

Economists and game theorists frequently enough decide from outside what are the values (to be) taken into account in a given decision or game (like to go to vote, or to persist in a previous investment), they have a 'prescriptive' attitude; and *they do not want to consider the hidden, personal, non-official values and rewards taken into account by real human subjects*. Relative to the external and official outcomes the subjects' decisions and conducts look 'irrational', but considering their real internal values, they are not at all.

Finally, it would be in its own right a form of irrational assumption, to ascribe to the other an irrational mind, while ascribing to ourselves rational capabilities; trusting the others assuming that they are different from us, and exploiting their presumed stupidity.[6] Thus, counting on those 'strange' motives (of *any* kind) seen in Y is neither irrational per se nor is counting on Y's irrationality.

8.2   Is Trust an 'Optimistic' and Irrational Attitude and Decision?

It is fairly commonplace for the attitude and decision to trust somebody and to rely on him is *per se'* to be considered rather irrational. Is this true and necessary? Is it irrational to decide to trust, to rely on others and take risks? Is trust too 'optimistic' and a decision or expectation that is ungrounded?

8.2.1   The Rose-Tinted Glasses of Trust

Frequently trust implies optimism, and optimists are prone to trust people. What is the basis of such a relationship and how 'irrational' is optimism? Trust frequently goes beyond evidence and proof, and doesn't even search for (sufficient) evidence and proof.[7] It can be a default attitude: *'Till I do not have negative evidence, I will assume that . . .'*; it can be a personality trait, an affective disposition (see Chapter 5). Is this necessarily an 'irrational' attitude and decision from an individual's perspective?

Optimism -1

As we have anticipated in Chapter 4, one feature defining *optimism* is precisely the fact that the subject, in the case of insufficient evidence, of a lack of knowledge, assumes all the uncertain cases in favor of her desired result. Her positive expectation is not restricted to the favorable evidence, but she assumes the positive outcome as 'plausible' (not impossible) (see Figure 8.1).

On the contrary, a prudent or pessimistic attitude, maintains the positive expectation within the limits of the evidence and data, while where there is uncertainty and ignorance the subject considers the negative eventuality (see Figure 8.2).

---

[6] Of course, it is also possible and perfectly rational in specific cases and circumstances, and on the basis of specific evidence, to trust Y precisely because he is naïve or even stupid.

[7] It should be clear that to us this is not a necessary, definitional trait of trust, but it is quite typical. It is more definitional of "faith" or a special form of trust: the 'blind' or 'faithful' one.

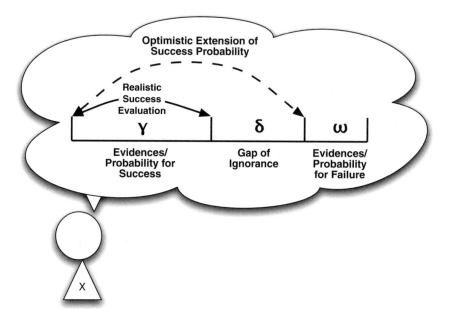

Figure 8.1   Strong Optimistic attitude

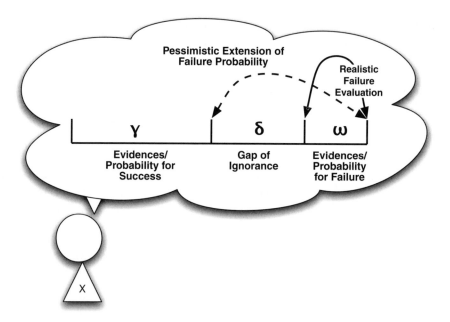

Figure 8.2   Extreme Pessimistic attitude

Extreme optimists and extreme pessimists have two opposite default assumptions and two opposite views of the 'onus of proof': does the onus lie on the prosecution side or on the defense side?

Given this (partial) characterization of optimism, when trust is not based on evidence it is an 'optimistic' attitude, since it assumes a favorable unproved outcome. Is this attitude necessarily counterproductive, too risky, or is it subjectively irrational? It seems that non-optimistic people (the more depressed ones) have a better (more realistic) perception of reality in some dimensions. For example, they have a more adequate (realistic) perception of the probability of the positive/negative events, and of their own control over events. While optimists would distort both the probability of favorable events, and the control they have on them. If something like this is true, optimism should be a disadvantage, since a distorted view of reality should lead to failures. However, a lot of studies on 'thinking positively' have shown that in several circumstances and tasks optimists fare better and that an optimistic attitude is an advantage (Scheier and Carver, 1985), (Scheier et al., 1986), (Taylor and Brown, 1988). How can we explain such an apparent contradiction: a distorted view of the world being more adaptive? We will answer this in the following paragraph.

Trust and Distrust as a Self-Fulfilling Prophecy

In sum, if the trustee knows that X trusts him so seriously that she is relying on him and exposing herself to risks – as we have explained – he can, for various reasons, become more reliable and even more capable, that is, more trustworthy. This is due to different possible psychological processes (Chapter 5).

The fact that X appreciates Y, has esteem for him, can make Y more proud and sure in his actions. He can either feel more confident, can increase his self-esteem or determination. Or Y can feel some sort of impulse in reciprocating this positive attitude (Cialdini, 2001). Or he can feel more responsible and obliged not to disappoint X's expectation and not to betray her trust. He would bring harm to X, which would be unfair and unmotivated. In other words, X's perceived trust in Y can both affect Y's motivation (making him more reliable, willing, persistent,..) or his effort and competence (attention, carefulness, investments, . . .) (see also Section 8.6 on Reciprocity).

Given this peculiar dynamic of trust one may say that trust can in fact be a *self-fulfilling prophecy*. Trust in fact is – as we know– an expectation, that is, a prediction. However, this (optimistic) prediction is not just the evaluation of an *a priori* probability that is independent and indifferent to the subject evaluation.

If X plays the roulette game and feels sure and trustful about the fact that the next result will be 'Red', this doesn't have any affect on the probability (50%) of it being Red. But there are a lot of 'games' in human life that are not like roulette, and where an optimistic attitude can actually help.

Consider, for example, courting a woman, or competing with another guy, or preparing for an exam; in this kind of 'game' the probability of success is not predetermined and independent from the participant's attitude and expectations. Self-esteem, self-confidence, trust, etc. that is, positive expectations, make the agent more determined, more persistent, less prone to give up; and also more keen to invest in terms of effort and resources. This does change the probability of its success. Moreover, a positive, confident and self-confident attitude as perceived by the others (*signaling*) can per se – in social interaction – be rewarding, can per se facilitate the success; for example of the courtship or of the examination.

This is why optimists are favored and more successful in many cases, compared with people with a more pessimistic and depressed attitude, although the latter might have a more realistic perception of their control (or lack of control) over events, or of the objective probabilities.

Having the 'prediction' (prophecy) is a factor that comes into the process which determines the result; thus it is a prediction that comes true, is fulfilled, thanks to its very existence. This is why trust is some sort of a 'self-fulfilling prophecy'.

However, not only can trust work like this, but also suspect and distrust can have self-realizing effects. For example, if X wants to be sure about Y, and wants to have proof and evidence about Y's trustworthiness, and creates repeated occasions for monitoring and testing Y's reliability (perhaps even exposing Y to 'temptations' in order to verify his fidelity), he is in fact altering the 'probability' of Y's betrayal.

Y might be irritated by X's controls and this irritation creates hostility; or Y can be disappointed and depressed by X's lack of trust and this might decrease either his confidence or his motivation. Moreover, in general, as a trusting attitude creates a positive feeling, and also a reciprocal trust, and a diffuse trust atmosphere; analogously, diffidence elicits diffidence, suspicion or mistrust produces offence or distance.

Our explanation is as follows. In life there are two quite different kinds of 'lotteries':

a) Lotteries where the probabilities of success are *fixed a priori and independent to the subject* and their mental attitudes. This is, for example, the case of playing dice or betting at roulette. The probability is given and the subject has no influence at all on the result; it doesn't depend on their prayers, or feelings, or on how they throw the dice (excluding possible tricks), and so on. In this kind of lottery there should be more dangers for optimists and some advantage (in the case of the risk of losing opportunities) for pessimists. However, there are other – more frequent and more important – lotteries in our social life.
b) Lotteries where the probabilities of the favorable result are not given, but are influenced by the attitude, the expectations, and even the feeling of the subject. Expectations in these cases are *self-fulfilling prophecies*. This is the case, for example, of an attempt to seduce a woman; an optimistic attitude – giving to the subject and exhibiting self-confidence, providing more persistence (less perplexity) and commitment, which also might mean to the other a stronger interest and motivation – can influence the probability of success which is not a priori given. The same holds for an interview for a job or for a contract; and in many social circumstances where our success depends both on our investment and persistence (and we invest and persist proportionally to the expected probability), and on the impression we give to the partners. So, optimism might be a *real* advantage. However, this makes optimism 'adaptive', 'effective', but not yet subjectively 'rational', although it paradoxically makes 'true' and realistic an expectation, which would be ungrounded (and subjectively irrational).

Optimism may be a 'subjectively rational' attitude if we assume some sort of awareness or knowledge of its self-fulfilling mechanism. And in fact we can assume a sort of implicit belief about this: the optimist has reinforced their position by practising their 'ungrounded' belief; has acquired by experience an implicit knowledge of such an effect. Thus their belief (expectation) is not so subjectively irrational. Unfortunately, the same reinforcement underlies a pessimistic attitude!

Optimism -2

Another fundamental feature of optimism is the fact that it focuses, makes explicit and throws attention onto the positive side of expectations. Actually expectations – including expectations on which trust is based – have a Janus nature, a double face.

In fact, since we introduce the idea of quantification of the degree of subjective certainty and reliability of *belief* about the future (the forecast), we get a hidden, strange, but nice consequence. There are other implicit opposite *beliefs* and thus *implicit expectations*.

For 'implicit' beliefs we mean in this case a belief that is not 'written', contained in any 'data base' (short term, working, or long term memory) but is only potentially known by the subject since it can be simply derived from actual beliefs. For example, while my knowledge that Buenos Aires is the capital city of Argentina is an explicit belief that I have in some memory and I have only to retrieve it, on the contrary my knowledge that Buenos Aires is not the capital city of Greece (or of Italy, or of India, and so on) is not in any memory, but can just be derived (when needed) from what I explicitly know. While it remains implicit, merely potential, until is not derived, it has *no effect* in my mind; for example, I cannot perceive possible contradictions: my mind is only potentially contradictory if I believe that p, I believe that q, and p implies *not q*, but I didn't derive that *not q*.

Now, a belief that '70% it is the case that p', logically (but not psycho-logically!) implies a belief that '30% it is the case that *not p*'.[8] This has interesting consequences on *expectations* and related emotions. The *positive expectation* that p entails an implicit (but sometimes even explicit and compatible) *negative expectation* (see Figure 8.3).

This means that any hope implicitly contains some fear, and that any worry implicitly preserves some hope. But also means that when one gets a 'sense of relief' because a serious threat that was expected does not arrive and the world is conforming to your desires, you also get (or can get) some exhaltation. It depends on your focus of attention and framing (Kahneman, 2000): are you focused on your worry and non existent threat, or on the unexpected achievement? Vice versa when you are satisfied about the actual expected realization of an important goal, you can also achieve some measure of relief while focusing on the implicit previous worry.

When one feels a given emotion (for example, fear), although not necessarily at the very moment of feeling it, one also feels the complementary emotion (hope) in a sort of oscillation or ambivalence and affective mixture. Only when the belief is explicitly represented and one can focus – at least for a moment – one's attention on it, can it generate the corresponding emotion.

Optimists do not think (elaborate) on the possibility of failure, the involved risks, or the negative parts of the outcome; or at least, they put them aside: they do not focus on it. In this way, they, for example, avoid the elicited feeling of worry, of avoidance, of prudence, or of non-enthusiasm (*'OK, it will be good, but . . . .; not so good'*).

---

[8] We are simplifying the argument. In fact it is possible that there is an interval of ignorance, some lack of evidence; that is that I 45% evaluate that p and 30% that *Not p*, having a gap of 25% neither in favor of *p* nor of *Not p* [29] [30].

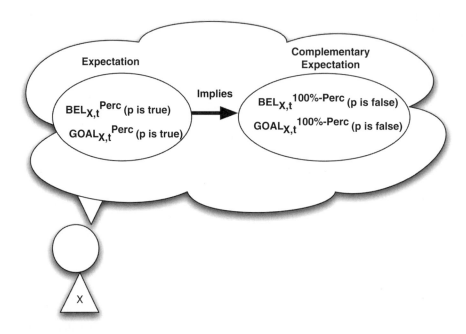

Figure 8.3   The expectation of an event (with a defined probability) implies also the expectation (with a correlated probability) of the denied event

Optimism -3

Moreover, 'optimism' in a sense is not only a prospective attitude, it is a more global attitude ('rose-tinted glasses'), even in an *a posteriori* evaluation of present results or of the past. Hence the saying: given a glass of water half-full, the optimist will see it as 'half full', while the pessimist will see it as 'half empty'. That is, the optimist will focus on the positive result, on what one has got, on the (even small) achievement or gain; and she will balance her disappointment with such a consolation. The pessimist, on the contrary, will focus on what has not been achieved, on what is lacking, and this will destroy, make marginal what has been obtained.

It is important to notice how these different perceptions of reality will reinforce the subjective prospective attitudes: the optimist will have their view confirmed that the positive expectation was not so wrong and that it is realistic to make good predictions; their attitude will be reinforced. The same for the pessimistic attitude! Both attitudes are 'self-fulfilling prophecies', self-realizing predictions; thus, both of them produce the expected results (or see the results as expected) and reinforce and reproduce themselves (Castelfranchi, 2000).

Is trust a form of optimism? An optimistic attitude? No; not necessarily. The fact is that without trust there is no society (Locke, Parsons, Garfinkel, Luhmann, etc.; see Chapter 9): so both optimistic people and pessimistic people have to trust in some way and under subjective limits and decisions. It is true: trust may be based on a very prudent and pessimistic attitude: 'I have really 'decided' – after a serious evaluation of the risks – on the basis of data and titles, to trust this medical doctor, but I continue to be anxious and to focus on risks and worries, and to be very pessimistic about the results'.

However, *optimists are obviously more trustful and more open to trust*. This is well predicted and explained in our model. They are all for the 'non-impossible' (plausible) eventualities; so they have better expectations; they perceive less risk or do not focus on risks and thus have a lower acceptance/trust threshold; they ascribe to people a pro-social attitude, common values, non-hostility; they appreciate the gains, what has been obtained, and do not focus on what they 'didn't achieve', so they are positively confirmed in their optimism and trust-decision.

When Trust Is Irrational

As we have seen in Section 3.7 it is possible to distinguish between rational and irrational trust. One particularly interesting case, also linked with the optimism concept, is the following one, where we really have an irrational form of trust; even 'subjectively' irrational. Let's call 'anti-rational' trust a decision to trust while subjectively going against our evidences. This is not a 'spontaneous' outcome of a trust evaluation and choice; this requires an act of will; it is a decision against our evaluation. This is when we say: *'I have decided to trust him anyway, although I know that . . ., although I'm sure he will betray me'*.

This decision and attitude is quite remarkable because it violates the usual relation between trust as attitude-evaluation (*e-T*) and trust as decision (*d-T*). The usual (and rational) relation is that: since I *e-trust Y* enough, I (decide to) *d-trust Y*. And if I *d-T Y* this means that I *e-T Y*. On the contrary, when I take that attitude and force myself to trust *Y*, I do not really trust *Y*. That is, I *d-trust* him, without really *e-trusting* him. These 'subjectively' not so rational trust attitudes are not necessarily 'objectively' irrational and dysfunctional. As we said, optimism can in fact be a self-fulfilling prophecy; it can influence the chances of success.

For example, as we said, even a decision to trust against evidence and beyond the current level of evaluation and expectation, can be objectively rational just because *X*'s act of trusting *Y* can influence *Y*'s trustworthiness, can increase the chances of success: Trust creates trust (see Chapter 6).

If *X* takes into account such an influence of her attitude and decision on *Y*, and predicts and calculates this effect (see Section 6.3.3. on the dynamic of trust) her decision will also become 'subjectively' rational, since her degree of e-trust and her certainty in expectation has been increased. Moreover, the 'decision' to trust can 'locally' be irrational, but it could be part of a more general or overall decision. In saying this, *X* should have some additional reason (goal) beyond the delegated one. For example, *X* wants to publicly demonstrate courage, or persuade someone to risk and invest (be a model), etc., and this higher goal also increases the need to cover and compensate the (probable) loss on τ. So at the global level the risky trust act is part of a rational decision.

Trust Versus Faith

As we have repeatedly explained, we accept the idea that, sometimes, trust is subjectively irrational, not based on justified evidences, not grounded (or going beyond evidences), or that it is just feeling-based and intuitive (but, perhaps, grounded on analogy and experience). However, we deny that trust is necessarily and by definition only this. There is also a trust based on evidence (observations, reports, reasoning, or learning). Trust is not contradictory to reasoning, argumentation, proof, demonstrations. In some sense some scholars mix up trust with faith.

Faith in the strict sense is not simply to believe something or *in* something without sufficient evidences. It is to believe *not* 'on the base of', 'in force of' evidence (even if evidence was there). Thus, in a sense it is renouncing evidence, refusing it; believing on a non-rational (reasons-based) ground. Frequently this is a meta-attitude; an aim. Faith rejects the need for evidence because the fact of desiring or searching for evidence or the attempt to ground our attitude on evidence is per se a 'sign' of doubt, a signal that we are doubting or might doubt. But real faith does not admit any doubt: either to doubt is prohibited ('dogma'); or we want to avoid or reject any doubt (which – notice – is some sort of meta-cognitive perception and activity).

The doubt invalidates the faith; it proves that the faith is no (longer) there. Trust on the contrary is evidence-based; not only is it not 'incompatible' with evidence (like faith), but it is inspired by evidence, signs, or experience. Not in the sense that it has always good and sufficient evidence, or that that evidence is always real 'reasons'. Actually trust can be based on different kinds of 'evidence', including feelings and emotions, intuition, practice, or mere plausibility ('not impossible'). But, in the sense that it is not *aimed at* being indifferent to evidence; evidence is very 'relevant' for trust, but not for real faith. Given this 'irrelevance' of evidence, not only is faith 'optimistic' and would also consider the 'plausible', but, it will even go against counter-evidence. It is indifferent to proof. Even if there was proof against what I believe, it is irrelevant to me, not taken into consideration. Faith aims at being non-rational (Occam).

We talk about 'faith' in a weak/broad sense, or more often 'faithful' trust, trust not justified or supported by the subjective evidences; blind; or trust not searching for evidence on the basis of some sort of meta-trust or default attitude. However, this is not the real, deep meaning of 'faith', and it is not the authentic, typical form of trust (as some authors claim).

8.2.2 Risk Perception

When we say that trust always implies some risk; that there is 'trust' and it is needed precisely because one has to assume some risk, we do not mean that this risk is necessarily explicit or focused in the mind of the trusting agent. Notice that this is not an *ad hoc* solution for an old controversial issue, it is just a general aspect of the theory of beliefs (Section 8.2.1) that we take for granted the grounding of trust attitudes (and thus decisions) on beliefs. Not only can beliefs be out of the focus of attention, or unconscious or even 'removed' in a psychoanalytic sense, but they can simply be 'implicit'. As we just said, one fundamental way in which beliefs are implicit is that they are just 'potential'; they are implied by the explicit data that we believe, but that has not yet been derived, not explicitly formulated or 'written' in some file or memory.

Subjectively speaking, for example, an agent can be fully trustful, not worrying at all, just because subjectively they don't perceive any risk and don't calculate the very small eventuality of a failure or harm. Subjectively their curve of probability is tending to a limit, is flat: 90, 95, 98% is equal to 100%, although this is actually impossible and realistically irrational (no prediction can be 100% certain about the future).[9]

---

[9] Even dead – contrary to moralistic *'memento mori'* – subjectively speaking is not sure: 'Who knows? Perhaps they will invent some miraculous drugs and interventions'; 'Who knows? Perhaps the water of immortality really exists; or perhaps there is resurrection and eternal life'.

So risk perception varies very much from one subject and context to another; and the risk can remain completely implicit in our mind.

Even trust beliefs can be implicit, not necessarily explicit and active. Not only in by default or affect-based forms of trust (see Chapters 4 and 5), but, for example, in a routine trust attitude, when I am used to relying on Y, and by experience I 'know' that he is reliable, and there is nothing to worry about.

8.3   Is Trust Just the Subjective Probability of the Favorable Event?

Our main disagreement with economists is about the following issue (as we have already seen in Section 8.1): Is trust simply reducible to *subjective probability* ?

This is in fact a dominant tradition in economics, game theory, part of sociology ((Gambetta, 1988); (Coleman, 1994)), and now in artificial intelligence and electronic commerce (Brainov and Sandholm, 1999). We argue in favor of *a cognitive view of trust* as a complex structure of beliefs and goals (in particular causal attributions, evaluations and expectations), even implying that the trustor must have a 'theory of the mind' of the trustee (see Chapter 2) ((Castelfranchi and Falcone, 1998), (Falcone and Castelfranchi, 2001)).

Such a structure of beliefs determines a 'degree of trust' and an estimation of risk, and then a decision whether to rely on the other, which is also based on a personal threshold of risk acceptance/avoidance (see Chapter 3).

In this chapter we use our cognitive model of trust to argue against probability reduction and the consequent *eliminative* behavior. We agree with Williamson (Williamson, 1985) that one can/should eliminate the redundant, vague, and humanistic notion of 'trust', *if* it simply covers the use of subjective probability in decisions. But we strongly argue against both this reduction and the consequent elimination. Trust cannot be reduced to a simple and opaque index of probability because agents' decisions and behaviors depend on the specific, *qualitative* evaluations and mental components. For example, internal or external attributions of risk/success, or a differential evaluation of trustee's competence vs. willingness, make very different predictions both about trustor's decisions and possible interventions and cautions. Let us extensively discuss some arguments against the reduction of trust to perceived probability, and eliminative behavior.

8.3.1   Is Trust Only about Predictability? A Very Bad Service but a Sure One

Very frequently an economic approach – in order to reduce trust to a well known notion and metrics (probability) – just eliminates an entire side of trust: a very typical one in common sense, in practice, and even in economic exchanges and in labor relationships: *trust as an expectation about the quality of the good or service; trust as belief about the competence, experience, skills of the trustee!*

It seems that the only concern of trust is money, and whether to be sure or not of receiving/cumulating it. But actually even money has a *quality*; not only can dollars be more reliable than euros (or vice versa), but money can be broken, forged, out of circulation, or just simulated!

Consider, for example, the already (see Chapter 1) cited definition by Gambetta: *Trust is the subjective probability by which an individual, X, expects that another individual, Y, performs a given action on which its welfare depends.*

As declared, we think that this definition stresses that trust is basically an estimation, an opinion, an expectation: a belief. It is also quite remarkable that there is no reference to exchange, cooperation, mutuality, Y's awareness.

However, it is also too restricted, since it just refers to one dimension of trust (predictability), while ignoring the 'competence/quality' dimension.[10]

Moreover, to express the idea of an uncertain prediction it uses the notion of 'subjective probability' and collapses trust in this notion and measure. This is quite risky since it might make the very notion of 'trust' superfluous (see below). Clearly enough, the reliability, the probability of the desired event, has nothing to do with its degree of quality, and we cannot renounce this second dimension of trust. When we trust a medical doctor we trust both his expertise, competence, skills and his taking care of us, his being reliable, trustworthy. Trust is an, at least, bi-dimensional notion (actually – as we have shown – is a multi-dimensional construct); *Agent1: 'Why don't you trust him? He is very reliable and well disposed'; Agent2: 'That's true; he is very willing, but is not expert in this domain, is not well prepared'.*

8.3.2 Probability Collapses Trust 'that' and 'in'

Trust is not simply the subjective probability of a favorable event: that is, trust 'that' the desired event and outcome will be realized. If trust is *'the subjective probability by which an individual, X, expects that another individual, Y, performs a given action on which its welfare depends'*, this does not only mean that the expected/desired event is an action of a given individual *Y*. What it does mean is that 'we trust *Y*', 'we trust *in Y*'. There is much more than a prediction of *Y*'s action (and the desire of that action or of its result). There is something 'about *Y*'; something we think of *Y*, or we feel towards *Y*. Is this just the estimated probability of his act or of the outcome? This definition does not capture some of the crucial kernel components of the very notion of trust: why we do not just trust 'that' *Y* will do a given favorable action, but we trust 'in' *Y*, and we see *Y* as endowed with some sort of qualities or virtues: trustworthiness, competence, reliability.

8.3.3 Probability Collapses Internal and External (Attributions of) Trust

Trust cannot be reduced to a simple and opaque index of probability because internal or external attribution of risk/success makes very different predictions about both the trustor's decisions and possible interventions and cautions.

As we saw in Section 2.7.2 one should distinguish between trust 'in' someone or something that has to act and produce a given performance thanks to its *internal characteristics*, and the

---

[10] In Chapter 1 we have added the following criticisms to this definition: it does not account for the meaning of 'I trust Y' where there is also the *decision* to rely on Y; and it doesn't explain what such an evaluation is made of and based on: the *subjective probability* includes too many important parameters and beliefs, which are very relevant in social reasoning. It also does not make explicit the 'evaluative' character of trust.

global trust in the global event or process and its result, which is also affected by external factors like *opportunities* and *interferences*.

Trust may be said to consist of or to (either implicitly or explicitly) imply the *subjective probability* of the successful performance of a given behavior α, and it is on the basis of this subjective perception/evaluation of risk and opportunity that the agent decides to rely or not, to bet or not on Y. However, the probability index is based on, derives from those beliefs and evaluations. In other words the global, final probability of the realization of the goal g, i.e. of the successful performance of α, should be decomposed into the probability of Y performing the action well (that derives from the probability of willingness, persistence, engagement, competence: *internal attribution*) and the probability of having the appropriate conditions (opportunities and resources *external attribution*) for the performance and for its success, and of not having interferences and adversities (*external attribution*).

Why is this decomposition important? Not only to cognitively ground such a probability (which after all is 'subjective' i.e. mentally elaborated) – and this cognitive embedding is fundamental for relying, influencing, persuading, etc., but because:

a) the agent trusting/delegating decision might be different with the *same global probability or risk*, depending on its composition;
b) trust composition (internal vs external) produces completely *different intervention strategies*: to manipulate the external variables (circumstances, infrastructures) is completely different from manipulating internal parameters.

Let's consider the first point (a). There might be different heuristics or different personalities with a different propensity to delegate or not in the case of a weak internal trust (subjective *trustworthiness*) even with the same global risk. For example, 'I completely trust him but he cannot succeed, it is too hard a task!', or 'the mission/task is not difficult, but I do not have enough trust in him'). The problem is that – given the same global expectation – one agent might decide to trust/rely in one case but not in the other, or vice versa! In fact, on those terms it is an irrational and psychological bias. But this bias might be adaptive, for example, perhaps useful for artificial agents. There could be logical and rational meta-considerations about a decision even in these apparently indistinguishable situations. Two possible examples of these meta-considerations are:

- to give trust (and then delegation) increases the experience of an agent (therefore comparing two different situations – one in which we attribute low trustworthiness to the agent and the other in which we attribute high trustworthiness to him; obviously, the same resulting probability – we have a criteria for deciding);
- the trustor can learn different things from the two possible situations; for example, with respect to the agents; or with respect to the environments.

As for point (b), the strategies to establish or increment trust are very different depending on the external or internal attribution of your diagnosis of lack of trust. If there are adverse environmental or situational conditions your intervention will be in establishing protection conditions and guarantees, in preventing interferences and obstacles, in establishing rules and infrastructures; while if you want to increase your *trust in* your trustee you should work on his motivation, beliefs and disposition towards you, or on his competence, self-confidence, etc.

We should also consider the *reciprocal influence between external and internal factors*. When *X* trusts the internal powers of *Y*, she also trusts his abilities to create positive opportunities for success, to perceive and react to the external problems. Vice versa, when *X* trusts the environmental opportunities, this evaluation could change the trust she has for *Y* (*X* could think that *Y* is not able to react to specific external problems).

Environmental, situational, and infrastructural trust,[11] are aspects of external trust. It is important to stress that *when the environment and the specific circumstances are safe and reliable, less trust in Y is necessary for delegation (for example for transactions).* Vice versa, when *X* strongly trusts *Y*, his capacities, willingness and faithfulness, *X* can accept a less safe and reliable environment (with less external monitoring and authority). We account for this 'complementarity' between the internal and the external components of trust in *Y* for *g* in given circumstances and a given environment.

8.3.4 Probability Misses the Active View of Trust

Reducing trust to subjective probability means also reducing trust to its 'dispositional' nature, just to a (partial) evaluation, to a belief, a forecast or an expectation (including motivational aspects). We will miss the other fundamental notions of trust as decision and act: trust as betting and taking a risk on somebody, as 'relying' and 'counting on' them. Probability has nothing to do with this; it is only one possible basis (factor) for such a decision (and the perceived risk); but is not 'trusting somebody for ...'. Moreover, the subjective probability says nothing about trusting or not *Y*; it depends on the threshold of the acceptable risk for *X*, and on the value of the foreseen gains and harms. Also after that decision and bet there is much more than a subjective probability: there is a 'positive expectation' and some worry.

8.3.5 Probability or Plausibility?

Moreover, to reduce trust to subjective probability to be taken into account in the 'subjective expected utility' guiding the decision to act or to 'delegate', is also reductive because we need a more sophisticated model of subjective evaluation of chances, including uncertainty and doubts, including 'plausibility' and gap of ignorance. We have just shown how important this is for the theory of trust, and for example the role of the 'plausibility' for characterizing optimism, or trust by default, or the real meaning of 'give credit' (see Chapter 4). Not only is a traditional probabilistic approach to trust reductive but it is also too elementary and obsolete.

8.3.6 Probability Reduction Exposes to Eliminative Behavior: Against Williamson

The traditional arrogance of economics and its attempt to colonize with its robust apparatus the social theory (political theory, theory of law, theory of organizations, theory of family, etc.[12]) coherently arrives – in the field of trust – to a 'collision' (Williamson, 1985) with the sociological view. The claim is that the notion of *trust* when applied in the economic

---

[11] They are claimed to be really crucial in *electronic commerce* and *computer mediated interaction*.

[12] In his section on 'Economics and the Contiguous Disciplines' ((Williamson, 1985) p. 251) Williamson himself gives example of this in law, political science, in sociology.

and organizational domain or, in general, in strategic interactions, is just a common sense, *empty term without any scientific added value*;[13] and that the traditional notions provided by transaction cost economics are more 'parsimonious' and completely sufficient to account for and explain all those situations where lay people (and sociologists) use the term 'trust' (except for very special and few personal and affective relationships[14]). The term trust is just for suggestion, for making the theory more 'user-friendly' and less cynical. It is just 'rhetoric' when applied to commerce[15] but does not explain anything about its nature which is and must be merely 'calculative' and 'cynical'.[16]

On the one hand, we should say that Williamson is pretty right: *if* trust is simply subjective probability, or if what is useful and interesting in trust is simply the (implicit) subjective probability (like in Gambetta's definition (Gambetta, 1988) and in the game-theoretic and rational decision use of trust), then the notion of trust is redundant, useless and even misleading. On the other hand, the fact is that trust is not simply this, and – more important – what (in the notion of trust) is useful in the theory of social interactions is not only subjective probability.

Not only is Williamson assuming a more prescriptive than scientific descriptive or explanatory attitude, but he is simply wrong in his elimination claims. And he is wrong even about the economic domain, which in fact is and must obviously be socially embedded. Socially embedded does not mean only – as Williamson claims – institutions, norms, culture, etc.; but also means that *the economic actors are fully social actors* and that they act in such a way also in economic transactions, i.e. with all their motives, ideas, relationships, etc. including the *trust* they have or not in their partners and in the institutions. The fact that they are unable to see what 'trust' adds to the economic analysis of risk[17] and that they consider those terms

---

[13] 'There is no obvious value added by describing a decision to accept a risk (. . .) as one of trust' ((Williamson, 1985), p. 265). 'Reference to trust adds nothing' ((Williamson, 1985) p. 265). 'I argue that it is *redundant* at best and can be *misleading* to use the term 'trust' to describe commercial exchange (. . .) Calculative trust is a contradiction in terms' ((Williamson, 1985) p. 256). Notice that he 'prescribes' the meaning of the word trust, and of its use in human sciences. Trust is only affective or moral; it cannot be based on evaluations and good evidence.

[14] '(. . .) trust, if obtained at all, is reserved for very special relations between family, friends, and lovers' ((Williamson, 1985), p. 273).

[15] 'I argue that it is *redundant* at best and can be *misleading* to use the term 'trust' to describe commercial exchange (. . .) Calculative trust is a contradiction in terms' ((Williamson, 1985) p. 256). '(. . .) the *rhetoric* of exchange often employs the language of promises, trust, favors, and cooperativeness. That is understandable, in that the *artful* use of language can produce deals that would be scuttled by abrasive calculation. If, however, the basic deal is shaped by objective factors, then calculation (credibility, hazard, safeguards, net benefits) is where the crucial action resides.' ((Williamson, 1985) p. 260). 'If calculative relations are best described in calculative terms, then the diffuse terms, of which trust in one, that have mixed meanings should be avoided when possible.' ((Williamson, 1985) p. 261). And this does not apply only to the economic examples but also to the apparent exception of 'the assault girl (. . .) I contend is not properly described as a condition of trust either' ((Williamson, 1985) p. 261). This example that is 'mainly explained by bounded rationality - the risk was taken because the girl did not get the calculus right or because she was not cleaver enough to devise a contrived but polite refusal on the spot - is not illuminated by appealing to trust'. ((Williamson, 1985) p. 267).

[16] 'Not only is 'calculated trust' a contradiction in term, but *user friendly terms,* of which 'trust' is one, have an additional cost. The world of commerce is reorganized in favor of the cynics, as against the innocents, when social scientists employ user-friendly language that is not descriptively accurate - since only the innocents are taken in' ((Williamson, 1985) p. 274). In other words, 'trust' terminology decorates and masks the *cynic reality* of commerce. Notice how Williamson is here quite prescriptive and neither normative nor descriptive even about the real nature of commerce and of the mental attitudes of real actors in it.

[17] Section 2 starts with 'My purpose in this and the next sections is to examine the (. . .) 'elusive notion of trust'. That will be facilitated by examining a series of examples in which *the terms trust and risk are used interchangeably*

as equivalent, simply shows how they are unable to take into account the interest and the contribution of cognitive theory.

Risk is just about the possible outcome of a choice, about an event and a result; trust is about somebody: it mainly consists of beliefs, evaluations, and expectations about the other actor, their capabilities, self-confidence, willingness, persistence, morality (and in general motivations), goals and beliefs, etc. Trust *in* somebody basically is (or better at least include and is based on) a rich and complex theory of them and of their mind. Conversely distrust or mistrust is not simply a pessimistic esteem of probability: it is diffidence, suspicion, negative evaluation *relative to* somebody.

From the traditional economic perspective all this is both superfluous and naive (non-scientific, rhetoric): common-sense notions. The economists do not want to admit the insufficiency of the economic theoretical apparatus and the opportunity of its cognitive completion. But they are wrong – even within the economic domain – not only because of the growing interest in economics towards a more realistic and psychologically-based model of the economic actor, but because mental representations of the economic agents and their images are, for example, precisely the topic of marketing and advertising (that we might well suppose has something to do with commerce).

8.3.7   Probability Mixes up Various Kinds of Beliefs, Evaluations, Expectations about the Trustee and Their Mind

We claim that the richness of the mental ingredients of trust cannot and should not be compressed simply in the subjective probability estimated by the actor for their decision. But why do we need an explicit account of the mental ingredients of trust (beliefs, evaluations, expectations, goals, motivations, model of the other), i.e. of the *mental background* of reliance and 'probability' and 'risk' components?

- First, *because otherwise we will neither be able to explain or to predict the agent's risk perception and decision.* Subjective probability is not a magic and arbitrary number; it is the consequence of the actor beliefs and theories about the world and the other agents. We do not arrive at a given expectation only on the basis of previous experiences or on the frequency of a series of events. We are able to make predictions based on other factors, like: analogical reasoning (based on a few examples, not on 'statistics'); other forms of reasoning like 'class-individual-class' (for example I can trust Y because he is a doctor and I trust

---

- which has come to be standard practice in the social science literature - (. . .)'. The title of section 2.1 is in fact 'Trust as Risk'. Williamson is right in the last claim. This emptying of the notion of trust is not only his own aim, it is quite traditional in sociological and game-theoretic approaches. For example in the conclusions of his famous book Gambetta says: *'When we say we trust someone or that someone is trustworthy, we implicitly mean that the probability that he will perform an action that is beneficial or at least not detrimental to us is high enough for us to consider engaging in some form of cooperation with him'* ((Gambetta, 1988) p. 217). What is dramatically not clear in this view is what 'trust' does *explicitly* mean! In fact the expression cited by Williamson (the 'elusive notion of trust') is from Gambetta. His objective is the elimination of the notion of trust from economic and social theory (it can perhaps survive in the social psychology of interpersonal relationships). 'The recent tendency for sociologists /the attack is mainly to Coleman and to Gambetta/ and economists alike to use the term 'trust' and 'risk' interchangeably is, on the arguments advanced here, ill-advised' ((Gambetta, 1988) p. 274).

doctors, even with no experience of them); a by default rule; an emotional activation; the existence of a norm and of an authority; etc.

* Second, because *without an explicit theory of the cognitive basis of trust any theory of persuasion/dissuasion, influence, signs and images for trust, deception, reputation, etc. is not 'parsimonious' but is simply empty.*

Let's suppose (referring to Williamson's example) that there is a girl walking in the park in the night with a guy; she is perceived by her father as under risk of assault. The father of that girl (Mr. Brown) is an anxious father; he also has a son from the same school as the guy *G* accompanying the girl. Will he ask his son *What is the probability that G will assault your sister?'* or *'How many times has G assaulted a girl?"* or *". . . Has some student of your college assauledt a girl?'*.

We do not think so. He will ask his son what he knows *about G*, if he has an evaluation/information about *G*'s education, his character, his morality, his family, etc. He's not asking rhetorical questions or simply being polite. He is searching for some specific and factual information upon which to found his prediction/expectation about risk. Coleman (Coleman, 1994) too stresses the importance of information, but he is not able to derive from this the right theoretical consequences: a view of trust also in terms of justified cognitive evaluations and expectations. In his theory one cannot explain or predict which information is pertinent and why. For example, why is the artistic talent of *G* or the color of his car irrelevant?

Now, why those questions? Which is the *relevance* of those data/beliefs *about Y* for the prediction about a possible violence? Which is the *relationship* between *Y*'s 'virtues', 'qualities' (*trustworthiness*) and the prediction or better positive *'expectation'* about *Y*'s behavior? 'Trust' is precisely this *relationship*. Trust is not just reducible to the strength of a prediction: to the 'subjective probability' of a favorable event. Trust is not just a belief, or worst the degree/strength of any belief. It is a grounded belief strength (either rationally justified or affect-based), and not of any belief, but of a belief about the action of an(other) agent, and a component of a 'positive expectation'. Is Williamson's theory able to explain and predict this relation? In his framework subjective probability and risk are *unprincipled and ungrounded notions*. What the notion of trust (its cognitive analysis) adds to this framework is precisely the explicit theory of the ground and (more or less rational) support of the actor's expectation, i.e. the theory of a specific set of beliefs and evaluations *about G* (the trustee) and about the environmental circumstances, and possibly even of the emotional appraisal of both, such that an actor makes a given estimation of probability of success or failure, and decides whether to rely and depend on *G* or not.

Analogously, what can one do within Williamson's framework to act upon the probability (either objective or subjective)? Is there any rational and principled way? He can just touch wood or use exorcism or self-suggestion to try to modify this magic number of the predicted probability. *Why and how* should, for example, information about 'honesty' change my perceived risk and my expected probability of an action of *G*? Why and how should, for example, training, friendship, promises, a contract, norms,[18] or control, and so on, affect (increase) the probability of a given successful action and my estimation of it? It remains unexplained.

---

[18] How and why 'regulation can serve to *infuse* trading *confidence* (i.e. trust) into otherwise problematic trading relations' as Williamson reminds us by citing Goldberg and Zucker ((Williamson, 1985) p. 268).

In the economic framework, first we can only account for a part of these factors; second this account is quite incomplete and unsatisfactory. We can account only for those factors that affect the rewards of the actor and then the probability that he will prefer one action to another. Honor, norms, friendship, promises, etc. *must be translated into positive or negative 'incentives' on choice* (for example to cooperate versus to defeat). This account is very reductive. In fact, we do not understand in the theory how and why a belief (information) about the existence of a given norm or control, or of a given threat, can generate a goal of G and eventually change his preferences. Notice, on the contrary, that our predictions and our actions of influencing are precisely based on a 'theory' like this, on a 'theory' of G's mind and mental processes beyond and underlying 'calculation'. Calculation is not only institutionally but also *cognitively embedded* and justified!

Other important aspects seem completely left out of the theory. For example, the ability and self-confidence of G, and the actions for improving them (for example training) and for modifying the probability of success, or the action for acquiring information about this and increasing the subjective estimated probability.

Trust is also about this: beliefs about G's competence and level of ability, and his self-confidence. And this is a very important basis for the prediction and esteem of the probability of success or the risk of failure.

Williamson is right and wrong. As we said, actually, we would agree with him (about the fact that one can/should eliminate the redundant, vague, and humanistic notion of 'trust'), but *if* and *only if* it simply covers the use of subjective probability in decisions. We have strongly argued against both this reduction and the consequent elimination. Since trust cannot be reduced to subjective probability, and needs a much richer and more complex model (and measure), it cannot be eliminated from economic decisions and models. Economics without an explicit notion and theory of trust cannot understand a lot of phenomena. For example: the real nature of the 'relational capital' (Chapter 10) and the importance of 'reputation'; that is, the role of the specific evaluations people have about me and why I invest in/for this, and the specific 'signals' my behavior sends out. The crucial role of trust (as positive attitude, as evaluation, as counting on you, as taking a risk, and so on) for eliciting a reciprocation attitude and behavior, for spreading trust, etc. (see later). The importance of trust, not only as subjective estimation/forecast, but as act and as social relation and link. And so on. In sum, reducing trust to subjective probability is a disservice to trust, and to economics.

8.4   Trust in Game Theory: From Opportunism to Reciprocity

Doubtless the most important tradition of studies on trust is the 'strategic' tradition, which builds upon the rational decision and Game Theories ((Luce and Raiffa, 1957) (Axelrod and Hamilton, 1981), (Shoham and Leyton-Brown, 2009)) to provide us a theory of trust in conflict resolution, diplomacy, etc. and also in commerce, agency, and in general in economy.

Let us start with our criticism of trust defined in terms of risk due to Y's temptation and opportunism; and as an irrational move in a strategic game. Then we will consider why (the act of) trust cannot be mixed up and identified with the act of 'cooperating' (in strategic terms). Finally we will discuss *Trust game* as a wrong model for trust; and why trust is not only and not necessarily related to 'reciprocity'.

We will discuss two positions, one strongly relating trust and cooperation in *Prisoner's or Social Dilemma* situations, and later in the so called *Trust game*.

8.4.1   Limiting Trust to the Danger of Opportunistic Behavior

Consider for example the view of trust that one can find in the conclusion of Gambetta's book [Gam-90] and in [Bac-99]: '*In general, we say that a person trusts someone to do A if she acts on the expectation that he will do A when two conditions obtain: both know that if he fails to do A she would have done better to act otherwise, and her acting in the way she does gives him a selfish reason not to do A.*'

In this definition we recognize the *Prisoner's Dilemma syndrome* that gives an artificially limited and quite pessimistic view of social interaction. In fact, by trusting the other (in term of decision not simply of evaluation!), X makes herself 'vulnerable', she gives to the other the *possibility* of damaging her, but: does she give to Y even the *temptation*, the *convenience* to do so? It is true that the act of trust exposes X by giving Y an opportunity, but it is not necessarily X or the game structure that gives him a *motive*, a reason for damaging her[19] (on the contrary, in some cases to trust someone represents an opportunity for the trustee to show his competencies, abilities, willingness, etc.).

It is not necessary for trust that trusting Y makes it convenient for him to disappoint the trustor's expectation. Perhaps the trustor's trusting in Y gives him (the trustee) a reason and a motive for *not* disappointing the trustor's expectation; perhaps the trustor's delegation makes the expected behavior of the trustee convenient for the trustee himself; it could create an opportunity for strict cooperation over a common goal.

Trust continues to be trust independently of making it convenient or not for the trustee to disappoint the trustor.

Of course, there could be always risks and uncertainty, but not necessarily *conflict* in the trustee's relationship between selfish interest and broader or collective interests. If it was true that there was no trust in strict cooperation based on a common goal, mutual dependence, a common interest to cooperate, and a joint plan to achieve the common goal (Conte and Castelfranchi, 1995). While on the contrary there is trust in any joint plan, since the success of the trustor depends on the action of the trustee, and vice versa, and the agents are relying on each other.

This strategic view of trust is not general; it is an arbitrary and unproductive restriction. It is a reasonable objective to study trust in those specific and advanced conditions; what is unacceptable is the pretense of defining not a sub-kind of trust, but trust *in se*, the general notion of trust in such a way, without caring at all about the common use and common sense, about other psychological, sociological, philosophical studies and notions. Let us analyze this problem in depth, taking into account more recent and important positions which mix up trust with cooperative attitudes or actions, or based on the *Trust game* and strictly relating trust with *reciprocity*.

8.4.2   'To Trust' Is not 'to Cooperate'

As we saw, *Trust* is (also) a decision and an intention, but this decision/intention is not about 'doing something for the other', to helping or to 'cooperating' with him (in our terminology 'adopting the other's goal'); the act of trusting is not a cooperative act *per se*. On the contrary, in a certain sense, the trustor (X) is expecting *from* the other some sort of 'help' (intentional or non-intentional): an action useful for the trustor.

---

[19] The 'reciprocity' view (see below) actually seems to reverse this claim.

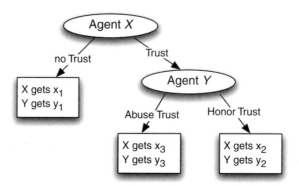

Figure 8.4   Example of Classical Trust Game

Of course, in specific cases, the decision to do something for the other (which is <u>not</u> a decision to trust him) can be joined with and even based on a decision to trust the other, when X is counting on an action of Y useful for herself as a consequence of her own action in favor of Y. An example is in fact when X does something for Y or favors Y while expecting some reciprocation from Y or for eliciting it.

This is not the only case: X might try to encourage an action in Y which would be of use to him (an action on which she decides to count and bet) not as a 'reciprocation' to 'helping' her, but simply as a behavioral consequence due to Y's independent aims and plans. For example, X might give Y a gun as a gift, because she knows that he hates Z and she wishes that Y would kill Z (not for X but for his own reasons).

Analogously, it is not the case that X always expects an adoptive act *from* Y and trusts him for this (decides to depend on him for achieving her goal), as 'reciprocation' of her own 'adoption'. However, this is certainly an important group of situations, with various sub-cases which are quite different from each other from the cognitive point of view.

In some cases, X counts on Y's feeling of gratitude, on a reciprocate motive of the affective kind. In other cases on the contrary she trusts Y's interest in future exchanges with her. In others, X relies just on Y's sense of honor and on his sensibility to promises and commitments. In yet other cases she knows that Y knows the law and worries about the authority and its sanctions.[20] In these cases the act of 'cooperating' (favoring the other and risking on it) is conceived as a (partial) means for obtaining Y's adoption and/or behavior. Either X wants to provide Y with *conditions* and instruments for his autonomous action based on independent motives, or she wants to provide Y *with motives* for doing the desired action.

For a detailed analysis of the Yamagishi approach to trust (Yamagishi & Yamagishi, 1994) (Yamagishi, 2003), and of our doubts about his mixing the two concepts of trust and cooperation, see Section 1.5.5.

8.5   Trust Game: A Procuste's Bed for Trust Theory

Figure 8.4 show the classical schema of a trust game.

---

[20] Notice that X might also adopt Y's goals, while expecting Y's 'cooperation', but not as a *means* for this. X might for example be an anticipatory reciprocator; since she knows that Y is doing an act in her favor, she wants to reciprocate and – in advance – does something for Y.

A trust game supposes this hypothesis: If $x_2>x_1>x_3$ agent X does not trust Y, make a decision with a guaranteed outcome x_1, or, trust Y and let him decide on action; If Y honors the trust, he will decide to benefit both, but if $y_3>y_2>y_1$ Y may abuse trust and maximize his own benefit.

Several authors use the trust game for characterizing trust. *'To isolate the basic elements involved in a trusting interaction we may use the Trust Game'* (Pelligra, 2005). On the contrary we argue that the trust game (as the great majority of game theoretic approaches and considerations about trust) gives us a biased and limited view of trust. It represents a Procuste's bed for the theory of trust.

The first two conditions for characterizing trust, as identified by Pelligra, are rather good:

i) *'potential positive consequences for the trustor'* from the trustee's behavior;
ii) *'potential negative consequences for the trustor'* from the trustee's behavior.

This means that X – as for her 'welfare', rewards, goal-achievement, depends on Y; she makes herself 'vulnerable' to Y, and Y gets some power over X. However, one should make it clear – as for condition (i) – the fact that X expects (knows and wishes) such consequences; and has decided to count on Y to realize them.

Condition (i) is only vague, insufficiently characterized (X might completely ignore that Y's behavior can produce good outcomes for him), but condition (iii) is definitely too restrictive for a general definition of the 'basic elements involved in a trusting interaction':

(iii) *'temptation for the trustee or risk of opportunism'.*

This is a wrong prototype for trusting interaction; too restrictive.

By relying and counting on Y (trusting him), X is exposing herself to risks: risks of failure, of non-realization of the goal for which she is counting on Y; and also risks of harm and damages due to her non-diffidence and vigilance towards Y. This is a well recognized aspect of trust (see Chapter 1). However, these risks (let us focus *in primis* on the failure; the non realization of the expected and 'delegated' action) are not necessarily due to Y's temptation (see also Section 2.8).

On the one hand, as we said, trust is also a belief (and a bet) on Y's competence, ability, intelligence, etc. X might be wrong about this, and can be disappointed because of this. Y might be unable or incompetent, and provide a very bad performance (service or product); Y can misunderstand X's request, expectation, or goals, and thus do something wrong or bad; Y can be absent minded or forgetful, and just disappoint and damage X for this.

On the other hand, when X trusts Y to carry out a given action which she is waiting for and counting on, Y is not necessarily aware of this. There are acts of trust not based on Y's agreement or even awareness. In these cases, Y can change his mind without any opportunism towards X; it is just X's reading of Y's mind and prediction that is wrong. Even if Y knows that X is relying on his behavior, he has no commitment at all towards X (especially if this is not common knowledge); he can change his mind as he likes. Even when there is a commitment and an explicit reliance (like in an exchange), Y changes his mind (and behavior) – violating X's expectations – just for selfish opportunism. He can change his mind, even for altruistic reasons, revising his intentions in X's interest.

8.6   Does Trust Presuppose Reciprocity?

Following the model proposed in this book, it is possible to contradict a typical unprincipled and arbitrary restriction of the notion and of the theory of trust, present in some of the economic-like approaches. It is based on a restriction of trust only to exchange relations, in contexts implying *reciprocity*. It is, of course, perfectly legitimated and acceptable to be interested in a sub-domain of the broader domain of trust (say 'trust in exchange relations'), and to propose and use a (sub)notion of trust limited to those contexts and cases (possibly coherent or at least compatible with a more general notion of trust). What would be less acceptable is to propose a restricted notion of something – fitting within a particular frame and specific issues – as the only one that is valid.

Consider, by way of an example, one of those limited kinds of definition, clearly game theory inspired, and proposed by R. Kurzban ((Kurzban, 2001), (Kurzban, 2003)): trust is *'the willingness to enter exchanges in which one incurs a cost without the other already having done so'*. As we have seen the most important and basic constituents of the mental attitude underlying trust behavior are already present (and more clear) in *non-exchange* situations.

Y can do an action to help *X* with many motives not including reciprocation; analogously, *X* can rely on *Y*'s action to have a broad set of different motives ascribed to *Y* (for instance, friendship, honesty, generosity, search for admiration, etc.) and the reasons active in cooperation, exchange, reciprocation situations, are only a subset of them.

It is simply not true that we either feel trust or not, and we have to decide to trust or not, *only in contexts of reciprocation*, when we do something for the other or give something to the other and expect (wish) that the other would reciprocate by doing his share. This notion of trust is arbitrarily restricted and it cannot be useful to describe in detail the case where *Y* simply and unilaterally offers and promises to *X* that he will do a given action α for her, and *X* decides to count on *Y*, does not commit herself to personally perform α, and *trusts Y* for accomplishing the task. The very notion of trust must include cases like this that describe real life situations. Should we even search just for a 'behavioral' notion? *Doing nothing* and counting on others is in fact a behavior.

Even cases based on an explicit agreement do not necessarily require reciprocation. Consider a real life situation where *X* asks *Y* *'Could you please say this to the Director, when you see her; I have no time; I'm leaving now'*. She is in fact trusting *Y* to really do the required action. *Y* is expected to do this not out of reciprocation (but, say, for courtesy, friendship, pity, altruism, etc.).

One might claim that *X* has given something to *Y*: his gentle *'Please'*; and *Y* has to do the required action in order to reciprocate the *'Please'*. But this is frequently not true since this is usually not doing enough: it is not the reason *X* expects of *Y* (in fact *X* has to be grateful after the action and she is in debt); it is not what *Y* feels or the reason why he does the action; he feels that his cost greatly exceeds the received homage. Moreover, there might be other kinds of requests, based on authority, hierarchy, etc. when *X* doesn't give anything at all to *Y* 'in exchange' for the required action which is simply 'due'. But, in these cases *X* also considers *Y* to be trustworthy if she is relying on him. In sum: *trust is not an expectation of reciprocation; and doesn't apply only to reciprocation situations*. Related to this misunderstanding is the fact that 'being vulnerable' is often considered as strictly connected with 'anticipating costs'.

This diffused view is quite complicated: it mixes up a correct idea (the fact that trust – as decision and action – *implies a bet, taking some risk, being vulnerable*) with the reductive idea of *an anticipated cost, a unilateral contribution*. But in fact, to contribute, to 'pay' something in anticipation while betting on some 'reciprocation', is just one case of taking risks. The expected beneficial action ('on which our welfare depends') from the other is not necessary 'in exchange'.

The risk we are exposed to and we accept when we decide to trust somebody, to rely and depend on them, is not always the risk of wasting our invested resources, our 'anticipated costs'. *The main risk is the risk of not achieving our goal*, of being disappointed over the entrusted/delegated and needed action, although perhaps our costs are very limited (just a verbal request) or nothing (just exploiting an independent action and coordinating our own behavior). Sometimes, there is the risk of frustrating our goal forever since our choice of Y makes inaccessible other alternatives that were present at the moment of our decision. We also risk the possible frustration of other goals: for example, our self-esteem as a good and prudent evaluator; or our social image; or other personal goods that we didn't protect from Y's access. Thus, it is very reductive to identify the risks of trust with the lack of reciprocation and thus wast our investment; risk, which is neither sufficient nor necessary.

In fact, in another article, Pelligra recognizes and criticizes the fact that *'most studies* [in economics and game theory] *consider trust merely as an expectation of reciprocal behavior'* while this is *'a very specific definition of trust'* (Pelligra, 2006).

However, Pelligra – as we saw – in his turn proposes a very interesting but rather restricted definition, which fits trust game and the previous conditions (especially (iii)). He defines trust as characterized by the fact that X counts on Y's *responsiveness* to X's act of trusting *('The responsive **nature** of Trust'* is the title of the article). We believe this is too strong.

When X trusts Y – even when agreeing on something – she can rely on Y's behavior not because Y will respond to her act of trusting him, but for many other reasons. X can count on the fact that there are norms and authorities (and (moral) sanctions) prescribing that behavior, independently of X's trust, and X assumes that Y is a worrying or respectful person. There might be a previous norm (independent of the fact that X is trusting Y), and X forecasts Y's behavior and is sure that Y will do as expected, just because the norm exists and Y knows it (A. Jones, 2002). The fact that X is trusting Y is *not* (in X's expectation) the reason behind Y's correct behavior.

However, let us assume that one wants to put aside, from a 'true'/'strict' notion of trust, any kind of reason external to the interpersonal relationship (no norms, no third parties, no contracts, etc.). There is some sort of 'genuine' trust ((Hardin, 2002), (K. Jones, 1996), (K. Jones, 2001), (Baier, 1986)) which would be merely 'interpersonal' (see Chapter 2). From this perspective, one might perhaps claim that 'genuine' trust is precisely based on responsiveness. But this vision also looks too strong and narrow. It might be the case that Y behaves as expected not because X trusts him but because X is dependent on him, for example for pity and help. He would do the same even if X wouldn't ask or expect anything. For example, X to Y: *'Please, please! Don't tell John what you saw. It would be a tragedy for me'*; Y to X: *'OK, be quiet!'*; later, Z to X: *'How can you trust him!?'*; X: *'I trust him because he is a sensible person, he understood my situation, he was moved'*.

Furthermore, in general, X may count upon feelings or bonds of benevolence, friendship, love: he counts on those motives that make Y do what is expected; not on Y's responsiveness to X's trust in him; like in the 'genuine' trust of a child towards his father.

8.7   The Varieties of Trust Responsiveness

As for the interesting idea that we respond to a trusting act (for example, by increasing our benevolence, reliability, efficacy, etc.) we acknowledge that this is a very important claim (see also (Falcone and Castelfranchi, 2001)); but it deserves some development.

As we have shown, trust has different components and aspects, so our claim is that we respond to trust in various (even divergent) ways, *since we can respond to different components or faces of the trusting act*, which can elicit a variety of emotions or behaviors.

For example, one thing is to react to the appreciation, the positive evaluation implicit in a decision to trust and manifested by the act of trust; or to respond to the *kindness* of not being suspicious, diffident; or to the exhibition of respect and consideration. For example, I might not feel grateful but guilty; suffering from low self-esteem and feeling that X's evaluation is too generous and misleading and her expectation could be betrayed.

I could also respond to the fact that the trustor is taking a risk on me, is counting on me, exposing her vulnerabilities to me by feeling 'responsible'. The trustor's manifestation of being powerless, dependent on me, could elicit two opposite reactions. On the one hand, the perceived lack of power and the appeal to me is the basis of possible feelings of pity, and of a helpful, benevolent disposition. On the other hand, this can elicit a sense of exploitation, of profiting, which will elicit anger and refusal of help: *'Clear! She knows that eventually there will be this stupid guy (me!) taking care of that! She counts on this'*.

We do not have a complete and explanatory theory of all the possible reasons why trust elicits a behavior corresponding to the expectations.

8.8   Trusting as Signaling

It is clear that in those cases where the act or attitude of trust is supposed to elicit the desired behavior, it is important that Y has to know (or at least to believe) X's disposition. This applies in both cases: when X just trusts and expects; when X is cooperating (doing something for Y) because she trusts Y and expects a given behavior. Since X plans to elicit an adoptive behavior from Y as a specific response to her act, she must ascertain that Y realizes her act toward him and understands its intentional nature (and – in case of cooperation – the consequent creation of some sort of 'debt'). This means that X's behavior is – towards Y – a 'signal' meaning something to him; in other and better words, it is a form of *implicit 'communication'* since it is *aimed to* be a signal for Y and to mean all that ((Schelling, 1960), (Cramerer, 1988), (Castelfranchi, 2004)).

X's cooperation in view of some form of intentional reciprocation (of any kind) needs to be a *behavioral implicit communication act* because Y's understanding of the act is crucial for providing the right motive for reciprocating. The same is for X's reliance on Y aimed at inducing Y's adoption. This doesn't mean that X necessarily intends that Y understands that she intends to communicate (Gricean meta-message): this case is possible and usual, but not inevitable. Let us suppose, for example, that X desires some favor from Y and, in order to elicit a reciprocating attitude, does something to help Y (say, offers a gift). It is not necessary (and sometimes is even counterproductive) that Y realizes the selfish plan of X, and thus the fact that she wants him to realize that she is doing something 'for' him and *intends him to recognize this*. It is sufficient and necessary that Y realizes that X is intentionally doing something just for him, and X's act is certainly also aimed at such recognition by Y: X's intention to favor

Y must be recognized, but *X*'s intention that *Y* recognizes this doesn't need to be recognized (Castelfranchi, 2004).

As we have already highlighted (see Chapter 2) the act of trusting is an ambiguous 'signal', conveying various messages, and different possible meanings. And a cognitive agent – obviously – reacts to the *meaning* of the event, which depends on his active interpretation of it.

8.9   Concluding Remarks

We have argued against *the idea that trust has necessarily to do with contexts that require 'reciprocation'; or that trust is trust in the other's reciprocation.* We have also implicitly adopted a distinction between, the concept of reciprocation/reciprocity *as behavior and behavioral relation* and the concept of reciprocation/reciprocity *as motive and reason for doing something beneficial for the other(s)* (Cialdini, 2001).

On the basis of this conceptual disambiguition and of our analytic model, it has been argued that we do not necessarily trust people because they will be willing to reciprocate; and that we do not necessarily reciprocate for reciprocating. Trusting people (also in strict social situations, with mutual awareness) means counting on their 'adopting' our needs, doing what we expect from them, out of many possible motives (from altruism to norms keeping, from fear of punishments to gratitude, from sexual attraction to reputation and social approval, etc.); reciprocating or obtaining reciprocation are just two of them. However, the theory of how trust elicits reciprocation and trust, and how reciprocation builds trust, is an important part of the theory of trust as personal and collective capital.

Trust certainly has an enormous importance in economy and thus in economics (for exchange, market and contracts, for agency, for money and finance, for organizations, for reducing negotiation costs, and so on.), as in politics (the foundational relations between citizens and government, laws, institutions), etc. However, this concerns all kinds and dimensions of trust; not only those aspects needed in strategic games.

References

Alloy, L.B. and Abramson, L.Y. Judgment of contingency in depressed and nondepressed students: sadder but wiser? *Journal of Experimental Psychology: general*, 108: 441–485, 1979.

Axelrod, R. and Hamilton, W. D. (1981) The evolution of cooperation. *Science*. 211: 1390–1396.

M. Bacharach and D. Gambetta (2001) Trust as type detection, in: C. Castelfranchi, Y.-H. Tan (eds.), *Trust and Deception in Virtual Societies*, Kluwer Academic Publishing, Dordrecht, The Netherlands, pp. 1–26.

Baier, A. (1986) Trust and antitrust, *Ethics*, 96: 231–260.

Barber, B. (1983) *The Logic and Limits of Trust*. New Brunswick, NJ: Rutgers University Press.

S. Brainov and T. Sandholm (1999) Contracting with uncertain level of trust, Proceedings of the AA'99 Workshop on 'Deception, Fraud and Trust in Agent Societies', Seattle, WA, 29–40.

Bratman, M. E. (1987) *Intentions, Plans and Practical Reason*. Harvard University Press: Cambridge, MA.

Cramerer, C. Gifts as economic signals and social symbols, *America Journal of Sociology*, 94, S180, 1988.

Castelfranchi, C. (1998) Modelling social action for AI agents. *Artificial Intelligence*, 103: 157–182, 1998.

Castelfranchi, C. (2000) Through the agents' minds: cognitive mediators of social action. In: *Mind and Society*. Torino, Rosembergh, pp. 109–140.

Castelfranchi, C. (2004) Silent agents. From observation to tacit communication. *Modeling Other Agents from Observations: MOO 2004* -WS at the International Joint Conference on Autonomous Agents and Multi-Agent Systems, July 19. URL: http://www.cs.biu.ac.il/~galk/moo2004/

Castelfranchi, C. (2009) Trust and reciprocity: misunderstandings. RISEC (Rivista Internazionale Scienze ECo-nomiche), University Bocconi.

Castelfranchi, C. and Falcone, R. (1998) Principles of trust for MAS: cognitive anatomy, social importance, and quantification, *Proceedings of the International Conference on Multi-Agent Systems (ICMAS'98),* Paris, July, pp.72–79.

Castelfranchi, C. and Falcone, R. 'Trust is much more than subjective probability: Mental components and sources of trust', *32nd Hawaii International Conference on System Sciences* - Track on *Software Agents,* Maui, Hawaii, 5–8 January 2000. Electronic Proceedings.

Castelfranchi, C. and Lorini, E. 'Cognitive Anatomy and Functions of Expectations'. In *Proceedings of IJCAI'03 Workshop on Cognitive Modeling of Agents and Multi-Agent Interactions, Acapulco, Mexico,* August 9–11, 2003.

Castelfranchi, C., Giardini, F., Marzo, M. (2006) Relationships between rationality, human motives, and emotions. *Mind & Society,* 5: 173–197.

Castelfranchi, C., Falcone, R., and Marzo, F., Being Trusted in a Social Network: Trust as Relational Capital. Proceedings of *iTrust 2006 - 4th International Conference on Trust Management,* Pisa, May, 2006, pp. 16–26.

Cialdini, R. B. (2001) *Influence: Science and practice* (4th ed.), Boston: Allyn & Bacon.

Coleman, J. S. (1994) *Foundations of Social Theory,* Harvard University Press, MA.

Conte, R. and Castelfranchi, C. (1995) *Cognitive and Social Action.* London: UCL Press.

Cramerer, C. (1988) Gifts as economic signals and social symbols, *America Journal of Sociology,* 94, S180.

Deutsch, M. (1985) *The Resolution of Conflict: Constructive and destructive processes.* New Haven, CT: Yale University Press.

Deutsch, M. (1958) Trust and suspicion. *Journal of Conflict Resolution.* 2: 265–279.

Epstein, S. Coping, ability, negative self-evaluation, and overgeneralization: experiment and theory. *Journal of Personality and Social Psychology,* 62: 826–836, 1992

Falcone, R. and Castelfranchi, C. (2001) Social trust: a cognitive approach, in *Trust and Deception in Virtual Societies* by Castelfranchi C. and Yao-Hua Tan (eds), Kluwer Academic Publishers, pp. 55–90.

Falcone, R. and Castelfranchi, C. (2001) The socio-cognitive dynamics of trust: does trust create trust?' In R. Falcone, M. Singh, Y.H. Tan, eds., *Trust in Cyber-societies. Integrating the Human and Artificial Perspectives,* Heidelberg, Springer, LNAI 2246, pp. 55–72.

Fukuyama, F. (1995) *Trust: The Social Virtues and the Creation of Prosperity,* New York: The Free Press.

Gambetta, D., ed. (1988) *Trust: Making and Breaking Cooperative Relations,* New York: Basil Blackwell.

Ganzaroli, A., Tan, Y.H., Thoen, W. (1999) The Social and Institutional Context of Trust in Electronic Commerce, *Autonomous Agents '99 Workshop on 'Deception, Fraud and Trust in Agent Societes',* Seattle, USA, May 1, 65–76.

Hardin, R. (2002) *Trust and Trustworthiness,* New York: Russel Sage Foundation.

Hart, K Kinship, contract and trust: the economic organization of migrants in an African city slum, in D. Gambetta, (1988) ed.

Henrich, Joseph, Robert Boyd, Samuel Bowles, Colin Camerer, Ernst Fehr, and Herbert Gintis (2004) *Foundations of Human Sociality: Economic Experiments and Ethnographic Evidence from Fifteen Small-Scale Societies.* Oxford University Press.

Holton, R. Deciding to trust, coming to believe, *Australasian Journal of Philosophy,* 72 (1): 63–76, 1994,

Daniel Kahneman and Amos Tversky (eds.) (2000) *Choices, Values, and Frames,* Cambridge University Press.

Kurzban, R. (2001) 'Are experimental economics behaviorists and is behaviorism for the birds?', *Behavioral and Brain Sciences,* 24: 420–41.

Kurzban, R. (2003) Biological foundation of reciprocity. In E. Omstrom and J. Walker, eds., *Trust, Reciprocity: Interdisciplinary Lessons from Experimental Research* (pp. 105–127), NY: Sage.

Jones, A. J. (2002) On the concept of trust, *Decision Support Systems,* 33 (3): 225–232 - Special issue: *Formal modeling and electronic commerce.*

Jones, K. (1996) Trust as an affective attitude, *Ethics,* 107: 4–25.

Jones, K. (2001) Trust: philosophical aspects in N. Smelser and P. Bates, eds., *International Encyclopedia of the Social and Behavioral Sciences;* Amsterdam: Elsevier Science, pp. 15917–15922.

Joyce, B., Dickhaut, J., and McCabe, K. (1995) Trust, reciprocity, and social history *Games and Economic Behavior,* (10): 122–142.

Luce, R. D. and, Raiffa, H. (1957) *Games and Decisions: introduction and critical survey,* New York: Wiley.

Luhmann, N. (1979) *Trust and Power,* Wiley, New York.

Luhmann, N. (1990) Familiarity, confidence, trust: Problems and alternatives, In D. Gambetta (ed.), *Trust* (Chapter 6, pp. 94–107). Oxford: Basil Blackwell.

Mashima, R., Yamagishi, T., and Macy, M. Trust and cooperation: A comparison between Americans and Japanese about in-group preference and trust behavior. *Japanese Journal of Psychology*, 75: 308–315, 2004.

Miceli, M. and Castelfranchi, C. The mind and the future: The (negative) power of expectations, *Theory & Psychology*, 12 (3): 335–366, 2002.

Pelligra, V. (2005) Under trusting eyes: the responsive nature of trust. In R. Sugden and B. Gui, eds., *Economics and Sociality: Accounting for the Interpersonal Relations*, Cambridge: Cambridge University Press.

Pelligra, V. (2006) *Trust Responsiveness: On the Dynamics of Fiduciary Interactions*, Working Paper CRENoS, 15.

Rousseau, D. M., Burt, R. S., and Camerer, C. Not so different after all: a cross-discipline view of trust. *Journal of Academy Management Review*, 23 (3): 393–404, 1998.

Scheier, M.F. and Carver, C.S. Optimisim, coping, and health: assessment and implications of generalized outcome expectancies. *Health Psychology*, 4: 219–247, 1985.

Scheier, M.F., Weintraub, J.K. and Carver, C.S. Coping with stress: divergent strategies of optimists and pessimists. *Journal of Personality and Social Psychology*, 51: 1257–1264, 1986.

Schelling, T. C. (1960) *The Strategy of Conflict*, Harvard University Press, Cambridge, MA.

Shoham, Yoav; Leyton-Brown, Kevin (2009) *Multiagent Systems: Algorithmic, Game-Theoretic, and Logical Foundations*, New York: Cambridge University Press.

Taylor, S. E., and Brown, J. D. Illusion and well-being: A social psychological perspective on mental health. *Psychological Bulletin*, 103: 193–210, 1998.

Tiger, L. (1979) *Optimism: The biology of hope*. New York: Simon & Schuster.

von Neumann, John; Morgenstern, Oskar (1944) *Theory of Games and Economic Behavior*, Princeton University Press.

Yamagishi, T. (2003) Cross-societal experimentation on trust: A comparison of the United States and Japan. In E. Omstrom and J. Walker, eds., *Trust, Reciprocity: Interdisciplinary Lessons from Experimental Research* NY, Sage, pp. 352–370.

Weinstein, N. D. (1980) Unrealistic optimism about future life events. *Journal of Personality and Social Psychology*, 39: 806–820.

Williamson, O. E. (1985) *The Economic Institutions of Capitalism: Firms, Markets, Relational Contracting*. New York: The Free Press.

Williamson, O. E. (1993) Calculativeness, trust, and economic organization, *Journal of Law and Economics*, 36: 453–486.

9

The Glue of Society

9.1 Why Trust Is the 'Glue of Society'

In the literature (and common sense) it is now commonplace to find the slogan that trust is the glue of society, it is crucial, vital in economics (financial activity, market, management, and so on), in social cooperation, in organization, for institutional effects and acts, in groups, etc. But why? What is the real reason that trust plays such an essential role in human social life in every dimension? The answer – for us – is not (simply) the need for reducing uncertainty, for feeling and acting in a more confident way; for relying on predictions: all necessary reasons it is true, but not sufficient to understand this phenomenon and all its implications.

For us, the most fundamental reason is 'sociality' per se, the Aristotelian view of the '*zoon politikon*'. Human beings are social in a basic and objective sense: *they depend on each other*; they live thanks to each other. More precisely (Castelfranchi, 1993) (Conte, 1995) human beings are different from each other, both in their skills and resources, and in their many desires and needs (and their subjective importance).

Moreover, they live in the same 'environment', that is, they *interfere* with each other: the realization of the goals of X is affected by the activity of Y in the same environment; Y can create favorable or unfavorable conditions. Each agent has seriously limited powers (competences, skills, resources) and cannot achieve all his/her (potential) goals; but, by exploiting the powers of others, they can satisfy and develop their goals. By exploiting others (for example, via cooperation over common goals, or via exchange, or via domination, etc.) human beings can multiply their powers and their achievement in an unbelievable way. Also, because there are powers that no single individual possesses (*co-powers*) and cannot just be 'exchanged' or unilaterally exploited, but depends on collaboration: only a multi-agent coordinated action can produce the desired outcome. However, in order for this transformation of limits and dependence into an explosion of powers be realized, X not only has to exploit Y (and possibly vice versa) but he has to 'count on' this, to 'rely' on Y, to 'delegate' the achievement of his own desire to Y's action; to Y. *This is precisely trust*. Dependence (and even awareness of dependence) without trust is nothing; it is an inaccessible resource (see Chapter 10).

Trust must be based on some experience of Y (or similar people), on some evaluation of Y's competences and features, on some expectation, and on the decision to bet on this, to take some risk while relying on Y. Moreover, X can rely on Y's understanding of this reliance, on

Y's 'adoption' of *X*'s goal and hope; on a positive (cooperative) attitude of *Y* (for whatever reason) towards *X*'s delegation. This is precisely trust; *the glue of society as the transition from passive and powerless 'dependence' to active and empowering interdependence relationships.*

As we saw, trust is not just an attitude, a passive disposition, a cognitive representation or an affective state towards another person. It is also an active and pragmatic phenomenon; because cognition (and affect) is pragmatic in general: *for* and *in* action, for realizing goals. So, trust is also an action (deciding and performing an action of betting and relying on another guy) and part of social actions: exchange, collaboration, obedience, etc.

9.2 Trust and Social Order

There is a special, substantial relationship between trust and social order; in both directions: on the one side, 'institutional', 'systemic' trust (to use sociological terms), builds upon the existence of shared rules, regularities, conventional practices, etc. and relies on this, in an automatic, non-explicit, mindless way; but, on the other side, spontaneous, informal social order (not the legal ones, with control roles and special authorities) exploits this form of trust and works thanks to it (Garfinkel, 1963). In particular, the *'stabilization'* of a given order of shared practices and common rules, creates trust (expectation and reliance about those behaviors), and this diffused and *self-confirming* trust (a self-fulfilling prophecy, as we know) stabilizes the emergent social order.

Garfinkel's theory is quite important, although partial, restricted only to some form of indirect trust, (he explicitly mentions 'trust' only a couple of time in quite a long paper), strongly inspired by Parsons and Schutz and joined with (based on) a rather strange ideological 'proclamation' against the need for psychological sub-foundations,[1] which is systematically contradicted one page later and throughout the entire paper.

The main thesis of Garfinkel is that social order and social structures of everyday life, emerge and stabilize and work thanks to our natural 'suspension' of doubts, of uncertainty, of worries; our by-default assumption is that what is coming will be normal, without surprises ('perceived normality'). We build our everyday life on such economic assumption of 'normality' and of shared, 'common' expectations about 'which game we are playing' and 'which are the well-known rules of this game'. And we react to the violation of this presupposed order and normality first of all by attempting to 'normalize' the event, to reinterpret it in another normal frame and game.[2]

Expectations about those rules and regularities, and the respect of them by the others, are *constitutive* of the game we play. *'The social structures consist of institutionalized patterns of normative culture; the stable features of the social structures as assemblies of concerted actions are guaranteed by motivated compliance with a legitimate order'* (p. 189).

Not only do the subjects suspend their possible vigilance and diffidence, but, they actively and internally 'adhere' to this order, by using those normal rules and game-definition as a *'frame'* for *'interpreting'* what is happening[3] (a rather psychological process!); and by *'accepting'* the events and the rules and the expectations themselves as 'natural', 'obvious'.

---

[1] 'Meaningful events [for a theory of trust] are entirely and exclusively events in a person's behavioral environment, ... Hence there is no reason to look under the skull since nothing of interest is to be found there but brains' (p. 190).

[2] Also because – we should add as cognitive scientist – for our need (mechanisms) of cognitive coherence and integration, and of social integration and coherence.

[3] 'To be clear, bridge players react to the others' actions as bridge events not just as behavioral events'.

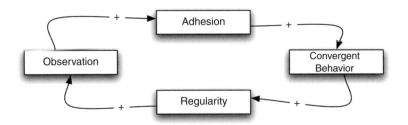

Figure 9.1   The self-reinforcing cycle of conformity

'The player **expects** that the fixed reference system be binding for himself and for the others.(..) The player **expects** that his adversary has the same expectations towards him [a rather psychological construct!]. We define these 'constitutive expectations''. Persons' treatments of their interpersonal environments are governed by constitutive expectancies, that is, they trust each other.

> 'The concept of trust is related to the concept of perceively normal environments as follows. To say that one person 'trusts' another means that (..) the player takes for granted the basic rules of the game as a definition of his situation, and that means of course as a definition of his relationship to others.' (p. 193)

The events in environments are *perceived as normal* (see discussion in Section 9.6 on trust and norms).

It is interesting that in Garfinkel's analysis those expectations play cognitive, pragmatic, and normative roles; they provide the 'frame' for interpreting what is happening; they guide the decision about what to do; they create not only predictions but entitled prescriptions and commitments (Castelfranchi, 2003).

One should model in a more systematic and explicit way (also by modeling the internal mechanisms of the agents) this *cycle of routines*; how 'social structures are typically maintained in a routinary way' (Figure 9.1).

X observes a given, seemingly regular practice and on such a basis interprets it as a rule of the game, what should be done, what to expect next time from the others; but also as what they have to do (in a given role). Thus, on one hand, they conform to that practice (also because they believe that the others expect something and prescribe this behavior); on the other hand, they create a pressure (through this behavior (*signaling*) and monitoring) on the others to conform. This actually reproduces that regularity; reinforces those beliefs and goals and makes them rule-based, routine mechanisms; confirms the validity of the 'frame' of interpretation; confirms that those are the rules of the game they are playing. Thus the cycle is self-maintaining and enforcing, and self-stabilizing (but only thanks to the *cognitive immergence* and *behavioral emergence*).

It is also important to notice and point out (Pendenza, 2000) that this kind of trust is 'natural', 'naive' (we would prefer to call it 'routine'), not based on specific reasons, reasoning, and assumptions, but just a basic reinforced attitude; just by-default, and rather automatic. It is a trust based on routines and habits. And the 'act' of trusting itself is not such a real 'decision' and 'intentional' act; it is more a functional rule-based act systematically reinforced (see Chapter 4).

9.2.1 Trust Routinization

There are various forms of trust not based on real deliberation, on intentional action, on conscious considerations of Y's virtues (trustworthiness) and of possible risks (Chapter 4). Some of them are non-reflective and automatic, rule-based; due to a basic 'natural' by-default attitude, that takes for granted the reliability of Y, or of infrastructures, or of social rules, etc. This is the trustful disposition analyzed in Chapter 4, or Garfinkel's basic trust in spontaneous and 'natural' social order.

However, there is another form of rule-based, routine trust; it *is routinized trust*. It is when trust is initially based on careful consideration, monitoring, hesitation, a serious evaluation of Y's willingness and competence. Consider, for example, a trapeze artist, starting her job with a new partner. She literally 'puts her life in the other's hands', and is very aware of his strength, attention to detail, ability, etc. But, after some months of perfect exercises and successes, she will fly in the air towards Y's hands, concentrating on her own acts, and automatically relying on Y's support. This holds for any "familiarization". Analogously, a blind man, who for the first time has a guide-dog and has to cross the street by following it, is very perplexed, careful and with a high perception of the risk; deciding over time to 'trust' his dog. But after they have been together for a long time, he will stop trying to control the dog all the time, and will follow it in a confident way. If this is the case in such risky relationships, a fortiori it is likely to be so in less dangerous stable reliance situations.

What, at the beginning, was an explicit belief, based on reasoning and observation, and a real decision to delegate and rely on; with successful repetitions, and exercises, becomes just a routine plan although involving (counting on) the actions of another agent; exactly as in a single-agent plan. What was a careful judgment has become some sort of reinforced classifier, and a feeling of safety and efficacy: a 'somatic marker' due to the repeated experience of success.

This routinized trust contains two form of trust: trust in the routine itself (see Chapter 4 on Routines implying trust), and procedural trust in Y implemented in a trusted routine.

This is also why a trust attitude and decision is not necessarily joined to an explicit consideration of risk. This idea can remain unformulated (just a logical consequence of a degree of certainty), implicit. Risks are there but not always psychologically present, although logically necessary (see Chapter 2).

9.3 How the Action of Trust Acquires the *Social Function* of Creating Trust

Trust (as attitude but especially as manifest action and signal) creates trust (see Chapter 6 and (Falcone and Castelfranchi, 2001)). A virtuous circle and thus in our model a 'function' can be created by the simple act of X trusts Y. In fact, in several ways this act can have effects that increase the probability of its reproduction (and spreading).

Since we define the 'function' of an action or of a feature an effect of it that is responsible for its reproduction without being intended, we consider these non accidental and sporadic effects of the action of trusting Y, which are responsible for the fact that it will be reproduced, to be its social 'function' (Castelfranchi, 2000).

There are different reasons and mechanisms responsible for this positive feedback; let's consider some of them.

a) It is true that generally speaking (but see also Chapter 6) if X's act of trusting somebody (Y) and relying on him is successful it will increase X's trust in Y (and also X's generalized trust and a trustful attitude). Assuming that there is a reasonable distribution of trustworthy agents, and that X bases his decision on some reasonable criteria or experience, there is a reasonable probability that the decision to trust brings some success.

In any case, when it brings success it will be reinforced, the probability of choosing it again will increase. While the decision not to try, not to risk cannot go in the same direction.[4] If you do not bet you can never win; if you bet you can either lose or win; if you lose (let's suppose) you will not bet again, thus being in the same situation as before.

b) The decision to trust Y – when known by Y (and frequently the act is in fact also an implicit message to Y 'I trust you') – may increase Y's trustworthiness (see Chapter 6); either, by increasing his commitment, attention, and effort, or by reinforcing his loyalty. This will increase the chance of a good result, and thus the probability of trusting Y again.

c) The fact that X trusts Y (has a good evaluation of him; is not diffident towards him, decides to make herself vulnerable to Y), can create in Y an analogous non-hostile disposition, a good-will towards X. Y will have reasonable trust in X in return (*trust reciprocation*); but this attitude and behavior will increase the probability that X trusts Y again.

d) The act of X of trusting Y can be observed by others, can be a *signal* for them:
    -   that Y is trustworthy (and increases the probability that they will rely on him too), or
    -   that Y is from a trustworthy group or role, or
    -   that this is a trustworthy context or community, where one can rely on the other without risk.

This will spread around trustful behaviors, that will be also perceived by X herself, and encourage again her trustful attitude.

Thus, in several independent ways the act of X trusting Y is responsible for its effects of increasing trust and thus of increasing the probability of its reproduction.

People are not usually aware of these effects or do not intend them, but via these effects in fact a trustful behavior has the *function* of creating trust, by spreading and reinforcing it. It is a virtuous circle, a loop (Figure 9.2).

Analogously, as trust creates and reproduces trust, distrust and diffidence enhance hostility and non-reliance and non-cooperation (*diffidence reciprocation and spreading*). This is a vicious circle in social life; what we call a 'kako-function' (Castelfranchi, 2000b). In fact it is a paradoxical 'function', since this behavior is also maintained and reproduced by its unintended effects.

---

[4] Except when transitory and based on the expectation of a better opportunity.

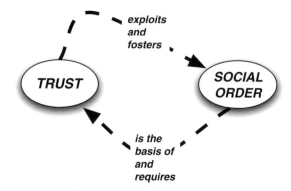

Figure 9.2   The virtuous loop between Trust and Social Order

9.4   From Micro to Macro: a Web of Trust

Let us also argue how the cognitive and interactive dynamics of trust produce the fact that trust-networks are real dynamic webs, with their own emergent macro-dynamics. They are not static, topological structures; only the sum of local relationships. Local events/changes have trans-local repercussions; the entire network can be changed; the diffused trust relationships can collapse, etc. As Annette Baier said ((Baier, 1994) p. 149): 'Trust comes in webs, not in single strands; and disrupting one strand often rips apart whole webs'.

For us Bayer is too extreme: also, merely interpersonal, dyadic (and even unilateral) trust attitudes, decisions, and relations (strands) exist. However, the phenomenon pointed out by Baier is important and must be modeled. We can also consider the individual or bilateral trust relations as very local, small and isolated webs. Actually it is true that trust attitudes and relations have a web nature.

How and why do these repercussions hold? Which are the mechanisms of this web-dynamics?

9.4.1   Local Repercussions

If X trusts Y, W, and Z, the fact that she revises her trust in Y can affect her trust in W and/or Z; there might be some repercussion. Not only because the trust in Y might have been also comparative: X has decided to actively trust Y, to choose Y and delegate to him, in comparison with W and Z. Thus, the success of this delegation (or the decision itself, thanks to Festinger's effect (Festinger, 1957)) can make more certain X's evaluation of Y and change the relative strength of trust in W and Z. While the failure of that delegation can comparatively increase trust in W and Z.

There might also be other repercussions. For example some common category (see Section 6.6) or some pertinent analogy between Y and Z, and thus the success or failure of Y can also change the intrinsic evaluation and expectation about Z. The trust in Y is betrayed or a disaster, but Y and Z belong to the same category/role (for example, layers); X generalizes his disappointments and negative expectations to the whole class; thus, also X's trust in Z collapses. And so on.

9.4.2 Trans-Local Repercussions

Also non-personal and non-local changes of the trust relationships can affect X's views and the entire network. Y can, for example, just observe J's delegation to K and on such a basis change her own trust disposition (and decision) towards Z. In fact, there might be analogies between J's delegation and X's potential delegation, or between K and Z. Or X can just imitate J, use his example as a model or cue, not only in a pseudo-transitive way (X's attitude towards K is derived from J's attitude), but in an analogical way: if J trusts K, X can trust Z (because X and J have similar needs, and K and Z similar properties) (see again Section 6.6).

Notice that the observation of the others' behavior and their evaluation (in relation to social conventions and norms, to fairness, honesty, etc.) is a basic fundamental social disposition and ability in human beings. We do not just observe the behaviors of agents that directly concern us (exchanging or cooperating with us, competing with us, etc.); we observe agents interacting with other agents not related with us. And we in a sense provide an altruistic 'surveillance' of each other (Gintis, 1957); we evaluate them, we spread around our blame or admiration, we spread the circulating reputation. This is the most powerful instrument for social control and social order in human societies, but, it will be equally important in virtual and in artificial societies. This is exactly why 'identification' of the agents matters in the first place.

Other trans-local mechanisms for trust repercussion apart from observation of distal events, are referrals (other agents report to me their evaluation or the success/failure of their delegation (Yu and Singh, 2003), and reputation: spreading around opinions and gossip about agents (Conte and Paolucci, 2002).

Particularly delicate conditions for web effects are *default* and *generalized forms of trust*. Some of them could collapse in an impressive way (like our trust in money). If, for example, the rule of my generalized trust is:

> 'Since and until everybody trusts everybody (or everybody is not diffident and suspicious towards the others) ===> I will trust anybody (I will not be diffident)'.

This rule, given one single case of personal bad experience, or of bad observation, or referral, or reputation, can invert its valence: I become suspicious in a generalized way. And if this rule is diffused (all the agents or many of them use it) the impact will be a generalized collapse of the general trust capital and atmosphere.

Analogously, if I follow an optimistic (non prudent) default rule: 'Except I have a negative example I will assume that agents in this community are trustworthy'.

The same can hold for affective trust disposition (Chapter 5) and a generalized mood in a given context that can be wasted by just one very negative personal experience or by the contagion of different moods of others.

Of course the network is not necessarily uniform or equally connected; it might be an archipelago of non well-connected 'islands' of very connected sub-communities. Thus the propagation might just have local effects.

9.5   Trust and Contracts

'A contract is not sufficient by itself, but is only possible because of the regulation of contracts, which is of social origin' E. Durkheim, (the Division of Labor in Society (1893), New York: The Free Press, 1997, p. 162).

Obviously, this *social* background includes trust, social conventions and trust in them, and in people respecting them, the authorities, the laws, the contracts.

9.5.1   Do Contracts Replace Trust?

A commonplace belief in trust theory is that we put contracts in place when and because there is no trust between the parties. Since there is no (not enough) trust, people want to be protected by the contract. The key in these cases is not trust but the ability of some authority to enforce contract application or to punish the violators. Analogously, in organizations people do not rely on trust but on authorization, permission, obligations and so forth.

As we have explained in Chapter 7 (on trust and third party relationships), for us this view is correct only if one adopts a very limited view of trust in terms of direct interaction and acquaintance, of friendliness, etc. But it is not true that 'trust', as a general category, is not there in contracts or in formal agreements and rules. In those cases (contracts, organizations) we just deal with *a more complex and specific kind of trust.* But trust is always crucial. A third party (and 'institutional') trust.

As we have said, we put a contract in place only because we believe that the agent will not violate the contract, and we count on that; and this is precisely 'trust'. We base this trust in the trustee (the belief that they will do what has been promised) either on the belief that they are a moral person and keeps their promises, or on the belief that they worry about law and punishment by the authorities (A). This expectation is the new level of X's trust in the trustee.

As we have explained, X relies on a form of paradoxical trust of Y in A: X believes that Y believes that A is able to control, to punish, etc. Of course, normally a contract is bilateral and symmetric, thus the point of view of Y should be added, and his trust in X and in A when it comes to monitoring X. Notice that Y's beliefs about A are precisely Y's trust in the authority when they are the client, while, when Y is the contractor, the same beliefs are the basis of their respect/fear toward A.

So contracts presuppose less informal/personal trust but require some more advanced (cultural, institutional) form of trust, in Y, and in the institution.

9.5.2   Increasing Trust: from Intentions to Contracts

What we have just described are not only different kinds and different bases of trust. They can also be conceived as different levels/degrees of social trust and *additional* supports for trust. We mean that one basis does not necessary eliminate the other but can supplement it or replace it when it is not sufficient. If I do not trust your personal persistence enough I can trust you to keep your promises, and if this is not enough (or is not there) I can trust you to respect the laws or to worry about punishments.

We consider these 'motivations' and these 'commitments' not all equivalent: some are stronger or more cogent than others. As we claimed in (Castelfranchi, 1995):

> *This more cogent and normative nature of S-Commitment explains why abandoning a Joint Intention or plan, a coalition or a team is not so simple as dropping a private Intention. This is not because the dropping agent must inform her partners -behaviour that sometimes is even irrational-, but precisely because Joint Intentions, team work, coalitions (and what we will call Collective-Commitments) imply S-Commitments among the members and between the member and her group. In fact, one cannot exit a S-Commitment in the same way one can exit an I-Commitment. Consequences (and thus utilities taken into account in the decision) are quite different because in exiting S-Commitments one violates obligations, frustrate expectations and rights she created. We could not trust in teams and coalitions and cooperate with each others if the stability of reciprocal and collective Commitments was just like the stability of I-Commitments (Intentions).*

Let us analyze this point in more detail by comparing *five* scenarios of delegation:

Intention Ascription

X is weakly delegating Y a task τ (let's say to raise his arm and stop the bus) on the basis of the hypothetical ascription to Y of an intention (he intends to stop the bus in order to take the bus).

There are two problems in this kind of situation:

- The *ascription of the intention* is just based on abduction and inferences, and to rely on this is quite risky (we can do this when the situation is very clear and very constrained by a script, like at the bus stop).
- This is just *a private intention and a personal commitment* to a given action; Y can change his private mind as he likes; he has no social obligations about this.

Intention Declaration

X is weakly delegating Y a task τ (to raise his arm and stop the bus) on the basis not only of Y's situation and behavior (the current script) but also or just on the basis of a declaration of intention by Y. In this case both the previous problems are a bit better:

- the *ascription of the intention* is safer and more reliable (excluding deception that on the other hand would introduce normative aspects that we deserve for more advanced scenarios);
- now Y *knows that X knows about his intention* and about his declaring his intention; there is no promise and no social commitment to X, but at least by changing his mind Y should care about X's evaluation of his coherence or sincerity or fickleness; thus he will be a bit more bound to his declared intention, and X can rely a bit more safely on it.

In other words, X's degree of trust can increase because of:

- either a larger number of evidences;
- or a larger number of motives and reasons for Y doing τ;
- or the stronger value of the involved goals/motives of Y.

Promises

Promises are stronger than a simple declaration or knowledge of the intention of another agent. Promises create what we called a social commitment, which is a right producing act, determining rights for Y and duties/obligations for X. We claim that this is independent of laws, authority, punishment. It is just at the micro level, as an inter-personal, direct relation (not mediated by a third party, be it a group, an authority, etc.).

The very act of committing oneself to someone else is a 'rights-producing' act: before the S-Commitment, before the 'promise', Y has no rights over X, Y is not entitled (by X) to exact this action. After the S-Commitment such a new and crucial social relation exists: Y has some rights on X, she is entitled by the very act of Commitment on X's part. So, the notion of S-Commitment is well defined only if it implies these other relations:

- Y is entitled (to control, to exact/require, to complain/protest);
- X is in debt to Y;
- X acknowledges being in debt to Y and Y's rights.

In other words, X cannot protest (or even better *he is committed to not protesting*) if Y protests (exacts, etc.).

One should introduce a relation of 'entitlement' between X and Y meaning that *Y has the rights* of controlling α, of exacting α, of protesting (and punishing), in other words, X is S-Committed to Y to not oppose these rights of Y (in such a way, X 'acknowledges' these rights of Y).

If Y changes his mind he is disappointing X's entitled expectations and frustrating X's rights. He must expect and undergo X's disappointment, hostility and protests. He is probably violating shared values (since he agreed about X's expectations and rights) and then is exposed to internal bad feelings like shame and guilt. Probably he does not like all this. This means that *there are additional goals/motives that create incentives for persisting in the intention. X* can reasonably have more trust.

Notice also that the declaration is more constraining in promises: to lie is worst.

Promises with Witness and Oaths

Even more binding is a promise in front of a witness, or an oath (which is in front of God). In fact, there are other bad consequences in the case of violation of the promise. Y would jeopardize his reputation (with very bad potential consequences; see [Cas11]) receiving a bad evaluation also from the witness; or if he'd behaved badly under oath he would elicit God's punishment.

Thus if I do not trust what you say I will ask you to promise this; and if I do not trust your promise I ask you to promise in front of other people or to take an oath on it. If I worry that you might break that promise I will ask for it in writing and signed. And so on.

Contracts

Even public promises might not be enough and we may proceed by adding binds to binds in order to make Y more predictable and more reliable. In particular we might exploit the third party more. We can have a group, an authority able to issue norms (defending rights and creating obligations), to control violation, to punish violators. Of course this authority or reference group must be shared and acknowledged, and as we said trusted by both X and Y. Thus we have got an additional problem of trust. However, now Y has additional reasons for keeping his commitment, and X's degree of trust is higher.

Notice that all these additional beliefs about Y are specific kinds or facets of trust in Y: X trusts that Y is respectful of norms, or that Y fears punishments; X trusts in Y's honesty or shame or ambition of good reputation, etc. To state this point even more clearly: the stronger Y's motive for doing τ and then the stronger his commitment to τ; the greater the number of those motives; and the stronger X's beliefs about this; the stronger will be X's trust in Y to do τ.

9.5.3   Negotiation and Pacts: Trust as Premise and Consequence

If X is negotiating with Y over a possible agreement and pact this means that she has some form and degree of trust in Y as (i) possible adequate and credible 'provider', and as (ii) willing and capable negotiator. She is not just wasting her time or doing this for fun. She is already betting and relying on Y to 'negotiate' and possibly achieve an agreement; and this is a possible bet and reliance on Y to do the delegated task (good, service).

Negotiation and pacts (contracts) *presuppose* some trust. However, pacts and contracts are there also to *create* some trust: the trust on which Y will, for example, give her money and confidently wait for a return. In fact, as we just said (Section 9.4.2), Y's promise (implicit or explicit), his 'commitment', gives X (additional) new bases for expecting that Y will act as desired. Y now has some 'duty' to do so; there are social and moral norms binding him; and X is entitled to her expectations; she has some 'right' over Y. Thus X has additional trust in Y; she feels more sure and safe.

With a 'contract' we have the added weight of legal authorities and guaranties to protect X and to oblige Y. So, negotiation and pacts are a social mechanism for producing trust by investing trust; for building new forms and bases of trust on previous ones. However, as we said, the new layer of trust is not always 'additional'; it is also completing or complementing some possible lack and insufficiency of trust. If X does not trust Y enough on a mere informal base, she would not rely on him. Then, a promise, pact, or contract gives her sufficient trust by binding Y's decision. Nevertheless, some trust must always be already there before the negotiation, or the promise, or the pact, or the contract. Why ask for a promise from Y if we do not believe that he is promise-sensible and bound? Why invoke the official signature of a contract if we perceive Y as indifferent to law, authority, sanctions, etc.?

9.6   Is Trust Based on Norms?

This is a quite diffused theory (especially in sociology; for example Garfinkel, 1963, and Giddens, 1984. See also Section 2.2.2 for our criticism on A. Jones' position).

1) On one hand, one should not over exploit the very dangerous ambiguity of the notion of 'norm' or of 'rule' in several languages, covering both a mere external descriptive 'regularity' (from Latin: 'regula', rule/norm), and a prescription or model aimed at inducing

a given conforming behavior in self-regulated systems, and the translation of this into some proximate *mechanism* affecting or producing the behavior of the system, conforming to that rule and thus 'regular'. It is not one and the same thing.

In particular it is quite different:

- saying that I trust and rely on a given predicted behavior (*based on* some perceived rule or regularity, or on some norm and conformity to it);
- saying that I trust the prediction; or that I trust the norm;
- saying that I trust the behavior in force of the explicit norm.

Is my prediction *based-on* such (perceived) regularity, on such a rule; or is that behavior *based on* that rule (affecting the mind of *Y*)? It is not at all the same thing.

2) However, even more important than this, one should be careful to preserve the very fundamental distinction made by Tommaso between: *'Id quod intelligitur'* and *'Id quo intelligitur'*: what (O) I'm thinking about, categorizing, recognizing, understanding, knowing, vs. what I'm using for thinking about O, for representing O in my mind (or externally): the representation, the scheme.[5] *I'm not thinking the representation; I'm thinking about my object of knowledge through the representation.*[6]

This clear distinction is fundamental for cognitive sciences (and semiotics).

Analogously, thanks to and through a given (implicit or explicit) 'rule' and learned regularity, I think that something *p* will happen (in the future). I do not believe – in the same way and sense – the rule (or *in* the rule). I believe *with/through/ thanks* to the rule ('Id quo'), not the rule ('Id quod').

If, for example, I believe that – since it starts raining – the ground will become wet, or if I believe that in springtime that tree in my garden will produce flowers (and I trust in this), I do *not* believe that the tree (or the rain) will follow/respect the norm. Not only do I not have some animistic, and 'intentional stance', but even less I believe that 'the rule will be respected'. I just *use* the rule (for inferring); its systematic use is a procedural, implicit assumption that it is true (reliable) and that 'it will be respected', but not an explicit belief and judgment, like my expectation about *p*.

If I strongly hope and even *trust* that she will accept my courtship this night, after my flowers, dinner, intimate atmosphere, wine, etc. as usual from my previous experiences, I do not 'trust' (believe) that 'she will respect the rule', or that 'the rule will be respected'.

Logicians seem rather insensible to this fundamental distinction between explicitly represented goals or beliefs, and merely procedural implementations. For example, one should not use the same predicate *(Bel x p)* to represent the *status/use/role* of 'being believed' of *p* in *X*'s mind, and the object 'belief'; object of various propositional attitudes: *(Goal Y (Bel X p)), (Bel*

---

[5] Quaestio 85; Prooemium Deinde considerandum est de modo et ordine intelligendi. Et circa hoc quaeruntur octo. Primo, utrum intellectus noster intelligat abstrahendo species a phantasmatibus. Secundo, utrum species intelligibiles abstractae a phantasmatibus, se habeant ad intellectum nostrum ut quod intelligitur, vel sicut id quo intelligitur (Thomas de Aquino, Summa Theologiae, I^a q. 84–89)

[6] Except I go to a meta-level, and take the representation itself (the 'significant') as my object of reflection.

Y (Bel X p)). These two *'Bel'* cannot be represented in the same way; I cannot use *(Bel X p)* to build a belief in *X*'s mind.

3) In sum, if it was true that any possible expectation, for its 'prediction' part, is based on some 'inference' and that an 'inference' is based on some 'rule' (about 'what can be derived from what'), it would be true that any trust is based on some 'rule' (but a cognitive one). However, even this is too strong; some activated expectations are not based on 'inferences' and 'rules' but just on associative reinforced links: I see *q* and this just activates, evokes the idea of *p*. This is not seriously a 'rule of inference' (like: *'If (A is greater than B) and (B is greater than C), then (A is greater than C)'*). So we would not agree that any expectation (and trust) is rule-based. However, one might expand the notion of 'rule' even to this simple and reactive 'mechanism' (mixing up the observed regularity that they produce, with a 'regula'/rule that should generate it). With such a broad and weak notion of rule, we might agree that any trust – being prediction based – is in some sense 'rule-based', it reflects some regularity and 'norm'. But not in the strict social or moral or cognitive sense; this holds only for social trust in its 'genuine' sense, based on goal-adoption and (implicit) commitments or on social norms and prescriptions.

4) Moreover, regularity is also about bad events; we also have 'negative' expectations (based on the same rules, 'norms' of any kind). Now, it is a real act of violence against the current notion of 'trust' that we are supposed to model, reducing it just to 'expectations based on perceived normality' (Garfinkel's claim).

We may have the expectation that the author of a horrible homicide will be condemned to die (given the laws of our states, and the practice of our government), both if we wish this to happen and expect it out of revenge, or if we are the killer condemned to die, or activists against the death sentence. However, if we are in favor of the death sentence, and we desire this, actually we 'trust' our authorities over this; if we are the condemned man, or the adversaries of the death sentence, we don't trust the authorities at all over this! This would be a serious distortion of the concept. This is why in our chapter about third party trust (Chapter 7) we say that this is a 'paradoxical', not true form of trust. There is a basic common mental ingredient (the belief about the future event), and this explains why the same belief becomes trust or not while just changing my role and goal. But it is not the right solution to reduce trust just to such a belief, and to call 'trust' fear and opposition.

Thus, in sum, *normality* and *regularity* are not sufficient for trust, and probably are not even necessary, if we do not extend conceptually the notion of 'rule' to cover any possible prediction device.

9.6.1   Does Trust Create Trust and does There Exist a Norm of Reciprocating Trust?

We have made it clear (Chapter 6) that it is not out of reciprocation that *Y* does the expected action after we have trusted him and decided to rely and depend on him; and also that trust is not always 'reciprocated' (even when *Y* performs the entrusted action). However, we acknowledge that there exist a property of trust to elicit trust, and we wonder about the idea that there might even exist *a norm of trust reciprocation. Since trust is not just a behavior, but a mental state and a feeling, it cannot really be 'prescribed', since is not really 'voluntary'.* Only the act, the intention can paradoxically be 'prescribed': 'Trust him! Rely on him!'; but not the real background disposition.[7]

However, moral (and religious) norms can impinge even on mere mental dispositions ('Do not desire . . .' 'Do not have this kind of thought'); thus there might be, and in fact it seems that there is, a social-moral *norm* about reciprocating trust: 'Since if *X* trusted you, you have to trust *X*'. To trust somebody seems to be a form of 'gentle' disposition or act, and it seems that we have to respond to a gentle act with a gentle act, to a smile with a smile.

There is a clear psychosocial phenomenon of trust propagation such that trust creates trust while diffidence creates hostility. If *X* trusts *Y*, this tends to elicit not only a 'benevolent' but also a 'trustful' attitude in *Y* towards *X*. However, we do not believe that it is mainly due to such a possible moral norm. We believe that it is mainly due to:

- The fact that while trusting *Y*, *X* makes himself dependent and vulnerable to *Y*, more exposed, and thus less dangerous, harmless.
- The fact that while trusting *Y*, *X* shows positive evaluations, esteem, thus a good disposition towards *Y*, which can be a good basis and a prognostic sign for 'benevolence' towards *Y*, that is, for adoption; (it is more probable that we help somebody who we perceive as competent and benevolent, although we do not currently intend to exchange with them).
- The fact that while trusting *Y*, *X* may even rely on common values, on sympathy (common feelings), on a sense of common membership, etc. and this makes him in his turn reliable, safe.

Nevertheless, we believe that such a norm of responding to trust with trust, exists. It is not responsible for eliciting trust in response to trust, but it is important for other functions. It is used for moral *evaluation*, and is responsible for blame, shame, etc.

9.7   Trust: The Catalyst of Institutions

As we said, trust is crucial for the whole of social life (exchange, cooperation, communication, rules of conflict, etc.), however *it is in particular fundamental* (or better, foundational) *for the 'institution'* (Searle, 1995).
   Together with:

- actors' recognition and assumption (acceptance) of the institutional act and effect, and with
- actors' 'as if' behavior (conforming to the assumption) (Tummolini, 2006), trust is the necessary ground on which our 'institutions' base themselves, their 'count-as' nature.

Actually, it is trust (and behavioral conformity) that 'institute' them and give them (make 'real') their artificial effects.
   In fact, the social 'representation', the collective *mise en scene* (Goffman, 1959), is strictly based on compliance and complicity, on collusion; that is, on the (prescribed) assumption that

---

[7] In those extreme cases trust as disposition wouldn't be enough for the intention, but we add independent, external, additional reasons which forces us to 'trust' in the sense of deciding to rely on Y.

everybody is doing their share (starting from such an assumption). If you do not believe – or better 'accept' – that C is equal to D, ('counts as' D) it doesn't actually count as D.

Only our cooperation and complicity creates the phenomenon, provides the needed 'power' and 'virtue' to those acts (like: signing, paying), roles (like: judges, policemen), objects (like: money, signatures). If C has to 'count as D' for us, then I have to pragmatically 'count on' its conventional effects, and thus I have to 'count on' you (us) for its 'counting as D' for you. I (have to) trust you to recognize and frame C as D, and *treat* C as D by behaving in front of C according to our convention.

Trust is not just the *glue*, but is the real *mediator*[8] *of the constructive process and mechanism* of the conventional actions/effects and the institutional building and maintenance.

9.7.1   The Radical Trust Crisis: Institutional Deconstruction

The most radical and serious economical-political crisis is in fact a *trust crisis*; when the 'doubt' corrodes the conventional, artificial, value, nature, and effect of *institutional* powers, actions, and objects.

I no longer believe that the court or the policeman has any authority over me (over us) since you and I do no longer recognize them. I do not believe that our money has a value, I'm not sure that the others will accept it, so why should I accept it or preserve it? I no longer believe that your act (signature, oath, declaration, etc.) has any value and effect, since to be effective as conventional-institutional act it presupposes the *acknowledgment* and *compliance* of other people, and I do not believe that the others believe in its validity and will be compliant.

This is a real institutional earthquake: I do not trust institutional authority, roles, acts, artifacts (such as money), because I do not believe that the others trust them. We move from a shared implicit trust in institutional artifacts and acts, to a *shared distrust*. But, since trust is the real foundation of their reification and effectiveness, this make them disappear: we realize, we see, that 'the king is naked!'.

Also the political 'representation' is an *institutional act*. We take X's (the 'representative') words or choices as 'representing' the preferences, the opinions, or at least the interests and values of his group (the people he 'represents'); and those people believe the same and rely on this. However, if the represented people's trust is in crisis, and thus there is no longer a real reliance and *'delegation'* to X, X no longer represents anybody. There is a serious detachment, a crisis of the relation between people and parties, voters and deputies, which essentially is a trust-delegation-reliance relation. If I no longer believe in you or 'count on' you, you no longer represent me. I don't necessarily perceive you as dishonest or selfish, but perhaps I perceive you as powerless or I perceive politics as distant and ineffective.

References

Baier, A. (1994) Trust and its vulnerabilities in *Moral Prejudices*, Cambridge, MA: Harvard University Press, pp. 130–151.

Castelfranchi, C., Cesta, A., Conte, R., Miceli, M. Foundations for interaction: the dependency theory, Lecture Notes in *Computer Science*; 728: 59–64, Springer, 1993.

---

[8] Trust is more than a *catalyst* for the institutional process; in fact it is also among the results of the 'reaction'.

Castelfranchi, C. (2000) Through the agents' minds: cognitive mediators of social action. In: *Mind and Society*. Torino, Rosembergh, pp. 109–140.

Castelfranchi, C. (2000b) Per una teoria pessimistica della mano invisibile e dell'ordine spontaneo. In Salvatore Rizzello (a cura di) *Organizzazione, informazione e conoscenza*. Saggi su F.A. von Hayek. Torino, UTET.

Castelfranchi, C. (1995) Social Commitment: from individual intentions to groups and organizations. In *ICMAS'95 First International Conference on Multi-Agent Systems*, AAAI-MIT Press, 41–49.

Castelfranchi, C., Giardini, F., Lorini, E., Tummolini, L. (2003) The prescriptive destiny of predictive attitudes: from expectations to norms via conventions, in R. Alterman, D. Kirsh (eds.) *Proceedings of the 25th Annual Meeting of the Cognitive Science Society*, Boston, MA.

Conte, R. and Castelfranchi, C. (1995) *Cognitive and Social Action*. London, UCL Press.

Conte, R. and Paolucci, M. (2002) *Reputation in Artificial Societies. Social Beliefs for Social Order*. Boston: Kluwer Academic Publishers.

Falcone, R. and Castelfranchi, C. (2001) Social trust: a cognitive approach, in *Trust and Deception in Virtual Societies* by Castelfranchi, C. and Yao-Hua, Tan (eds.), Kluwer Academic Publishers, pp. 55–90.

Festinger, L. (1957) *A Theory of Cognitive Dissonance.*, Stanford, CA: Stanford University Press.

Gintis, H. Strong reciprocity and human sociality, *Journal of Theoretical Biology*, 206: 169–179, 2000.

Yu, B. and Singh, M. P. (2003) Searching social networks. In *Proceedings of the Second International Joint Conference on Autonomous Agents and Multi-agent Systems (AAMAS)*. Pp. 65–72. ACM Press.

Garfinkel, H. (1963) A conception of, and experiments with, 'trust' as a condition of stable concerned actions. In O. J. Harvey (ed.) *Motivation and Social Interaction*. Ronald Press, NY, Ch. 7, pp. 187–238.

Giddens, A. (1984) *The Constitution of Society: Outline of the Theory of Structuration*. Cambridge, Cambridge University Press.

Goffman, E. (1959) *The Presentation of Self in Everyday Life*. Anchor Books.

Pendenza, M. *Introduzione* a Harold Garfinkel 'La fiducia', (Italian trnslation of H. Garfinkel, 1963), Armando, Roma, 2000.

Searle, J.R. (1995) *The Construction of Social Reality*, London: Allen Lane.

Tummolini, L. and Castelfranchi, C. The cognitive and behavioral mediation of institutions: Towards an account of institutional actions. *Cognitive Systems Research*, 7 (2–3), 2006.

Thomas de Aquino, Summa Theologiae (http://www.corpusthomisticum.org/sth1084.html)

10

On the Trustee's Side: Trust As Relational Capital

In most of the current approaches to trust, the focus of the analysis is on the trustor and on the ways to evaluate the trustworthiness of possible trustees. In fact, there are not many studies and analyses about the model of *being trusted*. But trust can be viewed at the same time as an instrument both *for an agent selecting the right partners in order to achieve its own goals* (the trustor's point of view), and *for an agent to be selected from other potential partners* (the point of view of the trustee) in order to establish a cooperation/collaboration with them and to take advantage of the accumulated trust. In the other chapters of this book we have focused our attention on the first point of view.

In this chapter[1] we will analyze trust as the agents' *relational capital*. Starting from the classical dependence network (in which *needs*, *goals*, *abilities* and *resources* are distributed among the agents) with potential partners, we introduce the analysis of what it means for an agent to be trusted and how this condition could be strategically used by him to achieve his own goals, that is, why it represents a form of power.

The idea of taking the trustee's point of view is especially important if we consider the amount of studies in social science that connect trust with *social capital* related issues. Our socio-cognitive model of trust (see previous chapters) is about the cognitive ingredients for trusting something or somebody, and how trust affects decisions, which are the sources and the basis for trusting, and so on; we do not model what it means to be trusted (with the exception of the work on trust dynamics (Chapter 6) in which the focus is also on the reciprocation and potential influences on the trustworthiness) and why it is important.

Here we address this point, analyzing what it means for trust to represent a strategic resource for agents who are trusted, proposing a model of '*trust as a capital*' for individuals.

---

[1] We thank Francesca Marzo for her precious contribution on the first reflections on this topic.

Our thesis is that *to be trusted*:

i) *Increases* the chance of being requested or accepted as a partner for exchange or cooperation.

ii) *Improves* the 'price', the contract that the trustee can obtain.

The reason for this new point of view derives directly from the fact that in human societies as well as in multi-agent systems it is strategically important not only to know who is trusted by whom and how much, but also to understand how being trusted can be used by several potential trustors. It has already been shown in the previous chapters that using different levels of trust represents an advantage when performing some tasks, such as allocating a task or choosing between partners. Therefore, having 'trust' as a cognitive parameter in agents' decision making can lead to better (more efficient, faster etc.) solutions than proceeding when driven by other kinds of calculation such as probabilistic or statistical ones. This study has already represented an innovation since trust has usually been studied as an effect rather than a factor that causes the development of a social network and its maintenance or structural changes.

In order to improve this approach and to understand dynamics of social networks better, we now propose a study of what happens on the other side of the two-way trust relationship, focusing on the trustee, in particular on a *cognitive trustee*. Our aim is an analytical study of what to be trusted means. In our view:

- To be trustworthy usually is an advantage for the trustee (agent Y); more precisely, received trust is a capital that can be invested, even if it requires choices and costs to be cumulated.
- It is possible to measure this capital, which is relational, that is depends on a position in a network of relationships.
- Trust has different sources: from personal experience that the other agents have had with Y; from circulating reputation of Y; from Y's belongingness to certain groups or categories; from the signs and the impressions that Y is able to produce.
- The value of this capital is context dependent (and market dependent) and dynamic.
- Received trust strongly affects the 'negotiation power' of Y that cannot simply be derived from the 'dependence bilateral relationships'.

Although there is a big interest in literature about 'social capital' and its powerful effects on the well being of both societies and individuals, often it is not clear enough what the object is that's under analysis. Individual trust capital (*relational capital*) and collective trust capital not only should be disentangled, but their relations are quite complicated and even conflicting. To overcome this gap, we propose a study that first attempts to understand what trust is as the competitive capital of individuals. How is it possible to say that 'trust' is a capital? How is this capital built, managed and saved? Then we aim to study the cognitive dynamics of this object analytically, with a particular focus on how they depend on beliefs and goals.

10.1   Trust and Relational Capital

Social capital ((Coleman, 1988), (Bourdieu, 1983), (Putnam, 1993), (Putnam, 2000),) can be seen as a multidimensional concept and can be studied in its relation both to social norms and shared values and to networks of interpersonal relations. While in the former case studies about conventions and collective attribution of meanings it was useful to study how social capital can be a capital for society, in the latter, one of the basic issues that needs to be studied is how it can happen that networks of relations can be built, how they develop, and how they can both influence individual behaviours and be considered as an individual capital.

We also would like to reiterate that social capital is an ambiguous concept. By 'social', a lot of scholars in fact mean 'collective', some richness, an advantage for the collective; something that favors cooperation, and so on. On the contrary, we assume here (as a first step) an individualistic perspective, considering the advantages of the trusted agent (deriving from his relationships with other agents), not the advantages for the collective, and distinguishing between 'relational capital' (Granovetter, 1973) and the more ambiguous and extended notion of 'social capital'. The individual (or organization) Y could use his capital of trust, for non-social or even anti-social purposes.

In economic literature the term 'capital' refers to a commodity itself used in the production of other goods and services: it is, then, seen as a man-made input created to permit increased production in the future. The adjective 'social' is instead used to claim that a particular capital not only exists in social relationships but also consists in some kind of relationship between economical subjects. It is clear that for the capital goods metaphor to be useful, the transformative ability of social relationships to become a capital must be taken seriously. This means that *we need to find out what is the competitive advantage not simply of being part of a network, but more precisely of being trusted in that network.*

In the other chapters in this book, the additional value of trusting is shown to be as a crucial argument in decision making and in particular in the choice of relying on somebody else for achieving specific goals included in the plans of the agents. Trust is analyzed as a valuation of the other and the expectations of him, and it is shown how these characteristics and mechanisms, being part of the decision process at the cognitive level, represent an advantage for society in terms of realizing cooperation among its actors and for the trustor in terms of efficiency of choices of delegation and reliance (Castelfranchi and Falcone, 1998).

Changing the point of view, we now want to focus on the trusted agent (the trustee). However, to account for this it is necessary to rethink the whole theory of negotiation power based on dependence ((Castelfranchi and Conte, 1996), (Sichman *et al.*, 1994), (Castelfranchi *et al.*, 1992), (Conte and Castelfranchi, 1996)).

Trying to build a theory of dependence including trust does not mean basing the theory of social capital on dependence, but to admit that the existing theory of dependence network and the consequent theory of social power is not enough without the consideration of trust. What we need, then, is a comprehensive theory of trust from the point of view of the trusted agent, in order to find out the elements that, once added to the theory of dependence, can explain the *individual social power in a network*, on the one hand, and, only in a second phase, the *social capital meant as a capital for the society.*[2]

---

[2] The advantage for a given community, group, organization or society of a diffuse trust atmosphere, of reciprocal trust attitudes and links in the social network, where (ideally) everybody trusts everybody (see Chapter 6), and trust is not monopolized by a few individuals who take advantage of that to seize negotiation power. This 'collective' meaning is mainly focused on the (quite confused) notion of 'trust capital', see http://www.socialcapitalgateway.org/eng-finland2007.html

Once a quantitative notion of the value of a given agent is formulated by calculating *how much the agent is valued by other agents in a given market for (in realizing) a given task*, we can say that this trust-dependent value is a real capital. It consists of all the relationships that are possible for the agent in a given market and, together with the possible relationships in other markets, it is the so-called *relational capital* of that agent. It differs from simple relationships in given networks, which are a bigger set, since it only consists of relationships the agent has with those who not only need him but have a good attitude toward him and, therefore, who are willing to have him as a partner. How much is he appreciated and requested? How many potential partners depend on Y and would search for Y as a partner? How many partners would be at Y's disposal for proposals of partnership, and what 'negotiation power' would Y have with them?

These relationships form a capital because (as with any other capital) it is the result of investments and it is costly cumulated it.

In a certain sense it represents a strategic tool to be competitive, and, also, as happens with other capitals such as the financial one, it is sometimes even more important that the good which is sold (be it either a service or a material good). For example, when Y decides to not keep a promise to X, he knows that X's trust in Y will decrease: is this convenient for future relationships with X? Will Y need to count on X in the future? Or, is this move convenient for reputation and other relationships?

For this reason it is very important to study how it is possible for the agent to cumulate this capital without deteriorating or wasting it: since the relational capital can make the agent win the competition even when the goods he offers is not the best compared with substitutive goods offered in the market. It should be shown quantitatively what this means and what kind of dynamic relationships exist between quality of offered good and relational capital.

10.2   Cognitive Model of Being Trusted

Before considering trust from this new perspective, let us underline a very important point, which will be useful for this work. The theory of trust and the theory of dependence are not independent from each other. Not only because – as we modelled ((Castelfranchi and Falcone, 1998), (Falcone and Castelfranchi, 2001)) before deciding to actively trust somebody, to rely on him (Y), one (X) has to be dependent on Y: X needs an action or a resource of Y (at least X has to believe so). But also because *objective* dependence relationships (Castelfranchi and Conte, 1996), that are the basis of adaptive social interactions, are not enough for predicting them. *Subjective* dependence is needed (that is, the dependence relationships that the agents know or at least believe), but is not sufficient; it is also necessary to add two relevant beliefs:

(i) the belief of being dependent, of needing the other;
(ii) the belief of the trustworthiness of the other, of the possibility of counting upon him.

If X does not feel dependent on Y, she could not rely on him.

It is important to remind ourselves (see Section 2.3) of a crucial clarification. X is (and feels) dependent on Y even if/when she is able to achieve her goal g, and to perform the (or an) appropriate action. X can trust Y and delegate and rely on him even when she has the alternative of 'doing it myself'. This is what one might call 'weak dependence': I would be

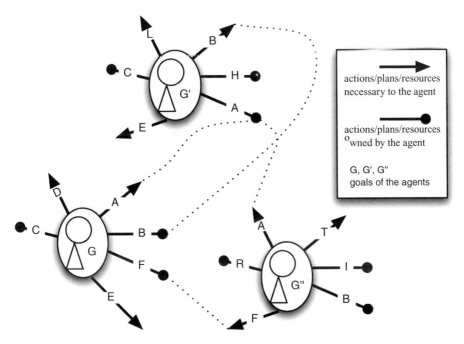

Figure 10.1   Objective dependence network

able to do the needed action myself. However, as we have explained 'weak dependence' is still 'dependence'. In fact, if X prefers and chooses to delegate to Y, this *necessarily* means that X sees some advantage, some convenience; that is: she not only achieves p – through Y's action – but also achieves some additional goal (say q; for example, a better quality, less costs or effort, etc.). Thus, necessarily the reliance is about realizing $p+q$, and relatively to this real global goal of X and of the delegated action, X is just and fully dependent on Y.

10.2.1   Objective and Subjective Dependence

The theory of dependence includes in fact two types of dependence:

(1) The *objective dependence*, which says who needs whom for what in a given society (although perhaps it also ignores this). This dependence already has the power to establish certain asymmetric relationships in a potential market, and it determines the actual success or failure of the reliance and transaction (see Figure 10.1).

(2) The *subjective (believed) dependence*, which says who is believed to be needed by who. This dependence is what determines relationships in a real market and settles on the negotiation power; but it might be illusory and wrong, and one might rely upon unsuitable agents, even if one could actually do the action oneself. For example, in Figures 10.2A and 10.2B the dependence relationships as believed by X and Y respectively are shown: they are different from the objective dependence shown in Figure 10.1, but in fact it is on these beliefs that the agents make decisions.

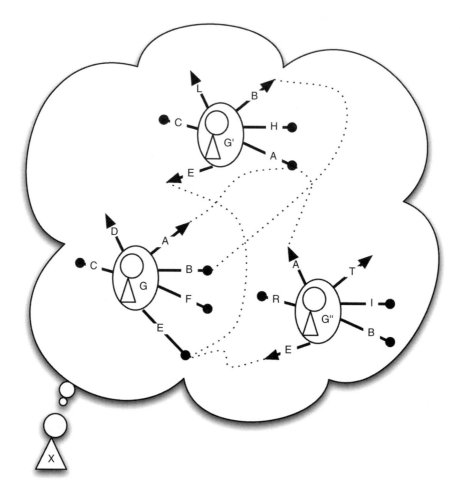

Figure 10.2A   Subjective dependence network (believed by X)

More formally, let $Agt = \{Ag_1,.., Ag_n\}$ a set of *agents*; we can associate to each agent $Ag_i \in Agt$:

- a set of *goals* $G_i = \{g_{i1},..,g_{iq}\}$;
- a set of *actions* $Az_i = \{\alpha_{i1},..., \alpha_{iz}\}$; these are the elementary actions that Ag_i is able to perform;
- a set of *plans* $\Pi = \{p_{i1},..,p_{is}\}$; Ag_i's plan library: the set of rules/prescriptions for aggregating the actions; and
- a set of *resources* $R_i = \{r_{i1},..,r_{im}\}$.

Each goal needs a set of actions/plans/resources.

Then, we can define the *dependence relationship* between two agents (Ag_j and Ag_i) with respect to a goal g_{jk}, as *Obj-Dependence* (Ag_j, Ag_i, g_{jk}) and say that: *An agent Ag_j has an Objective Dependence Relationship with agent Ag_i with respect to a goal g_{jk} if there are*

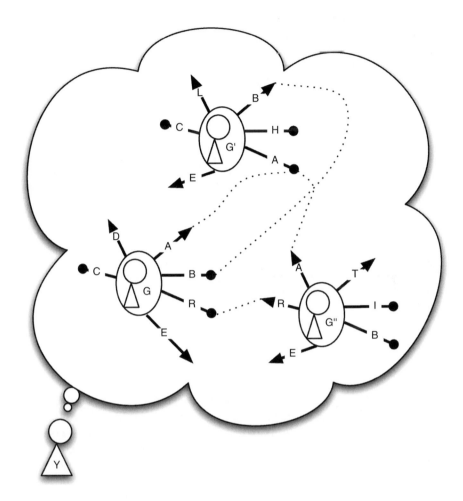

Figure 10.2B   Subjective dependence network (believed by Y)

necessary actions, plans and/or resources that are owned by Ag_i and not owned by Ag_j.in order to achieve g_{jk}.

In general, Ag_j has an Objective Dependence Relationship with Ag_i if in order to achieve at least one of its goals $g_{jk} \in G_j$, there are necessary actions, plans and/or resources that are owned by Ag_i and not owned by Ag_j (or, that is the same, they are owned by Ag_j but not usable by it for several reasons).

Ag_j has not got the 'power of' achieving g_{jk}, while Ag_i has this 'power of'.

As in (Castelfranchi *et al.*, 1992) we can introduce the *unilateral, reciprocal, mutual* and *indirect* dependence (see Figure 10.3). In very short and simplified terms, we can say that the difference between reciprocal and mutual is that the first is on different goals while the second is on the same goal.

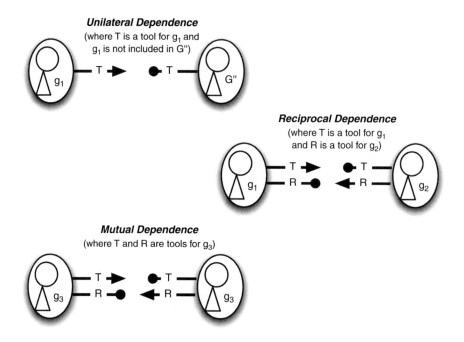

Figure 10.3   Unilateral, Reciprocal and Mutual Dependence

If the world knowledge were perfect for all the agents, the above described objective dependence would be a common belief about the real state of the world (Figure 10.1). In fact, the important relationship is the network of dependence *believed by each agent* (see Figures 10.2A and 10.2B). In other words, we cannot only *associate* a set of goals, actions, plans and resources with each agent, but we have to evaluate these sets as believed by each agent (the subjective point of view) and also take into consideration that they would be partial, different from each other, sometimes wrong, and so on. In more practical terms, each agent will have a different (subjective) representation of the dependence network as exemplified in Figures 10.1, 10.2A, and 10.2B.

For this reason we introduce the formula $Bel_k G_z$ that represents Ag_z's goal set as believed by Ag_k. The same for $Bel_k Az_z$, $Bel_k \Pi_z$, and $Bel_k R_z$, respectively, for actions, plans and resources. In practice, the dependence relationships should be re-modulated on the basis of the agents' subjective interpretation. The really operative part of the resulting interactions among the agents is due to their beliefs about the reciprocal dependences rather than the objective dependences; although, the final results of those interactions are also due to the 'objective' powers and dependence relations, even if ignored by the agents.

We call *Subj-Dependence(Ag_j, Ag_i, g_{jk})* when representing Ag_j's point of view with respect its dependence relationships with Ag_i about its k-th goal g_{jk}. Analogously, we call *Obj-Dependence(Ag_j, Ag_i, g_{jk})* for representing the objective dependence relationship of Ag_j with Ag_i about its k-th goal g_{jk}. In the first place, each agent should correctly believe what is true of their own goals, actions, plans, resources; while they could mismatch the sets of other agents.

We define *Dependence-Network(Agt,t)* the set of dependence relationships (both subjective and objective) among the agents included in *Agt* set at the time *t*. Each agent $Ag_j \in Agt$ must have at least one dependence relation with another agent in *Agt*.

More formally, a dependence network of a set of agents *Agt* at the time *t* can be written:

$$Dependence\text{-}Network(Agt, t) = Obj\text{-}Dependence(Ag_j, Ag_i, g_{jk}) \qquad (10.1)$$
$$\cup \; Subj\text{-}Dependence(Ag_j, Ag_i \; g_{jk})$$

with $Ag_j, Ag_j \in Agt$.

10.2.2   Dependence and Negotiation Power

Given a *Dependence-Network(Agt,t)*, we define

Objective Potential for Negotiation of $Ag_j \in Agt$ about its own goal g_{jk} – and call it *OPN(Ag_j, g_{jk})* – the following function:

$$OPN(Ag_j, g_{jk}) = f \left(\sum_{i=1}^{l} \frac{1}{1 + p_{ki}} \right) \qquad (10.2)$$

Where:

- f is in general a function that preserves monotonicity (we will omit this kind of function in the next formulas);
- l represents the number of agents in the set *Agt* that have an objective dependence relation with Ag_j with respect to g_{jk} (this dependence relation should be either reciprocal or mutual: in other words, there should also be an action, plan, or resource owned by Ag_j that is necessary for the satisfaction of any of Ag_i's goals);
- p_{ki} is the number of agents in *Agt* that are objectively requiring (there is an analogous dependence relation) the same actions/plans/resources (as useful for g_{jk}) to Ag_i on which is based the dependence relation between Ag_j and Ag_i and that in consequence are competitors with Ag_j actions/plans/resources in an incompatible way (Ag_i is not able to satisfy all the agents at the same time: there is a saturation effect). See Figure 10.4 for an example.

So, in case there are no competitors with Ag_j ($p_{ki}=0$ for each $i \in \{1, \ldots, l\}$) we have:

$$OPN(Ag_j, g_{jk}) = f \left(\sum_{i=1}^{l} \frac{1}{1 + p_{ki}} \right) = l \qquad (10.3)$$

More precisely, this *Objective Potential for Negotiation* should be normalized and evaluated with respect to each of the potential required tasks (actions, plans, resources) for the goal in object (g_{jk}): in fact, the achievement of this goal could require different performances of the dependent agents (see, for example, Figure 10.5: Ag_j needs *A*, *B* and *C* to achieve its goal g_{jk}). In the dependence network there are three agents. Ag_1 can offer *A* and *B* and can exploit *N* by Ag_j; Ag_2 can offer *A* and *C* and can exploit *L* by Ag_j; Ag_3 can offer *B* and can exploit *N* by Ag_j. Finally, Ag_2 is concurrent with Ag_j on *B* with both Ag_1 and Ag_3 (see also Figure 10.4).

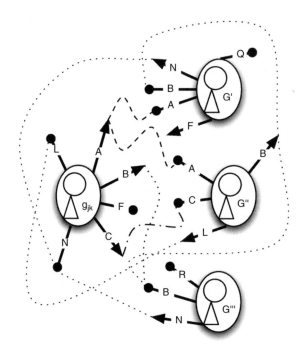

Figure 10.4   Ag$_j$ Ag$_2$ and Ag$_3$ as competitors on a resource owned by Ag$_1$

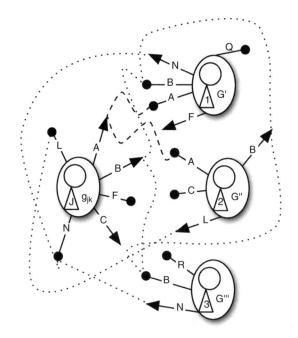

Figure 10.5   Example of *Objective Potential for Negotiation*

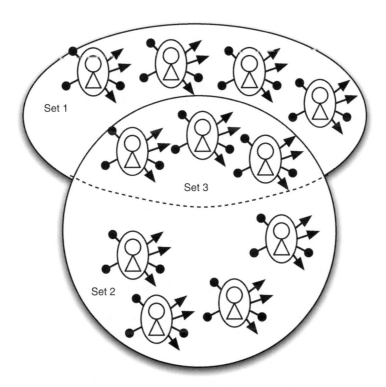

Figure 10.6 Matching the agents who depend on Ag_j for something and those on which Ag_j depends for its goal g

In general, we can represent the objective dependence of Ag_j as shown in Figure 10.6: *set1* represents the set of agents who depend on Ag_j for something (actions, plans, resources), *set2* represents the set of agents on which Ag_j depends for achieving their own specific goal g_{jk}. The intersection between *set1* and *set2* (part *set3*) is the set of agents with whom Ag_j could potentially negotiate for achieving g_{jk}. The greater the overlap the greater the *negotiation power* of Ag_j in that context.[3]

However, the negotiation power of Ag_j also depends on the possible alternatives that its potential partners have: the fewer alternatives to Ag_j they have, the greater its negotiation power (see Figure 10.4). We can define the *Subjective Potential for Negotiation* of $Ag_j \in Agt$ about its own goal g_{jk} – and call it $SPN(Ag_j, g_{jk})$ – the following function:

$$SPN(Ag_j, g_{jk}) = \sum_{i=1}^{I^{Bj}} \frac{1}{1 + p_{ki}^{Bj}} \qquad (10.4)$$

---

[3] Even if increasing the number of agents in the overlap doesn't necessarily increase the probability of achieving Ag_j's goal (maybe one (or more) of the needed resources is not owned by an increasing number of agents).

where the apex B_j means 'believed by Ag_j'; in fact in this new formula Ag_j both *believes* the number of potential collaborative agents (l) and the number of competitors (p_{ki}) for each of them.

It is clear how, on the basis of these parameters (l^{Bj} and $p_{ki}{}^{Bj}$), the negotiation power of Ag_j is determined. And, at the same time, his own decisions will be strongly influenced. Analogously, we can interpret Figure 10.5 as the set of relationships among the agents, believed by Ag_j. In this case we take the subjective point of view.

10.2.3 Trust Role in Dependence Networks

We would like to introduce into the dependence network the trust relationship. In fact, the dependence network alone is not sufficient for a real allocation of tasks among the agents. It is true that Ag_i should be able and willing to realize the action α_k: But how? And, will it be sufficient given my expectations? Would it be more or less trustworthy than Ag_z? To answer these questions the agents in the dependence network have to establish among themselves the reciprocal trust about the different tasks they can allocate to each other.

Indeed, *although it is important to consider the dependence relationship between agents in society, there will be not an exchange in the market if there is not the trust to strengthen these connections.* Considering the analogy with Figure 10.4, we will now look at a representation as given in Figure 10.7 (where *Set 4* includes the set of agents that Ag_j considers trustworthy for achieving g_{jk}).

We have now a new subset (the dark agents in Figure 10.7) containing the potential agents for negotiation. By introducing the basic beliefs about trust in the *Subjective Potential for Negotiation* (of $Ag_j \in Agt$ and its own goal g_{jk}) we also introduce the superscript index T to differentiate it from the *SPN* without trust and we have:

$$SPN^T(Ag_j, g_{jk}) = \sum_{i=1}^{l^{Bj}} \frac{DoA_{ik}{}^{Bj} * DoW_{ik}{}^{Bj}}{1 + p_{ki}{}^{Bj}}) \tag{10.5}$$

with $1 \geq DoA_{ik}{}^{Bj}, DoW_{ik}{}^{Bj} \geq 0$.

where $DoA_{ik}{}^{Bj}$ and $DoW_{ik}{}^{Bj}$ are, respectively, the degree of ability and willingness (with respect to the goal g_{jk}) of the agent Ag_i as believed by Ag_j (see Chapter 3). We do not consider here the potential relations between the values of $DoA_{ik}{}^{Bj}$ and $DoW_{ik}{}^{Bj}$ with the variable $p_{ki}{}^{Bj}$.

On analyzing Figure 10.7, we can see that there are two other agents (medium dark) that are trustworthy according to Ag_j on the goal g_{jk} but they do not depend on Ag_j for something. In fact the dependence and trust relationships are strongly interwined and not simply sequential as shown above. Not only does the decision to trust presuppose a belief of being dependent, but notice that a dependence belief (*BelDep*) implies on the other hand an element of trust. In fact to believe oneself to be dependent means:

- (*BelDep-1*) to believe not to be able to perform action α and to achieve goal g; and
- (*BelDep-2*) to believe that Ag_i is *able* and in condition to achieve g, performing α.

Notice that (*BelDep-2*) is precisely one component of trust in our analysis: the *positive evaluation* of Ag_i as competent, able, skilled, and so on. However, the other fundamental

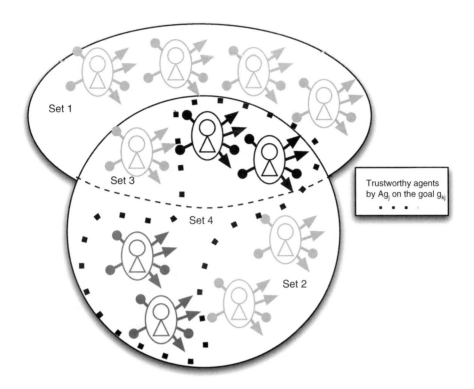

Figure 10.7 Subset of Agents selected by introducing also the trustworthiness of the agents (in Ag_j's point of view) in the dependence network

component of trust as evaluation is lacking: reliability, trustworthiness: Ag_i really intends to do, is persistent, is loyal, is benevolent, etc. Thus he will really do what Ag_j needs.

Given the basic role played by 'believed networks of dependence', established by a believed relationship of dependence based on a belief of dependence, and given that this latter is one of the basic ingredients of trust as a mental object, we can claim that this overlap between theories is the crucial issue and our aim is namely to study it in great depth.

Analogously, but less relevant in this case, we can introduce the *Objective Potential for Negotiation* (of $Ag_j{\in}Agt$ about its own goal g_{jk}), we have:

$$OPN^T(Ag_j, g_{jk}) = \sum_{i=1}^{l} \frac{DoA_{ik} * DoW_{ik}}{1 + p_{ki}} \qquad (10.6)$$

where DoA_{ik} and DoW_{ik} respectively represent *objective Ag_i's* ability and willingness to use actions/plans/resources for the goal g_{jk}.

When a cognitive agent trusts another cognitive agent, we talk about social trust. We consider here the set of actions, plans and resources owned/available by an agent that can be useful in achieving a set of tasks ($\tau_1, \ldots, \tau_r$).

We *now take the point of view of the trustee in the dependence network*: so we present a cognitive theory of trust as a capital. That is to say that if somebody is (potentially) strongly useful to other agents, but is not trusted, its negotiation power is not high.

As showed in Chapter 3 we call *Degree of Trust* of the agent Ag_j for the agent Ag_i about the task τ_k $(DoT(Ag_j Ag_i \tau_k))$:

$$DoT(Ag_j, Ag_i, \tau_k)^{Bj} = DoA_{ik}^{Bj} * DoW_{ik}^{Bj} \qquad (10.7)$$

In the same way we can also define the *self-trust* of the agent Ag_i about the task τ_k:

$$ST(Ag_i, \tau_k) = DoA_{ik}^{Bi} * DoW_{ik}^{Bi} \qquad (10.8)$$

We call the *Objective Trust Capital* of $Ag_i \in Agt$ about a potential delegable task τ_k the function:

$$OTC(Ag_i, \tau_k) = \sum_{j=1}^{l} DoA_{ik}^{Bj} * DoW_{ik}^{Bj} = \sum_{j=1}^{l} DoT(Ag_j, Ag_i, \tau_k)^{Bj} \qquad (10.9)$$

Where l is the number of agents (included in the dependence network) who need to delegate the task τ_k. Note that we are calling as objective trust capital the sum of the trustworthiness that the other agents in the *DN* attribute to Ag_i rather than the capital Ag_i could deserve on the basis of his own objective relationships: in other words, *it is referred to the partial (subjective) points of view of the other agents*.

In words, the cumulated trust capital of an agent Ag_i with respect to a specific delegable task τ_k, is the sum (all the agents need that specific task in the network dependence) of the corresponding abilities and willingness believed by each potentially dependent agent.[4]

We call the *Subjective Trust Capital* of $Ag_i \in Agt$ for a potential delegable task τ_k the function:

$$STC(Ag_i, \tau_k) = \sum_{j=1}^{l^{Bi}} DoA_{ik}^{Bi Bj} * DoW_{ik}{}^{Bi Bj} = \sum_{j=1}^{l^{Bi}} DoT(Ag_j, Ag_i, \tau_k)^{Bi Bj} \qquad (10.10)$$

Where the apex $B_i B_j$ means 'as Ag_i believes is believed by Ag_j'. Subjectivity means that both the network dependence and the believed abilities and willingness are believed by (the point of view of) the agent Ag_i. The subjectivity consists in the fact that both the network dependence and the believed abilities and willingness are believed by (the point of view of) the agent Ag_i (see Figure 10.8A and 10.8B).

Starting from Trust Capital we would like to evaluate its *usable part*. In this sense, we introduce the *Subjective Usable Trust Capital* of $Ag_i \in Agt$ for a potential delegable task τ_k as:

$$SUTC(Ag_i, \tau_k) = \sum_{j=1}^{l^{Bi}} \frac{DoT(Ag_j, Ag_i, \tau_k)^{Bi Bj}}{1 + p_{kj}{}^{Bi}} \qquad (10.11)$$

Where the apex $B_i B_j$ means 'what Ag_i believes is believed by Ag_j' and where $p_{kj}{}^{Bi}$ is (following Ag_i's belief about the beliefs of Ag_j) the number of other agents in the dependence network

---

[4] We might consider in an even more objective way the capital based on the real trustworthiness of Ag_i (with respect to the task) rather than based on the $DoT(Ag_j Ag_i \tau_k)$ of the various agents Ag_j. But the trust capital is not interesting if it is not in the mind of the potential 'users'.

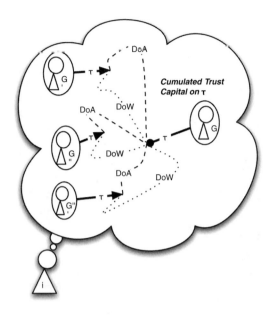

Figure 10.8A   *Ag_I* believes the specific dependence network in which he has a cumulated Trust Capital (about a specific task)

Figure 10.8B   *Ag_I* believes how other agents evaluate his own abilities and willingness about a specific task

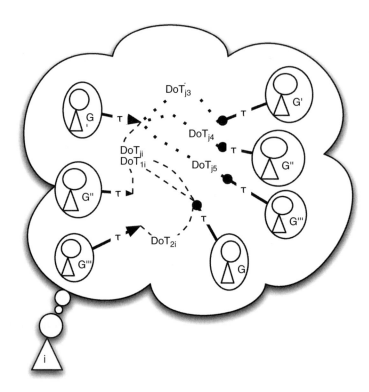

Figure 10.9 Example of positive (Ag_j, Ag_1, Ag_2) and negative (Ag_3, Ag_4, Ag_5) contributors to Ag_i's capital of trust (in Ag_i's mind)

that can realize and achieve the same task (with a trust value comparable with the one of Ag_i) to whom Ag_j can delegate the task τ_k (see Figure 10.9).

We say that there are two *comparable trust values* when the difference between them is in a range under a given threshold that could be considered meaningless with respect to the achievement of the task.

In Figure 10.9, Ag_1 and Ag_2 strengthen the trust capital of Ag_i (they are competitors with Ag_j about the task τ); while Ag_3, Ag_4 and Ag_5 weaken the trust capital of Ag_i because they are competitors with Ag_i in offering (at the same trustworthy value) the task τ. As shown in Figure 10.9, it is possible that Ag_i believes in potential competitors (jeopardizing his trust capital), but they are not really competitors because there are no links with his potential clients/delegating (see Ag_3, Ag_4 and Ag_5 that are not linked with Ag_1 and Ag_2 but only with Ag_j).

Of course, we can analogously introduce the *Objective Usable Trust Capital* of $Ag_i \in Agt$ about a potential delegable task τ_k as:

$$OUTC(Ag_i, \tau_k) = \sum_{j=1}^{l} \frac{DoT(Ag_j, Ag_i, \tau_k)}{1 + p_{kj}} \qquad (10.12)$$

In this paragraph we have introduced in the dependence network (that establishes, objectively or subjectively, how each agent can potentially depend on other agents to solve its own tasks) the trust relationships (that introduce an additional dimension, again evaluated both objectively and subjectively, in a potential partner selection for achieving tasks). In general, we can say that the introduction of trust relationships reduces the set of potential partners for each agent and for each task, with respect to the situation with the dependence relationships alone: more formally $OPN>OPN^T$, and $SPN>SPN^T$. Even if exceptions are possible: may be an agent trusts other agents on a specific task with respect to whom he really depends on.

From the comparison between $OUTC(Ag_i, \tau_k)$, $SUTC(Ag_i, \tau_k)$, $OTC(Ag_i, \tau_k)$, $STC(Ag_i, \tau_k)$, and $ST(Ag_i, \tau_k)$ a set of interesting actions and decisions are taken from the agents (we will see in the next paragraph).

10.3 Dynamics of Relational Capital

What has not been considered enough in organization theory is the fact that the *relational capital* is peculiar in its being crucially based on beliefs: again, what makes relationships become a capital is not simply the structure of the networks (who 'sees' whom and how clearly) but the evaluations and the levels of trust which characterize the links in the networks (who trusts whom and how much). Since trust is based on beliefs – including, as we said, the believed dependence (who needs whom) – it should be clear that *relational capital is a form of capital, which can be manipulated by manipulating beliefs.*

Thanks to a structural theory of what kind of beliefs are involved, it is possible not only to answer some very important questions about agents' power in networks, but also to understand the dynamical aspects of relational capital. In addition, it is possible to study what a difference between trustee's beliefs and others' expectations of him implies in terms of both reactive and strategic actions performed by the trustee.

10.3.1 Increasing, Decreasing and Transferring

As far as the dynamic aspects of this kind of capital are concerned, it is possible to make hypotheses on how it can increase or how it can be wasted, depending on how each of the basic beliefs involved in trust might be manipulated. In general, starting from the analysis of the previous paragraph, we can see how matching the different terms we have different interesting situations.

First of all, even if $OTC(Ag_i, \tau_k)$ is a relevant factor for the agent Ag_i (it shows in absolute terms how the trustworthiness of Ag_i is recognized), in fact the really important thing for an agent cumulating trust capital is $OUTC(Ag_i, \tau_k)$ that indicates not only the trustworthiness cumulated in the dependent agents, but also the number of possible other concurrent agents on that offered task. So, for example, it may be more important to have competence on tasks which are not highly required, but with a low number of concurrents, than viceversa.

Again it is interesting to consider the $SUTC(Ag_i, \tau_k)$ factor (in which a relevant role is played by the beliefs of the involved trustee) and its relationships with $OUTC(Ag_i, \tau_k)$, $SPN^T(Ag_j, g_{jk})$, and $OPN^T(Ag_j, g_{jk})$ factors. As we have seen in the previous paragraph, these factors are constituted by the beliefs of trustee or trustor, so it can be interesting to analyze the different situations matching them and evaluating the consequences of their coherence or incoherence.

A general rule (that could be easily translated into an algorithm) regards the fact that the trust capital of an agent (say Ag_i) increases when:

- the number of other trusted agents (competitors) in the DN offering the solution to the given task (or classes of tasks) decreases; and/or
- the number of agents (delegators/clients) in the DN requiring the solution to the given task (or classes of tasks) increases.

Following this analysis, the trustee should work to decrease the number of competitors (for example, disconnecting the links in the network, reducing their reputation, and so on) and/or he should work to increase the delegators (for example, connecting new ones, changing the needs of the connected ones, and so on).

Let us consider what kind of strategies can be performed to enforce the other's dependence beliefs and his beliefs about *agent's competence*. If Ag_i is the potential trustee (the collector of the trust capital) and Ag_j is the potential trustor we can say:

i) Ag_i can make Ag_j dependent on him by making Ag_j lack some resource or skill (or at least inducing Ag_j to *believe* so). He has to work on $SPN^T(Ag_j, g_{jk})$.

ii) Ag_i can make Ag_j dependent on him by activating or inducing in them a given goal (need, desire) in which Ag_j is not autonomous (Castelfranchi and Falcone, 2003) but is dependent on Ag_i (or in any case they believe so). In this case they have to find a way to include in G_j an additional g_{jk} such that Ag_j is dependent on Ag_i for that goal (and they believe that).

iii) Since dependence beliefs are strictly related to the possibility of others (for example Ag_j) being able to see the agent (for example Ag_i) in the network and to know their ability to perform useful tasks, the goal of the agent who wants to improve their own relational capital will be to *signal* their presence and their skills ((Schelling, 1960), (Spece, 1973), (Bird and Smith, 2005)). While to show his presence he might have to shift his position (either physically or figuratively, for instance, by changing his field), to communicate his skills he might have to hold and show something that can be used as a signal (such as an exhibition, certificate, social status, proved experience, and so on). This implies, in the plan of actions of the trustee, several necessary sub-goals to provide a signal. These sub-goals are costly to achieve and the cost the agent has to pay to achieve them has to prove the signals to be credible (of course without considering cheating by building signals). It is important to underline that using these signals often implies the participation of a third party in the process of building trust as a capital: a third party which must be trusted (Falcone and Castelfranchi, 2001). We would say the more the third part is trusted in society, the more expensive will it be for the agent to acquire signals to show, and the more successful these signals will be at increasing the agent's relational capital. Later we will see how this is related to the process of transferring trust from one agent to another (building reputation). Obviously Ag_i's *previous performances* are also 'signals' of trustworthiness. And this information is also provided by the circulating *reputation* of Ag_i ((Conte and Paolucci, 2002), (Jøsang and Ismail, 2002)).

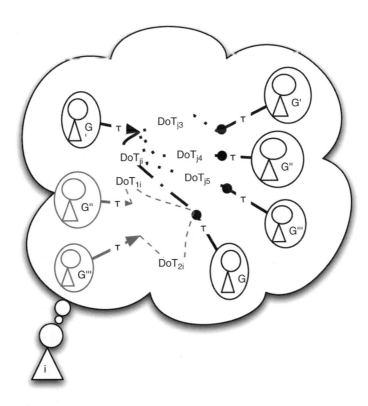

Figure 10.10   Example in which Ag_i might increase his own trust capital reducing in some way Ag_j's trust in his competitors (Ag_3, Ag_4, Ag_5)

iv) Alternatively, Ag_i could work to reduce the believed (by Ag_j) value of the ability of each of the possible competitors of Ag_i (in a number of p_{kj}) on that specific taskτ_k, See Figure 10.10: he has to work $SPN^T(Ag_j, g_{jk})$.

Let us now consider how *willingness beliefs* can be manipulated. In order to do so, consider the particular strategy that needs to be performed to gain the other's good attitude through gifts (Cialdini, 1990). It is true that the expected reaction will be of reciprocation, but this is not enough. While giving a gift Ag_i knows that Ag_j will be more inclined to reciprocate, but Ag_i also knows that his action can be interpreted as a sign of the good will he has: since he has given something without being asked, Ag_j is driven to believe that Ag_i will not cheat on her. Then, the real strategy can be played on trust, sometimes totally and sometimes only partially – this will basically depend on the specific roles of agents involved.

Again in formal terms, we can say that Ag_i has to work to increase his DoW_i as believed by Ag_j $(Bel_j(DoA_i))$.

Alternatively, it could work to reduce the believed (by Ag_j) value of willingness of each of the possible competitors of Ag_i (in number of p_{kj}) on that specific taskτ_k, See again Figure 10.10.

An important consideration we have to take is that a dependence network is mainly based on the set of actions, plans and resources owned by the agents and necessary for achieving the agents' goals (we considered a set of tasks each agent is able to achieve; its 'power of'). The interesting thing is that the dependence network is modified by the dynamics of the agents' goals: from their variations (as they evolve in time), from the emergency of new ones, from the disappearance of old ones, from the increasing request of a subset of them, and so on (Pollack, 1990). On this basis, the role of each agent in the dependence network changes, which in fact changes the trust capital of the involved agents.

Relational capital can also be circulated inside a given society. If somebody has a good reputation and is trusted by somebody else, they can be sure this reputation will be passed on and transfered to other agents – and this is always considered a good marketing strategy, word of mouth. What is not clear yet is how these phenomena work. But when trust in an agent circulates, it is strategically important for the agent to know how this happens and which paths (not only figuratively) trust follows.

In fact, not all the ways are the same: it is possible that being trusted by a particular agent could mean that he just has one more agent in his relational capital, but gaining the trust of another agent can be very useful to him and exponentially increase his capital thanks to the strategic role or position of this other agent. That said, the importance of understanding if and how much an agent is able to manage this potentiality of his capital should be clear.

Basically, here also, a crucial part is played by the involved agents: for this reason it is necessary for an agent to know *the multiplicative factors* represented by the recognized and trusted evaluator in society. It is not necessarily true, in fact, that when somebody trusts somebody else and they in turn trust a third one, the first one will trust the third one: the crucial question is 'which role does the first ascribe to the second'. If the second one is trusted as an evaluator by X, then X can trust the third one to achieve specific goals (see Chapter 6 for the analysis of the *trust transitivity*). Usually how well these transitive processes work depends on what kind of broadcasting and how many links the evaluator has and how much he is trusted in each of those links, so, basically, it depends on the evaluator's relational capital.

10.3.2 Strategic Behavior of the Trustee

Up until now we have just considered trust as something quantitatively changeable, but we did not talk about subjective difference in the way trust is perceived by the two parts of the relationship. Nevertheless, to be realistic, we must take into account the fact that there is often a difference between how the others actually trust an agent ($OTC(Ag_i, \tau_k)$) and what the agent believes about ($STC(Ag_i, \tau_k)$); but also between this and the level of trustworthiness that an agent perceives in themsel (we can refer to the $ST(Ag_i, \tau_k)$ factor for this). Since being able is not necessarily the reason for trust: it can be a diffuse atmosphere that makes the others trust the agent, although the agent doesn't possess all the characteristics required to be trusted.

In fact, these subjective aspects of trust are fundamental to the process of managing this capital, since it can be possible that the capital is there but the agent does not know how to reach it. Can it be possible to use the relational capital even if the person who uses it is not aware of having it?

At the basis of the possible discrepancy in the subjective assessment of trustworthiness there is the perception of how much an agent feels themselves to be trustworthy in a given task and the assessment that they do of how much the other agents trust them in the same task.

In addition, this perception can change and become closer to the objective level while the task is performed: the agent can either find out that they are being more or less trustworthy than they had believed, or realize that the others' perception was wrong (either positively or negatively). All these factors must be taken into account and studied together with the different components of trust, in order to build hypotheses on strategic actions that the agent can perform to cope with his relational capital.

We must consider what can be implied by these discrepancies in terms of strategic actions: how can they be individuated and valued? How will the trusted agent react when he becomes aware of that? He can either try to acquire competences in order to reduce the gap between others' valuation and his own, or exploit the existence of this discrepancy, taking economic advantage of the reputation over his capability and counting on the others' scarce ability to monitor and test his real skills.

10.4   From Trust Relational Capital to Reputational Capital

However, there is another 'evolutionary' step in this path from dependence and interpersonal trust relationships, to a personal, competitive 'relational capital' and the consequent 'negotiation power' and role in the 'market'. The 'relational capital' of the individual is not just the sum of the evaluations of the other members, and a simple interpersonal complex relation. This is just the basic, inter-personal layer. But the agents communicate about the features, the reliability, the trustworthiness of the others; and they not only communicate their own opinion, but they also report and spread around – without personal commitment – what they have heard about Y: Y's reputation

On such a basis a complex phenomenon emerges: Y's circulating reputation in that community; which is represented in nobody's mind in particular. However, this circulating phantom determines the individual perception of Y's reputation, then his trustworthiness in that community. In other words, beyond Y's ' trust-relational capital' there is an additional, emerging 'capital': Y's reputation. This capital in many contexts – in particular in open, anonymous 'markets', where individuals do not know each other – is the really fundamental one to determine Y's value in that market and his negotiation power.

This view of 'reputation' (rather close to Conte and Paolucci's theory-(Conte and Paolucci, 2002)) gives the right role to this important phenomenon and is less reductive than the view we have used before. We have (correctly, but in a rather reductive way) presented 'reputation' as one of the possible bases and sources of our evaluations and trust in Y. Apart from *personal experience* of X about Y, apart from *reasoning and instantiation from general categories*, *roles*, etc., apart from *various forms of 'transfer'* of trust, *reputation* (that is the circulating fame, voice, gossip about Y in a given community) can be the base of X 's opinion and trust in Y.

However, 'reputation' is not the single-agent 'opinion' about Y, or the communicated personal evaluation of Y: it is an emerging, anonymous phenomenon – to which nobody responds – which is self-organizing over the various implicit or explicit *messages* 'about' Y's virtues, competence and trustworthiness.

So, the relationship between trust and reputation is more dialectic: trust of community members in Y indirectly contributes to Y's reputation; and reputation contributes to their individual and diffuse trust in Y. Moreover, reputation is not just a mental object, a piece of information we use for evaluating Y, but is an emergent sociological phenomenon, beyond the individual mind.

10.5 Conclusions

As we said, individual trust capital (relational capital) and collective trust capital should not only be disentangled, but their relations are quite complicated and even conflicting. In fact, since the individual is in competition with the other individuals, he is in a better position when trust is not uniformly distributed (everybody trusts everybody), but when he enjoys some form of concentration of trust (an oligopoly position in the trust network); while the collective social capital could do better with a generalized trust among the members of the collectivity. Agents compete and invest to cumulate their individual 'trust capital' (or 'relational capital'), even by showing their superiority and the low trustworthiness and ability of the competitors, or even by propagating false information about the others and a bad reputation (Conte and Paolucci, 2002).

References

Bird, R. B. and Smith, E. Alden (2005) Signaling theory, strategic interaction, and symbolic capital, *Current Anthropology*, 46 (2), April.

Bourdieu, P. (1983) Forms of capital. In: Richards, J. C. ed. *Handbook of Theory and Research for the Sociology of Education*, New York, Greenwood Press.

Castelfranchi, C. and Falcone, R. (1998) Principles of trust for MAS: cognitive anatomy, social importance, and quantification, *Proceedings of the International Conference of Multi-Agent Systems (ICMAS'98)* , pp. 72–79, Paris, July.

Castelfranchi, C. and Falcone, R. (2003) From automaticity to autonomy: the frontier of artificial agents, in Hexmoor, H, Castelfranchi, C., and Falcone, R. (eds.), *Agent Autonomy*, Kluwer Publisher, pp. 103–136.

Castelfranchi, C., and Conte, R. (1996) The Dynamics of Dependence Networks and Power Relations in Open Multi-Agent Systems. In Proc. COOP'96 – Second International Conference on the Design of Cooperative Systems, Juan-les-Pins, France, June, 12–14. INRIA Sophia-Antipolis. pp.125–137.

Castelfranchi, C., Miceli, M. and Cesta, A. (1992) Dependence relations among autonomous agents. In E. Werner, Y. Demazeau (eds.), *Decentralized A. I.*, 3: 215–227, North Holland, Amsterdam.

Cialdini, R. B. (1990) *Influence et manipulation*, Paris, First.

Coleman, J. C. Social capital in the creation of human capital. *American Journal of Sociology* 94: S95–S120, 1988.

Conte, R. and Castelfranchi, C. (1996) Simulating multi-agent interdependencies. A two-way approach to the micro-macro link. In U. Mueller and K. Troitzsch (eds.) *Microsimulation and the Social Science*. Berlin, Springer Verlag, Lecture Notes in Economics.

Conte, R. and Paolucci, M. (2002) *Reputation in Artificial Societies. Social Beliefs for Social Order*. Kluwer.

Falcone, R. and Castelfranchi, C. (2001) Social Trust: A Cognitive Approach, in *Trust and Deception in Virtual Societies*, Castelfranchi, C. and Yao-Hua, Tan (eds.), Kluwer Academic Publishers, pp. 55–90.

Granovetter, M. The strength of weak ties. *American Journal of Sociology*, 78: 1360–1380, 1973.

Jøsang, A. and Ismail, R. (2002) *The Beta Reputation System*. In the proceedings of the 15th Bled Conference on Electronic Commerce, Bled, Slovenia, 17–19 June 2002.

Pollack, M. (1990) Plans as complex mental attitudes in Cohen, P. R., Morgan, J. and Pollack, M. E. (eds.), *Intentions in Communication*, MIT Press, USA, pp. 77–103.

Putnam, R. D. (1993) *Making Democracy Work. Civic traditions in modern Italy.* Princeton NJ, Princeton University Press.

Putnam, R. D. (2000) *Bowling Alone. The collapse and revival of American community,* New York, Simon and Schuster.

Schelling, T. (1960) *The Strategy of Conflict.* Cambridge, Harvard University Press.

Sichman, J, R., Conte, C., Castelfranchi, Y. Demazeau (1994) A social reasoning mechanism based on dependence networks. In *Proceedings of the 11th ECAI,*

Spece, M. Job market signaling. *Quarterly Journal of Economics,* 87: 296–332, 1973.

11

A Fuzzy Implementation for the Socio-Cognitive Approach to Trust

In this chapter[1] we will show a possible implementation of the socio-cognitive model of trust developed in the other chapters of the book. This implementation (Falcone *et al.*, 2005) uses a fuzzy approach (in particular, it uses the so-called Fuzzy Cognitive Maps –FCM (Kosko, 1986). In particular our attempt is to show, using a specific implementation, how relevant a trust model is based on beliefs and their credibility.

As previously described, our model introduced a degree of trust instead of a simple probability factor since it permits trustfulness to be evaluated in a rational way: Trust can be said to consist of, or even better (either implicitly or explicitly) imply, the *subjective probability* (in the sense of a subjective evaluation and perception of the risks and opportunities) of the successful performance of a given behavior, and it is on the basis of this subjective perception/evaluation that the agent decides to rely or not, to bet or not on the trustee. In any case this probability index is based on (derives from) those beliefs and evaluations. In other words, the global, final probability of the realization of the goal *g* (i.e. of the successful performance of an action α) should be *decomposed* into the probability of the trustee performing the action well (that derives from the probability of its willingness, persistence, engagement, competence: *internal attributions*) and the probability of having the appropriate conditions (opportunities and resources: *external attributions*) for the performance and for its success, and of not having interferences and adversities (*external attributions*).

In such a way we understand how the attribution of trust is a very complex task, and that the decision making among different alternative scenarios is based on a complex evaluation of the basic beliefs and of their own relationships. And again, how the (even minimal) change of the credibility value of any (very relevant) belief might influence the resulting decision (and thus the trustworthiness attributed to the trustee); or vice versa, how significant changes in the credibility value of any unimportant belief does not significantly modify the final trust.

---

[1] We would like to thank Giovanni Pezzulo for his precious contribution on this chapter.

Trust Theory Cristiano Castelfranchi and Rino Falcone
© 2010 John Wiley & Sons, Ltd

11.1 Using a Fuzzy Approach

Given our purpose of modelling a graded phenomenon like trust (that is difficult to estimate experimentally) we have chosen a Fuzzy Logic Approach (FLA). A clear advantage with FLA is the possibility of using natural language labels (like: '*this doctor is very skilled*') to represent a specific real situation. In this way, it is more direct and simple to use intervals rather than exact values.

In addition, the behavior of these systems (e.g. their combinatorial properties) seems to be good at modeling several cognitive dynamics (Dubois and Prade, 1980), even if finding 'the real function' for a mental operation and estimating the contribution of convergent and divergent belief sources remain ongoing problems.

We have used an implementation based on a special kind of fuzzy system called Fuzzy Cognitive Maps (FCM); they allow the value of the trustfulness to be computed, starting from belief sources that refer to trust features. The values of those features are also computed, allowing us to perform some cognitive operations that lead to the effective decision to trust or not to trust (e.g. impose an additional threshold on a factor, for example risks). Using this approach we describe beliefs and trust features as approximate (mental) objects with a strength and a causal power over one another.

11.2 Scenarios

The scenario we are going to study is *medical house assistance* and we will look at it in two particular instances:

a) A *doctor* (a human operator) visiting a patient at home, and
b) A *medical automatic system* used to support the patient (without direct human intervention).

The case studies under analysis are:

- An *emergency situation*, in which there is a need to identify what is happening (for example, a heart attack) as soon as possible, to cope with it; we consider in this case the fact that the (first) therapy to be applied is quite simple (perhaps just an injection).
- A *routine situation*, in which there is a systematic and specialist therapy which needs to be applied (using quite a complex procedure) but in which there is no immediate danger to cope with.

We will show how the following factors can produce the final trust for each possible trustee who is dependent on it:

- The initial strength of the different beliefs (on which trust is based); but also
- How much a specific belief impacts on the final trust (the causality power of a belief).

It is through this second kind of factor that we are able to characterize some *personality traits* of the agents (Castelfranchi *et al.*, 1998).

11.3   Belief Sources

As shown in Chapter 2, our trust model is essentially based on specific beliefs which are both a *basis* of trust and also its *sub-components* or *parts*. These beliefs are the analytical account and the components of trust, and we derive *the degree of trust* directly from the *strength* of its componential and supporting beliefs (see Chapter 3): *the quantitative dimensions of trust are based on the quantitative dimensions of its cognitive constituents.*

However, what is the origin and the justification of the strength of beliefs? Our answer is: Just their sources. In our model, depending on the nature, the number, the convergence/divergence, and the credibility of its sources a given belief is more or less strong (certain, credible).

Several models propose a quantification of the degree of trust and make it dynamic, i.e. they can change and update such a degree (Jonker & Treur, 1999), (Schilloet *et al.*, 1999). But they only consider direct interaction (experience) or reputation as sources. In this implementation we have considered four possible types of belief sources:

- *direct experience* (how the personal – positive or negative – experience of the trustor contributes to that belief);
- *categorization* (how the properties of a class are transferred to their members);
- *reasoning* (more general than just categorization); and
- *reputation* (how the other's experience and opinion influences the trustor beliefs).

We do not consider learning in the model's dynamic. We are just modeling the resulting effects that a set of trustor's basic beliefs (based on various sources) have on the final evaluation of the trustee's trusfulness about a given task and in a specific situation. At present we do not consider how these effects feed back on the basic beliefs.

11.4   Building Belief Sources

Agents act depending on what they believe, i.e. *relying on* their beliefs. And they act on the basis of the degree of reliability and certainty they attribute to their beliefs. In other words, trust/confidence in an action or plan (reasons to choose it and expectations of success) is grounded on and derives from trust/confidence in the related beliefs.

For each kind of source we have to consider the impact it produces on trustor's beliefs about trustee's features. These impacts result from the composition of the value of the content (property) of that specific belief (the belief's object) with a subjective modulation introduced by some epistemic evaluations about that specific source. In fact when we have a belief we have to evaluate:

- the *value of the content* of that belief;
- *what the source* is (another agent, my own inference process, a perceptive sense of mine, etc.);
- *how this source evaluates* the belief (the subjective certainty of the source itself);
- how the *trustor evaluates this source* (with respect to this belief).

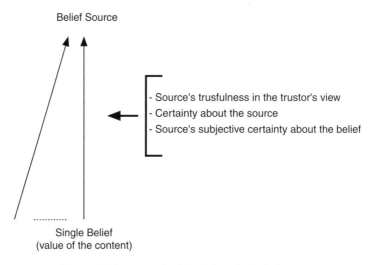

Figure 11.1   From single beliefs to the belief source

Those beliefs are not all at the same level. Clearly some of them are meta-beliefs, and some of them tune, modulate the value and the impact of the lower beliefs. The general schema could be described as a cascade having two levels (see *Figure* 11.1); at the bottom level there is the single belief (in particular, the value of the content of that specific belief; this value should be used (have a part) in the trustor's evaluation of some trustee's feature); at the top level there is the composition of the previous value with the epistemic evaluations of the trustor. At this level all the contributions of the various sources of the same type are integrated.

Let us consider as an example the belief source of the kind '*Reputation*' about a *doctor's ability* (see *Figure* 11.2). In order to have a value, we have to consider many opinions about

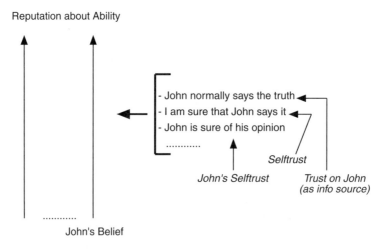

Figure 11.2   Case of belief source of Reputation

the ability of that doctor. For example, John may have an opinion: *I think that the doctor is quite good at his work*. In this case we have the belief's content: '*the doctor is quite good at his work*' and the belief source: '*John*'. Considering, in this specific case, the four factors above described, we have:

- the value of the content (doctor is *quite good at his work*);
- the degree of certainty that the trustor has about the fact that John has expressed this opinion (*I am sure* that John told me (thinks) that, etc.);
- how good John considers his own belief (when John says: 'I think', he could mean: *I am sure/ I am quite sure/ I am not so sure* and so on);
- the credibility of John's opinion (from the trustor's point of view).

The first factor represents a property, a belief and the value of its content (for example, ability); it is a source's belief that becomes an object of the trustor's mental world. The second factor represents a trustor's degree of certainty that the source expressed (communicated) that belief (it is also linked with the trustor's self-trust). The third factor represents an epistemic evaluation that the source makes on the communicated belief. Finally, the fourth factor represents a degree of trust in the source's opinion, and it depends on a set of trustor's beliefs about source's credibility, ability to judge and so on.

The second, third and the fourth factor are not objects of the same level, but rather meta-beliefs: they represent a *modulation* of the beliefs. In our networks, this can be better represented as impact factors. So, in our network we have two main nodes: 'John's belief' and 'Reputation about ability'. The first factor sets the value of the first node. The second, third and fourth factors set the value of the edge from the first to the second node.

The impact factors are not evaluation beliefs, but rather epistemic ones: they describe the way to see the other beliefs and their degree of certainty. So, at the level of building belief sources, evaluation and epistemic factors are separated; from the belief sources level up, in our FCM representation, they are combined in a unique numerical value.[2]

11.4.1 A Note on Self-Trust

Self-trust is a belief that relies on many beliefs, as, in general, trust is: their belief-FCM can be built in the same manner. As for trustfulness, self-trust is specific of a task or of a context. Among belief sources there can be, as usual, personal opinions and others' ones – i.e. reputation.

In self-trust computation there is also a set of motivational factors: self-image, auto-deceiving, and so on. Since our implementation does not represent motivational factors, at this moment we are not able to take into account these factors; so we calculate self-trust in the same way trust is calculated.

---

[2] Even quantitative information (how much I know about) is combined; for example, a low value about the ability of a doctor can derive from: low evaluation; low confidence in my information sources, little information.

11.5   Implementation with Nested FCMs

In order to understand the following parts of the work, we need to describe how a belief source value is computed starting from many different opinions[3] of different sources (e.g. in a MAS system, each agent is a source and can communicate an opinion about something). In an FCM this situation is modelled with a set of nodes representing single beliefs that have an edge ending on the (final) belief source node of the FCM. For each of these nodes, two values are relevant: the value of the node itself and the value of the edge.

The value of the node, as usual in this kind of model, corresponds directly to the fuzzy label of the belief; for example, *John says that the doctor is quite good at his work* can be considered as a belief about the doctor's ability with value 0.5 that impacts over the others/reputation belief source of a doctor's ability.[4]

Computing the *impact factor* of this belief (i.e. the value of the edge in the FCM) is more difficult. We claim that the impact represents not a cognitive primitive; rather, it has to be computed by a *nested FCM*, that takes into account mainly epistemic elements about the opinion itself and its source.

In our experiments with FCMs evaluation and epistemic issues are mixed up in a single value; this was a methodological choice, because we wanted to obtain one single final value for trustfulness. But this is the place where the two different kinds of information can be kept separate because they have a different role. Figure 11.3 shows many elements involved in this FCM: mainly beliefs about the source of the belief, grouped into three main epistemic features. Here we give an example of such an FCM in the medical domain. This FCM has single beliefs that impact on these features; the resulting value represents the final impact of a single belief over the belief source node.

This nested FCM was filled in with many nodes in order to show the richness of the elements that can intervene in the analysis. A similar FCM can be built for each single belief that impacts into the belief sources nodes; some of those nested FCMs have overlapping nodes, but in general each belief can have a different impact, depending on epistemic considerations.

It is possible to assign different impacts to the three different epistemic features; in this case we wanted to give them the same importance, but it depends from both the contingent situation, from personality factors and even from trust: for example, my own opinions can be tuned by self-trust (e.g. sureness about my senses and my understanding ability), and Mary's opinions can be tuned by trust about Mary. This leads to a very complex structure that involves trust analysis about all the sources (and about the sources' opinions about the other sources). For the sake of simplicity in the example we use all maximal values for impacts.

In general, it is important to notice that the 'flat' heuristic (same weights) we use in order to mix the different factors is not a cognitive claim, but a need derived from simplicity and lack of empirical data. In the following paragraph we investigate a very similar problem that pertains to how to sum up the different belief sources.

---

[3] We call this information *opinions* and not *beliefs* because they are not into the knowledge structure of an agent; an agent can only have a belief about another agent's beliefs (*John says that the doctor is good* is a belief of mine, not of John). This belief sharing process is mediated by opinions referred by John, but it can even be false, misleading or misinterpreted. What is important, however, is that beliefs are in the agent's cognitive structure, whether they correspond or not to other agent's opinions, beliefs or even to reality.

[4] It is important to notice that this node does not represent an opinion of John; it represents a belief of the evaluator, that can be very different from the original John's opinion (for example, it can derive from a misunderstanding).

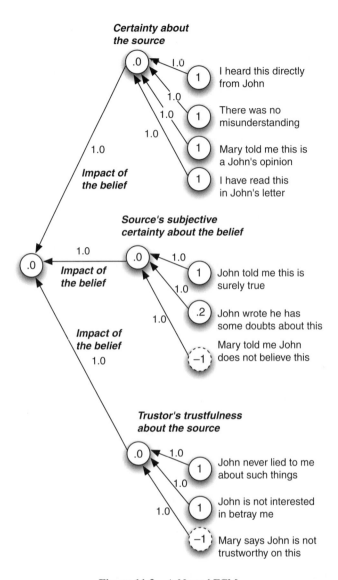

Figure 11.3 A Nested FCM

11.6 Converging and Diverging Belief Sources

In order to consider the contribution of different sources we need a theory of how they combine. The combination of different information sources is a classical complex problem (Dragoni, 1992), (Castelfranchi, 1996). It is in particular an evident problem in the case in which we are going to model human behaviors. In fact, humans use very different strategies and mechanisms, essentially based on their personalities and experiences.

This problem is very relevant in the case in which there are diverging opinions (beliefs). In these cases humans could use various heuristics for combining the opposite values simply because the elements which should be combined could produce an incoherent picture: if someone says that Mary dresses a hat and another one says that she does not dress a hat, I cannot infer that Mary dresses half a hat; or again if there are two persons that both say that Dr Smith is not too good a doctor and also not too bad a doctor while two other persons give us two diverging evaluations on Dr White (one says that he is an excellent doctor and another says that he is a really bad doctor) we would not have an equivalent evaluation of Dr White and Dr Smith, and our decision would be guided by other criteria. *These criteria are linked with context, emotions, personality factors.* We could have people who, in the presence of diverging opinions, decide to suspend judgment (they become unable to decide), or people who take into consideration the best opinion (*optimistic personality*), or, on the contrary, people who take into consideration the worst opinion (*pessimistic personality*). And so on. A good model should be able to implement different heuristics. For the moment, in our model, we simply sum up all the contributions and we squash the result with a threshold function. In fact, the exact heuristics that humans choose depend on the situation and eventually the exact threshold functions can be the object of empirical analysis or simulations. The model itself is independent to those heuristics, that is they can be easily substituted.

11.7   Homogeneous and Heterogeneous Sources

We have the problem of summing up the contribution of many different sources. We have already discussed the case of homogenous sources (e.g. different opinions about a feature/person/thing, etc.), when an heuristic has to be chosen.

The same problem occurs when we want to sum up the contribution of heterogeneous fonts (e.g. direct experience and reputation about the ability of a doctor). Even in this case, many heuristics are possible. For example, which is more relevant, our own personal experience or the reputation about a specific ability of a person? There is not a definitive answer to this question: are we able to evaluate that ability in a good way? Or is it better to rely on the evaluation of others? And vice-versa. Our analysis is limited to a plain estimation of all the relevant factors, but many other strategies are possible, as in the case of homogenous sources. Also, in this case, some strategies depend on personality factors.

We have described how it is possible to model belief sources starting from the single beliefs; now we describe how trust is computed starting from the belief sources.

11.8   Modeling Beliefs and Sources

Following a belief-based model of trust we can distinguish between trust *in the trustee* (be it either someone, e.g. the doctor, or something, e.g. the automated medical system) which has to act and produce a given performance thanks to its internal characteristics, and trust *in the* (positive and/or negative) *environmental conditions* (like opportunities and interferences) affecting the trustee's performance, which we call 'external factors'. In this work we take into account:

- three main beliefs regarding the trustee: *an ability/competence belief*; a *disposition/ availability belief*, and *an unharmfulness belief;*
- two main beliefs regarding the contextual factors: *opportunity beliefs* and *danger beliefs.*

Table 11.1 Internal and external factors for the *automated medical system*

Internal factors	*Ability or Competence beliefs*	They concern the efficacy and efficiency of the machine; its capability to successfully apply the right procedure in the case of correct/proper use of it. Possibly also its ability to recover from an inappropriate use
	Disposition or Availability beliefs	They are linked to the reliability of the machine, its regular functioning, its ease of use; possibly, its adaptability to new and unpredictable uses
	Unharmfulness beliefs	They concern the lack of internal/ intrinsic risks of the machine: the dangers implied in the use of that machine (for example side effects for the trustor's health), the possibility of breaking and so on
External factors	*Opportunity beliefs*	Concerning the opportunity of using the machine, independently of the machine itself, from the basic condition to have the room for allocating the machine to the possibility of optimal external conditions in using it (regularity of electric power, availability of an expert person in the house who might support its use, etc.)
	Danger beliefs	They are connected with the absence of the systemic risks and dangers external to the machine that could harm the user: consider for example the risk for the trustor's privacy: in fact we are supposing that the machine is networked in an information net and the data are also available to other people in the medical structure

What are the meanings of our basic beliefs in the case of the doctor and in the case of the automated medical system? For both the latter and former, the internal and external factors are shown in Table 11.1 and 11.2.

Each of the above mentioned beliefs may be generated through different sources; such as: direct experience, categorization, reasoning, and reputation. So, for example, ability/competence beliefs about the doctor may be generated by the direct knowledge of a specific doctor, and/or by the generalized knowledge about the class of doctors and so on.

11.9 Overview of the Implementation

An FCM is an additive fuzzy system with feedback; it is well suited to the representation of a dynamic system with cause-effect relations. An FCM has several nodes, representing causal concepts (belief sources, trust features and so on), and edges, representing the causal power of a node over another one. The values of the nodes representing the belief sources and the values of all the edges are assigned by a human; these values propagate in the FCM until a stable state is reached; so the values of the other nodes (in particular the value of the node named trustfulness) are computed.

Table 11.2   Internal and external factors for the *doctor*

Internal factors	*Ability or Competence beliefs*	They concern the (physical and mental) skills of the doctor; his/her ability to make a diagnosis and to solve problems
	Disposition or Availability beliefs	They concern both the willingness of the doctor to commit to that specific task (subjective of the specific person or objective of the category), and also his/her availability (in the sense of the possibility to be reached/informed about his/her intervention).
	Unharmfulness beliefs	They concern the absence (lack) of the risk of being treated by a doctor; namely the dangers of a wrong diagnosis or intervention (for example, for the health of the trustor).
External factors	*Opportunity beliefs*	Concerning the opportunities not depending on the doctor but on conditions external to his/her intervention. Consider for example the case in which the trustor is very close to a hospital in which there is an efficient service of fast intervention; or again, even if the trustor is not very close to a hospital he/she knows about new health policies for increasing the number of doctors for quick intervention; and so on. Conversely, imagine a health service not efficient, unable to provide a doctor in a short time; or, again, a particularly chaotic town (with heavy traffic, frequent strikes) that could hamper the mobility of the doctors and of their immediate transfer to the site where the patient is.
	Danger beliefs	These beliefs concern the absence (lack) of the risks and dangers which do not depend directly on the doctor but on the conditions for his/her intervention: for instance, supposing that the trustor's house is poor and not too clean, the trustor could see the visit of a person (the doctor in this case) as a risk for his/her reputation.

In order to design the FCM and to assign a value to its nodes we need to answer four questions:

1) Which value do I assign to this concept?
2) How sure am I of my assignment?
3) What are the reasons for my assignment?
4) How much does concept impact on another linked concept?

We address the *first and the second question* above by assigning numeric values to the nodes representing the belief sources. The nodes are *causal concepts*; their value varies from −1 (true negative) to +1 (true positive). This number represents the value/degree of each single trust feature (say *ability*) by combining together both the credibility value of a belief (degree of credibility) and the estimated level of that feature. Initial values are set using adjectives from natural language; for example, 'I believe that the ability of this doctor is *quite good* (in his work)' can be represented using a node labeled 'ability' with a little positive value (e.g.

+0.4). For example, the value +0.4 of ability either means that the trustor is *pretty sure* that the trustee is *rather good* or that he/she is *rather sure* that the trustee is *really excellent*, etc. In this implementation we do not address how the degree of credibility/certainty of the belief combines with the degree of the content dimension (even if this analysis is quite relevant); we just use a single resulting measure.

We address the *third question* above designing the graph. Some nodes receive input values from other nodes; these links represent the reasons on which their values are grounded. Direct edges stand for fuzzy rules or the partial causal flow between the concepts. The sign (+ or −) of an edge stands for causal increase or decrease. For example, the Ability value of a doctor influences positively (e.g. with weight +0.6) his Trustfulness: if ability has a positive value, Trustfulness increases; otherwise it decreases.

We address the *fourth question* above by assigning values to the edges: they represent the impact that a concept has over another concept. The various features of the trustee, the various components of trust evolution do not have the same impact, and importance. Perhaps, for a specific trustee in a specific context, ability is more important than disposition. We represent the different quantitative contributions to the global value of trust through these weights on the edges. The possibility of introducing different impacts for different beliefs surely represents an improvement with respect to the basic trust model.

FCMs allow causal inference to be quantified in a simple way; they model both the strength of the concepts and their relevance for the overall analysis. For example, the statement: 'Doctors are *not very accessible* and this is an *important factor* (for determining their trustfulness) in an emergency situation' is easily modeled as a (strong) positive causal inference between the two concepts of accessibility and trustfulness. FCMs also allow the influence of different causal relations to be summed up. For example, adding another statement: 'Doctors are *very good* in their ability, but this is a *minor factor* in an emergency situation' means adding a new input about ability, with a (weak) positive causal influence over trustfulness. Both accessibility and ability, each with its strength and its causal power, contribute to establish the value of trustfulness.

11.9.1   A Note on Fuzzy Values

Normally in fuzzy logic some labels (mainly adjectives) from natural language are used for assigning values; each label represents a range of possible values. There is not a single universal translation between adjectives and the exact numerical values in the range.

FCM is different from standard fuzzy techniques, in that it requires the use of crisp input values; we have used the average of the usual ranges, obtaining the following labels, both for positive and negative values: *quite*; *middle*; *good;* etc. However, as our experiments show, even with little variation of these values in the same range, the FCMs are stable and give similar results.

As Figure 11.4 shows, the ranges we have used do not divide the whole range {−1,1} into equal intervals; in particular, near the center (value zero) the ranges are larger, while near the two extremities they are smaller. This implies that a little change of a value near the center normally does not lead to a 'range jump' (e.g. from *some* to *quite*), while the same little change near the extremities can (e.g. from *very* to *really*).

This topology is modeled in the FCM choosing the threshold function; in fact, it is possible to choose different kinds of functions, the only constraint is that this choice must be coherent with the final convergence of the algorithm. With the function chosen in our implementation,

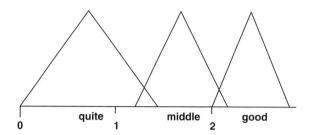

Figure 11.4   Fuzzy Intervals. (Reproduced with kind permission of Springer Science+Business Media © 2003)

changes in big (positive or negative) values have more impact on the FCM, this is a tolerable result even if it does not correspond with a general cognitive model.

11.10   Description of the Model

Even if FCMs are graphs, ours can be seen as having four layers. The first layer models the influence of the 'beliefs sources': *Direct Experience* (e.g. 'In my experience . . .'), *Categorization* (e.g. 'Usually doctors . . .'), *Reasoning* (e.g. 'I can infer that . . .'), *Reputation* (e.g. 'A friend says that . . .'). Their value is meant to be stable (i.e. it does not change during computation), because these nodes could be assumed as being the result of an 'inner FCM' where each single belief is represented (e.g. direct experience about ability results from many nodes like: 'I was visited *many times* by this doctor and he was *really good* at his work', '*Once* he made a *wrong* diagnosis', . . .). So their value not only represents the strength of the feature expressed in the related beliefs, but also their number and their perceived importance, because belief sources represent the synthesis of many beliefs.

The second layer shows the five relevant basic beliefs: *Ability, Accessibility, Harmfulness, Opportunities* and *Danger*. These basic beliefs are distinguished in the third layer into *Internal Factors* and *External Factors*. *Ability, Accessibility* and *Harmfulness* are classified as *Internal Factors*; *Opportunities* and *Danger* are classified as *External Factors*. *Internal* and *External Factors* both influence *Trustfulness*, which is the only node in the fourth layer. For the sake of simplicity no crossing-layer edges are used, but this could be easily done since FCM can compute cycles and feedback, too.

11.11   Running the Model

Once the initial values for the first layer (i.e. belief sources) are set, the FCM starts running. The state of a node N at each step s is computed taking the sum of all the inputs, i.e., the current values at step s-1 of nodes with edges coming into N multiplied by the corresponding *edge weights*. The value is then *squashed* (into the $-1,1$ interval) using a *threshold function*. The FCM run ends when an *equilibrium* is reached, i.e., when the state of all nodes at step s is the same as that at step s-1.

At this point we have a resulting value for trustfulness, that is the main goal of the computational model. However, the resulting values of the other nodes are also shown: they are useful for further analysis, where thresholds for each feature are considered.

11.12 Experimental Setting

Our experiments show the choice between a doctor and a medical apparatus in the medical field. We assume that the choice is mainly driven by trustfulness. We have considered two situations: a 'Routine Visit' and an 'Emergency Visit'. We have built four FCMs representing trustfulness for doctors and machines in those two situations. Even if the structure of the nets is always the same, the values of the nodes and the weights of the edges change in order to reflect the different situations. For example, in the 'Routine Visit' scenario, *Ability* has a great causal power, while in the 'Emergency Visit' one the most important factors is *Accessibility*.

It is also possible to alter some values in order to reflect the impact of *different trustor personalities* in the choice. For example, somebody who is very concerned with *Danger* can set its causal power to *very high* even in the 'Routine Visit' scenario, where its importance is generally low. In the present work we do not consider those additional factors; however, they can be easily added without modifying the computational framework.

11.12.1 Routine Visit Scenario

The first scenario represents many possible routine visits; there is the choice between a *doctor* and a *medical apparatus*. In this scenario we have set the initial values (i.e. the beliefs sources) for the *doctor* hypothesizing some direct experience and common sense beliefs about doctors and the environment.

Most values are set to zero; the others are:

- Ability – Direct Experience: *quite (+0.3)*;
- Ability – Categorization: *very (+0.7)*;
- Avaialability – categorization: *quite negative (−0.3)*;
- Unharmfulness – categorization: *some negative (−0.2)*;
- Opportunity – Reasoning: *some (+ 0.2)*;
- Danger – Reasoning: *some negative (−0.2)*

For the *machine* we have hypothesized no direct experience. These are the values:

- Efficacy – Categorization: *good (+0.6)*;
- Accessibility – Categorization: *good (+0.6)*;
- Unharmfulness – Categorization: *quite negative (−0.3)*;
- Opportunity – Reasoning: *some (+0.2)*;
- Danger – Categorization: *quite negative (−0.3)*;
- Danger – Reasoning: *quite negative (-0.3)*

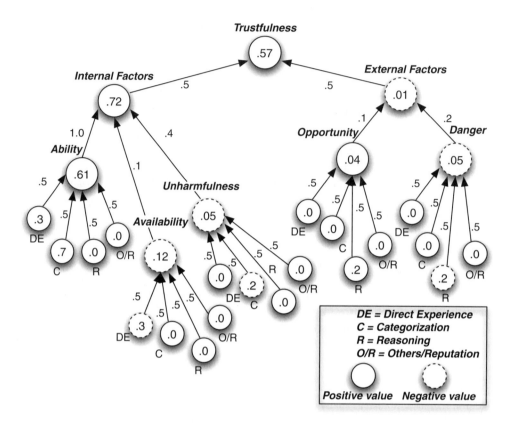

Figure 11.5 Routine Visit FCMs for the Doctor. (Reproduced with kind permission of Springer Science+Business Media © 2003)

We have also considered the causal power of each feature. These values are the same both for the *doctor* and the *machine*. Most values are set to *mildly relevant (+0.5)*; the others are:

- Ability: *total causation (+1)*;
- Accessibility: *only little causation (+0.1)*;
- Unharmfulness: *middle negative causation (−0.4)*;
- Opportunity: *only little causation (+0.1)*;
- Danger: *little negative causation (-0.2)*

The results of this FCM are shown in Figure 11.5 and 11.6: trustfulness for the doctor results *good (+0.57)* while trustfulness for the machine results only *almost good (+0.22)*.

The FCMs are quite stable with respect to minor value changes; setting Machine's 'Accessibility – Direct Experience' to *good (+0.6)*, 'Accessibility – Categorization' to *really good (+0.8)* and 'Danger – Categorization' to *little danger (−0.5)* results in a non dramatic change in the final value, that changes from *almost good (+0.23)* to *quite good (+0.47)* but does not overcome the doctor's 'trustfulness'. This is mainly due to the high causal power of ability with respect to the other features.

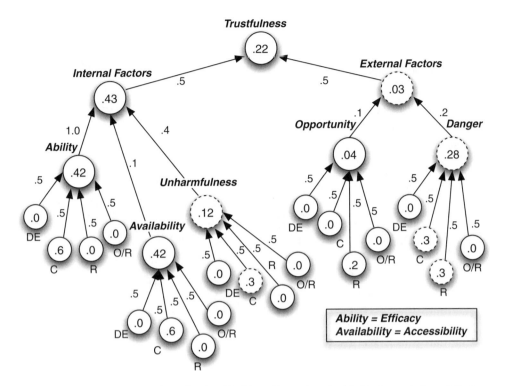

Figure 11.6   Routine Visit FCMs for the Machine. (Reproduced with kind permission of Springer Science+Business Media © 2003)

We can also see the influence of different personalities. For example, if we assume that doctors are supposed to involve high external risks ('Danger − Reputation': +1), with the usual values, the trustor's 'trustfulness' does not change very much (*good (+0.47)*). But if the patient is somebody who gives high importance to danger (danger: *total causality (−1)*), the doctor's trustfulness decreases to *negative (−0.42)*.

11.12.2   Emergency Visit Scenario

We have here hypothesized an emergency situation where somebody needs a quick visit for an easy task (e.g. a injection). In this scenario the values for the nodes are the same as before, but some edges drastically change: *Reliability* becomes very important and *Ability* much less. The values for the edges are:

- Ability: little causation (+0.2);
- Willingness: very strong causation (+1);
- Unharmfulness: strong negative causation (−0.8);
- Opportunity: middle causation (+0.5);
- Danger: quite strong causation (+0.6).

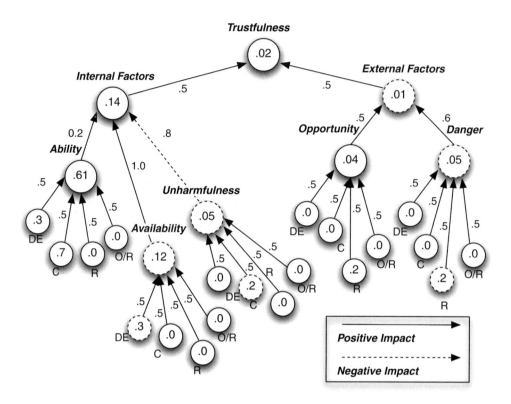

Figure 11.7 Emergency Visit FCMs for the Doctor. (Reproduced with kind permission of Springer Science+Business Media © 2003)

The results also change drastically: trustfulness for the doctor is *only slightly positive (+0.02)* and for the machine it is *quite good (+0.29)* (see Figure 11.7 and 11.8).

The FCMs are very stable; altering some settings for the doctor (Ability – Direct Experience: *very good* and Danger – Categorization: *only little danger*) results in a change in the trustfulness value that becomes *almost good* but does not overcome the machine's one. We obtain the same results if we suppose that Doctor's Ability - Direct Experience: *perfect* and Ability's Causal Power: *very strong*.

On the contrary, if we introduce a big danger (+1) either internal (*harmfulness*) or external (*danger*) in each FCM the trustfulness values fall to *negative* in both cases (respectively −0.59 and −0.74 for the doctor; and −0.52 and −0.67 for the machine).

11.12.3 Trustfulness and Decision

We consider three steps: evaluation (i.e. how much trust do I have); decision (to assign or not assign a task); delegation (make the decision operative). Obtaining the trustfulness values is only the first step. In order to make the final choice (e.g. between a doctor and a machine in our scenarios) we have to take into account other factors, mainly costs and possible saturation thresholds for the various features.

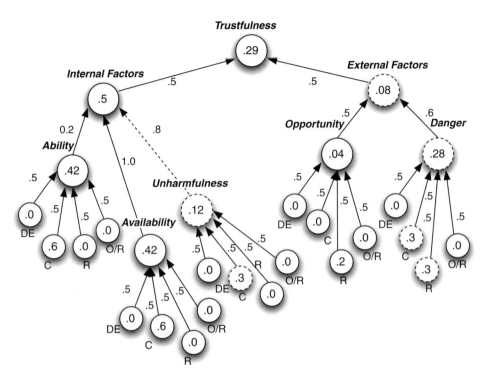

Figure 11.8 Emergence Visit FCMs for the Machine. (Reproduced with kind permission of Springer Science+Business Media © 2003)

FCMs not only show the overall trustfulness value, but also the values of each belief. We can fix a threshold for one or more features and inhibit a choice even if trustfulness is acceptable (i.e. 'I trust him, but the danger is too high'). In addition, the final function for decision has to also take into account the costs for each decision choice. In the present analysis we do not consider here these additional factors.

11.12.4 Experimental Discussion

The two scenarios try to take into account all the relevant factors for trustfulness: beliefs sources, basic beliefs and their causal power. Moreover, FCMs allow experimentation of changes in values due to different personalities.

As already specified, belief sources are figured values, possibly derived from inner FCMs where many beliefs play their role. We have assumed four types of beliefs sources, but for many of them, we give no values. We have set all their causal power to *middle causality* (+0.5) in order to let them be 'neutral' in the experiments. Some different personalities can augment or reduce the values (e.g.: somebody who cares only about his own experience may assign a strong causal power to the corresponding edges).

Basic beliefs, both internal and external, are the core of the analysis; we have expanded our model (see Chapters 2 and 3 in this book) by representing and quantifying the different importance of trust components/determinants (for different personalities or different situations). Our experiments show that the relative importance assigned to each feature may drastically change the results. Most of the differences in FCM's behavior are due to the strong causal power assigned to ability (routine visit scenario) and accessibility (emergency visit scenario), even if the basic beliefs values are the same.

11.12.5   Evaluating the Behavior of the FCMs

We have conducted several experiments modifying some minor and major beliefs' sources in the FCM of routine visit scenario for the doctor. This allows us to evaluate their impact on the overall results.

We can see that the FCMs are quite stable: changing minor factors does not lead to catastrophic results. However, modifying the values of some major factors can lead to significant modifications; it is very important to have a set of coherent parameters and to select the most important factors very accurately.

However, our first aim is not to obtain an exact value for trustfulness for each FCM; on the contrary, even if we consider the whole system as a qualitative approach, it has to be useful in order to make comparisons among competitors (i.e. the doctor and the machine in our scenarios). So, an important question about our system is: *how much can I change the values (make errors in evaluations) and conserve the advantage of a competitor over the other?*

In the routine visit scenario the two trustfulness values are far removed from one another (0.57 for the doctor vs. 0.23 for the machine). Even if we change several factors in the machine's FCM its trustfulness does not overcome its competitor's one.

11.12.6   Personality Factors

Given the way in which the network is designed, it is clear that the weights of the edges and some parameters of the functions for evaluating the values of the nodes are directly expressing some of the personality factors. It is true that some of these weights should be learned on the basis of the experience. On the other hand, some other weights or structural behaviours of the network (given by the integrating functions) should be directly connected with personality factors. For example, somebody who particularly cares about their safety can overestimate the impact of danger and unharmfulness, or even impose a threshold on the final decision. Each personality factor can lead to different trust values even with the same set of initial values for the beliefs sources. Many personalities are possible, each with its consequences for the FCM; for example: *Prudent*: high danger and unharmfulness impact; *Too Prudent*: high danger and unharmfulness impact, additional threshold on danger and unharmfulness for decision; *Auto*: high direct experience impact, low impact for the other beliefs sources; *Focused on Reputation*: high reputation impact, low impact for the other beliefs sources.

Some personality factors imply emotional components, too. They can lead to important modifications of the dynamics of the FCM, for example modifying the choice of the heuristic for combining homogenous and heterogeneous fonts.

To summarise, we can say that our experiments aim to describe the dynamics of trust and to capture its variations due to belief sources variation, and the different importance given to the causal links and personality factors. The scenarios presented here fail to capture many factors; in addition, we have assigned values and weights more as a matter of taste than through experimental results. More, the results of the experiments are shown as an attempt to describe the behavior of this kind of system; for example, its additive properties or the consequences of the choice of the threshold function. The adequacy of such a behavior to describe cognitive phenomena is an ongoing problem.

However, the experimental results show that it is possible to mimic many commonsense assumptions about how trust varies while some features are altered; our aim was in fact to capture trust variations more than assign absolute values to it. In our view, this experiment confirms the importance of an analytic approach to trust and of its determinants, not simply reduced to a single and obscure probability measure or to some sort of reinforcement learning.

In the next two paragraphs we introduce:

- some learning mechanisms with respect to the reliability of the belief sources; and
- some comparison experiments among different strategies for trusting other agents using a *Contract Net* protocol: we will show how our cognitive approach presents some advantages.

11.13   Learning Mechanisms

In the previous paragraphs we have considered the belief sources as a static knowledge of the agents. Briefly in this part we show how it could be possible to extend this approach by trying to model some of the dynamics generated by a learning process and by trust itself.

We give an agent in a MAS system the capacity to evaluate its 'sources of opinions', i.e. the other agents, according to a specific advising attitude parameter: *trust in Y as an information source*, (that is different from trust in Y to perform differently).

In order to build a belief source, we considered a node representing a single belief and an edge representing the impact of this single belief: the value of the edge (the impact factor) represents *the validity of the source* with respect to this single communicative episode. Some elements of this episode are unique (e.g. certainty about the source) but others are shared between all the other communicative episodes with the same source: the trustfulness of the source is applicable to all the class of the possible beliefs about the opinions of a source about an argument. These values can be learned and applied to future cases; for example, if it results, from some interactions with John, that he systematically lies, the impact of (my belief about) his opinion (e.g. *John says p...*) will drastically diminish or even become negative, and this value can be used in further interactions.

As shown the FCM computes its results until it stabilizes. This leads to a stable result for each node involved in the FCM. Here we propose a second phase: *the FCM 'evaluates its sources', i.e. modifies the impact of each single belief source according to the final value of its belief source node.*

For example, many nodes $n_1, \ldots, n_n$ representing single beliefs (opinions given by the source) can contribute to the value of a belief source node N. Each node n_j (with $1 \leq j \leq n$) has an impact $i_1, \ldots, i_n$ over N; the impact value is calculated by the inner FCM previously described. After the FCM stabilization, the difference between the (final) value of the belief source and the value of each single belief (of the source) can be seen in information terms as an *error*.

The *learning phase* consists in trying to minimize errors; in our terms, the impact of a bad opinion (and the importance of the corresponding source), has to be lowered; the reverse is also true. In order to achieve this result, we have two different strategies: *implicit* and *explicit* *revision*.

11.13.1   Implicit Revision

In an *implicit revision* procedure, the *error* (i.e. the difference between N and n_j) can be back-propagated and modify the value of i_j. The rationale is that the FCM adjusts the evaluation of its sources, the impact of their opinions (as mediated by our beliefs); for example, in the case of a source that systematically lies (gives as output the opposite of the final value of the target node), step to step its impact will be nearer to -1.

The revision process has two steps. The first step is the change of the impact factor: a standard back-propagation algorithm can be used in order to achieve this result. The second step leads to a feedback of this revised value over the nested FCM: some low-weight edges are assumed that back-propagate the value until the nested FCM stabilizes. For example, if the impact factor value was lowered, all the nodes in the nested FCM will be lowered a bit, until it stabilizes.

This procedure has to be better explained. Since the value of the edge (the impact factor) represents the validity of the source, this value changes as I have a feedback between an opinion and my final belief. For example, in evaluating the ability of a doctor, a significant difference between a value furnished by a source and the final value I assume, means that the source was not totally valid. This can result from different reasons: the source is not trustworthy, a misunderstanding, poor information and so on. With regard to these problems the implicit revision strategy is blind: it revises the value of the impact successively to all the nodes in the nested FCM, without caring *what nodes* are responsible for the error and so need to be changed.

This process adjusts the evaluation of a single belief, but it can be in part shared with other belief impact evaluations with respect to the same source: some nodes of the nested FCM apply only to the current situation (e.g. certainty about the source) but others are related to all the interactions with the same source (e.g. trustfulness about the source). So, this form of learning from a single episode generalizes for the following episodes. This kind of learning is non specific. We cansider a better one: explicit revision.

11.13.2   Explicit Revision

The *explicit revision* consists of the revision of some beliefs about the source; since these beliefs are part of the *inner FCM*, the (indirect) result is a modification of the impact. So, *explicit revision* means revising the values of some nodes of the inner FCM (or building new ones); the revised *inner FCM* computes a different impact value f.

In order to obtain explicit revision, the first important issue is to decide *where* to operate in the inner FCM. In some cases it would be useful to insert a new node representing a bad or good 'past experience' under the 'trustfulness' feature; in this case the value is easily set according to the usual set of fuzzy labels; an example of such a node can be *This time John was untrustworthy*.

Even if it is unrealistic to think that a single revision strategy is universal, there can be many heuristics: for example, a wrong opinion can be evaluated in different ways if I am sure

Table 11.3 Source evaluation

Willingness Episodes	The source reveals itself not to be trustworthy (*I know John intentionally lied to me*)
Competence Episodes	The source reveals itself not to be competent (*John did not lie but he was not able to give an useful information*)
Self trust Episodes	The evaluator was responsible for a misunderstanding (*I misunderstood*)
Accidental Problems	There were contingent problems and no revision is necessary (*there was an unpredictable circumstance*)

that somebody intended exactly this (there was no misunderstanding) or if I do not remember exactly what he said (or even if I am not sure that it was his opinion). In the latter cases, if I am not certain of the source, it can be better to assign the error to my evaluation rather than to ignorance of the source or even worse to his intention to deceive me. Such a change has no impact on other interactions with the same sources (but it can lead to change my self trust value).

11.13.3 A Taxonomy of Possible Revisions

There are many possible ways to evaluate an episode of interaction in order to learn from it and to decide to change one's beliefs. As we have shown, not only the sources' opinions, but also the full set of interaction episodes have to be categorized; from this kind of categorization the following belief revision process depends. For example, in order to comprehend the motivation of the discrepancy between the source evaluation and my evaluation ('*John says that this doctor is pretty good, but it results to me to be not so good...*'). The first thing to consider is 'what was the main factor' from which this discrepancy depends.

Obviously this decision process pertains to a cognitive apparatus and it is impossible at a pure-belief level; so a cognitive agent needs some revision strategies that individuate the error source and try to minimize it for the future. In Table 11.3 we propose a crude taxonomy of the problems that can intervene in an episode of interaction.

Implicit revision performs better with regard to computational speed. However, explicit revision has many advantages. First of all, taking into account single cognitive components allows a better granularity; this can make the difference where fine-grained distinctions are needed, for example in order to distinguish between trust somebody as an information source and as a specialist, or to distinguish a deceiver from a not informed source. Also, a single belief can be shared among many different FCMs, so this operation leads to the generalization and reuse of the obtained results.

In general, explicit revision takes into account the single cognitive components of trust, and this feature is one of our main desiderata. We derive trust from its cognitive components, i.e. from single agent's beliefs. So it is better to store information learned by experience into the same representation form (i.e. beliefs) rather than using compounded values (as an impact is), in order to integrate them into the representational and reasoning system of the agent.[5]

---

[5] However, it has to be noticed that since we have a fine-grained distinction between different belief sources, even the implicit mechanism results in being sufficiently accurate and specific for many purposes, even if it loses part of

Keeping them separate can lead to building different graphs and applying different revision heuristics, too.

The process of evaluation of the sources is described as a step that follows the stabilization of the FCM. In computational terms, thanks to the characteristics of the FCMs, these feedback processes can even be made in a parallel way. We see this process as a cognitive updating of beliefs: the parallel options are better to model gradual opinion shifts, while the two phases division allows the mimicry of counterfactual phenomena and drastic after decision changes.

When is it good for the system to evaluate its sources? In our experiments we assume, in a conservative way, that the mechanism only starts after a decision; the rationale is that a decision taken is a sufficient condition to assume that the stabilization reached is sufficient; less conservative criteria are possible, of course: this choice can be considered an heuristic rather than a part in the way the system works.

An interesting 'side-effect' of this source evaluation mechanism is that revision has less effect on well established sources (i.e. agents that have many interactions with us); we are less inclined to revise the stronger ones, mainly for economic reasons. The process described takes into account all the past experiences, so introducing a new example (or a counterexample) has less impact on the case of many interactions.

11.14   Contract Nets for Evaluating Agent Trustworthiness[6]

In this paragraph we show a first significant set of results coming from a comparison among different strategies for trusting other agents using a contract net protocol (Smith, 1980). We introduced three classes of trustors: a *random trustor*, a *statistical trustor*, a *cognitive trustor*. All the simulations were performed and analyzed using the cognitive architecture AKIRA (Pezzullo & Calvi, 2004).

The results show the relevance of using a *cognitive representation* for a correct trust attribution. In fact, a cognitive trustor performs better than a statistical trustor even when it has only an approximate knowledge of the other agents' properties.

11.14.1   Experimental Setting

We implemented a contract net with a number of trustors who delegate and perform tasks in a variable environment. Each agent has to achieve a *set of tasks* and is defined by a set of features: *ability set, willingness, delegation strategy*.

- The *Task set* contains the tasks an agent has to achieve; it is able either to directly perform these tasks, or to delegate them to some other agent.
- The *Ability set* contains the information about the agent's skills for the different tasks: each agent has a single ability value for each possible task; it is a real number that ranges in *(0,*

---

the expressiveness of the explicit one. Sometimes the trade off between computational power and expressiveness can lead to the adoption of the implicit mechanism.

[6] We would like to thank Giovanni Pezzulo and Gianguglielmo Calvi for their relevant contribution to the implementation and analysis of the model discussed in this paragraph.

1). At the beginning of the experiment these values are randomly assigned to each agent on each possible task.

- The *Willingness* represents how much the agent will be involved in performing tasks (e.g. how many resources, will or amount of time it will use); this modulates the global performance of the agent in the sense that even a very skilled agent can fail if it does not use enough resources. Each agent has a single willingness value that is the same for all the tasks it tries to perform; it is a real number that ranges in *(0, 1)*.
- The *Delegation strategy* is the rule an agent uses for choosing which agent to delegate the task to (e.g. random, cognitive, statistical). It is the variable we want to control in the experiments for evaluating which trustor performs better.

Agents reside in an *environment* that changes and makes the tasks harder or simpler to perform. Changes are specific for each agent and for each task: in a given moment, some agents can be in a favorable environment for a given task, some others in an unfavorable one. For example, two different agents, performing the same task, could be differently influenced by the same environment; or, the same agent performing different tasks in the same environment could be differently influenced by it in performing the different tasks. Influences range in *(−1, 1)* for each agent for each task; they are fixed at random for each simulation. The environment changes randomly during the simulations: this simulates the fact that agents can move and the environment can change. However, for all experiments, if a task is delegated in an environment, it will be performed in the same one.

11.14.2   Delegation Strategies

In the contract net, on the basis of the offers of the other agents, each agent decides to whom to delegate (Castelfranchi and Falcone, 1998) depending on their delegation strategy. We have implemented a number of different agents, having different *delegation strategies*:

- a *random trustor*: who randomly chooses the trustee to whom to delegate the task. This kind of trustor has no a priori knowledge about: the other agents, the environment in which they operate, their previous performances. There is no learning. This is used as a base line.
- a *statistical trustor*: inspired by a number of works, including (Jonker and Treur, 1999), assigns a major role to learning from direct interaction. They build the trustworthiness of other agents only on the basis of their previous performances, without considering specific features of these agents and without considering the environment in which they performed. It is one of the most important cases of trust attribution; it uses the previous experience of each agent with the different trustees (failures and successes) by attributing to them a degree of trustworthiness that will be used to select the trustee in a future interaction. There is a training phase during which this kind of trustor learns the trustworthiness of each agent through a mean value of their performances (number of failures and successes) on the different tasks in the different environments; during the experimental phase the statistical trustor delegates the most trustful agent (and continues learning, too). There is no trustor's ability to distinguish how the properties of the trustee or the environment may influence the final performance.

- a *cognitive trustor*: this kind of trustor takes into account both the specific features of the actual trustee and the impact of the environment on their performance. In this implementation there is no learning for this kind of agent but an a priori knowledge of the specific properties of the other agents and of the environment. It is clear that in a realistic model of this kind of agent, the a priori knowledge about both the internal properties of the trustees and the environmental impact on the global performance will not be perfect. We did not introduce a learning mechanism for this kind of agent (even if in Section 11.13 we discussed this problem and showed potential solutions) but we introduced different degrees of errors in the knowledge of the trustor that corrupted their perfect interpretation of the world. The cognitive model is built using Fuzzy Cognitive Maps. In particular, two special kind of agents will be analyzed:
- *best ability trustor*: who chooses the agent with the best ability score.
- *best willingness trustor*: who chooses the agent with the best willingness score.

These two kind of cognitive agents can be viewed as having different 'personalities'.

11.14.3   The Contract Net Structure

We have performed some experiments in a *turn world*, others in a *real time world*. In the turn world the sequence is always the same. The first agent (randomly chosen) posts their first task (*Who can perform the task τ?*) and they collect all the replies from the other agents (*I can perform the task τ in the environment w*). All data given from the offering agents are true (there is no deception) and in particular the cognitive trustors know the values of ability and willingness for each agent (as we will see later, with different approximations).

Depending on their delegation strategy, the trustor delegates the task to one of the offering agents (in this case, even to themselves: self-delegation). The delegated agent tries to perform the task; if it is successful, the delegating agent gains one *Credit*; otherwise it gains none. The initiative passes to the second agent and so on, repeating the same schema for all the tasks for all the agents. At the end of each simulation, each agent has collected a number of *Credits* that correspond to the number of tasks that the delegated agents have successfully performed.

We have introduced no external costs or gains; we assumed that each delegation costs the same and the gain of each performed task is the same. Since the agents have the same structure and the same tasks to perform, gained credits are the measure of success of their delegation strategy.

In the *real time world* we have disabled the turn structure; the delegation script is the same, except for no explicit synchronization of operations. This means that another parameter was implicitly introduced: *time* to execute an operation. Collecting and analyzing messages has a time cost; agents who have more requests need more time in order to fulfill them. In the same way, agents who do more attempts in performing a task, as well as agents who reason more, spend more time. In *real time world* time optimization is another performance parameter (alternative or together with *credits*), and some alternative trust strategies become interesting: in real time experiments we introduced another strategy:

- the *first trustful trustor*: it is a variant of the cognitive trustor and it has the same FCM structure; but it delegates to the first agent whose trust exceeds a certain threshold: this is less accurate but saves the time of analyzing all the incoming messages. Furthermore, if some busy agent accepts only a limited number of tasks, or if there is a limited time span for performing the maximum number of tasks, it is important to be quick to delegate them.

11.14.4   Performing a Task

When an agent receives a delegation, it tries to perform the assigned task. Performing a task involves three elements: two are features of the agent (task specific ability and willingness). The third is the (possible) external influence of the environment. In order to be successful, an agent has to score a certain number of hits (e.g. *3*); a hit is scored if a random real number in *(0, 1)* is rolled that is less than its ability score. The agent has a number of tries that is equal to ten times its willingness value, rounded up (i.e. from 1 to 10 essays). The *environment can interfere* with an agent's activity giving a positive or negative modifier to each roll (so it interferes with ability but not with willingness). If the number of scored hits is sufficient, the task is performed; otherwise it is not.

The rationale of this method of task resolution is that, even if the tasks are abstract in our simulations, they semantically represent concrete ones: they involve a certain (controllable) amount of time; they are 'cumulative', in the sense that the total success depends on the success of its components; they can be achieved at different degrees (in our experiments the number of hits is used as a threshold of minimum performance and after being successful the agent can skip the other essays). Moreover, and most importantly for our theoretical model, the contribute of willingness (persistence) is clearly separated from ability; for each task 'attempting' is a prerequisite of 'doing' and an agent can fail in either. The contribution (positive or negative) of the environment is limited to the second phase, representing a favorable or unfavorable location for executing the task. The duration of the task is used to introduce another crucial factor that is *monitoring*: a delegator has to be able to control the activity of the delegee and possibly retire the delegation if it is performing badly; this aspect will be introduced in one of the experiments.

In the simulations we used *3 hits as a default*; however, all the effects are stable and do not depend on the number of hits. We have performed experiments with a different number of hits, obtaining similar results: choosing higher values leads to less tasks performed on average, but the effects remain the same.

This kind of task performing highlights both the role of ability and of willingness (persistence).

11.14.5   FCMs for Trust

For the sake of simplicity we have assumed that each cognitive agent has access to all true data of the other agents and of the environment; these data involve their task specific *ability*, their *willingness* and their current *environment*. All these data are useful for many strategies: for example, the *best ability trustor* always delegates to the agent with the higher ability value for that task.

The *cognitive trustor*, following our socio-cognitive model of trust, builds an elaborated mind model of all the agents; this is done with Fuzzy Cognitive Maps, as described in

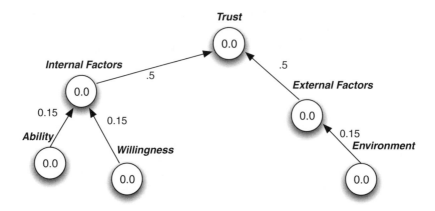

Figure 11.9 The FCM used by the *cognitive trustor*

(Falcone *et al.*, 2003).[7] The values of three nodes (ability and willingness as internal factors and environment as external factor) were set according to agent knowledge. The values of the edges reflect the impact of these factors and are always the same in the simulations. It has to be noticed that we never tried to optimize those factors: the results are always significant with different values. An additional point: while in the experiments the environment modifies the ability, in the 'mental representation' of FCMs this is not the case: this is not a piece of information that an agent is meant to know; what it knows is that there is an external (positive or negative) influence and it aggregates it by fulfilling the cognitive model. Figure 11.9 shows an (un-initialized) FCM.

11.14.6   Experiments Description

The first aim of our experiments is to compare the *cognitive trustor* and the *statistical trustor* in different situations: their delegation strategy represents two models of trust: derived from direct experience versus experience built upon a number of cognitive features. The random strategy was added as a baseline for the difficulty of the setting. The *best ability* and *best willingness* strategies are added in order to verify, in different settings, which are the single most influential factors; as it emerges from the experiments that their importance may vary, depending on some parameters.

In all our experiments we used exactly six agents (even if their delegation strategies may vary); it is important to always use the same number of agents, otherwise the different sets of experiments would not be comparable. In each experiment the task set by all agents is always the same; their ability set and willingness, as well as the environment influence, are randomly initialized.

The experiments are performed in a variable environment that influences (positively or negatively) the performance of some agents in some tasks, as previously explained.

---

[7] With respect to the general model, for the sake of simplicity we assume that unharmfulness and danger nodes are always 0, since these concepts have no semantic in our simulations.

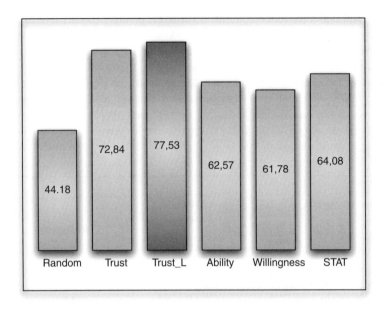

Figure 11.10 Experiment 1: comparison between many delegation strategies, 3 hits; (measuring the success rates). (Reproduced with kind permission of Springer Science+Business Media © 2005)

In order to allow the *statistical trustor* to learn from the experience, all the simulation sets were divided in two phases (two halves). The first phase is meant for training only: the statistical trustor delegates several times the tasks to all the agents and collects data from successful or unsuccessful performance. It uses these data in order to choose to whom to delegate in the second phase (in fact, it continues to learn even in the second phase). The delegation mechanism is always the same: it chooses the agent who has the best ratio between performed and delegated tasks; this number is updated after each result following a delegation. In order to measure the performance of this strategy, we analyzed only experimental data from the second phases.

The first experiment (***EXP1***) compares the *random trustor* (*RANDOM*), the best *ability trustor* (*ABILITY*), the *best willingness trustor* (*WILLINGNESS*), the *statistical trustor* (*STAT*), and two other cognitive strategies that differ only because of how much they weight the environmental factor: *no impact* (*TRUST*) does not consider the environment, while *low impact* (*TRUST_L*) gives it a low impact (comparable to the other factors). In Figure 11.10 we show 250 simulations for 100 tasks.

We can see that the cognitive strategies always beat the statistical one.[8] Moreover, it is important to notice that recognizing and modeling the *external components* of trust (the environment) leads to a very high performance: the cognitive trustor who does not consider the environment (*TRUST*) beats the statistical one (*STAT*), but performs worse than the cognitive trustor who gives a role to the environment (*TRUST_L*).

---

[8] The results are similar e.g. with five hits (250 simulations, 100 tasks): *RANDOM*: 26,24; *TRUST*: 57,08; *TRUST_L*: 61,43; *ABILITY*: 40,58; *WILLINGNESS*: 48,0; *STAT*: 49,86.

We have performed three more experiments in order to verify another interesting condition about learning (250 simulations, 100 tasks). Sometimes it is not possible to learn data in the same environment where they should be applied. For this reason, we have tested the *statistical trustor* by letting them learn without environment and applying data in a normal environment (*EXP2* – positive and negative influences as usual), in an always positive environment (*EXP3* – only positive influences), and in an always negative environment (*EXP4* – always negative influences). As easily foreseeable, the mean performance increases in an always positive environment and decreases in an always negative environment; while this is true for all strategies, the statistical strategy has more troubles in difficult environments. Figure 11.11

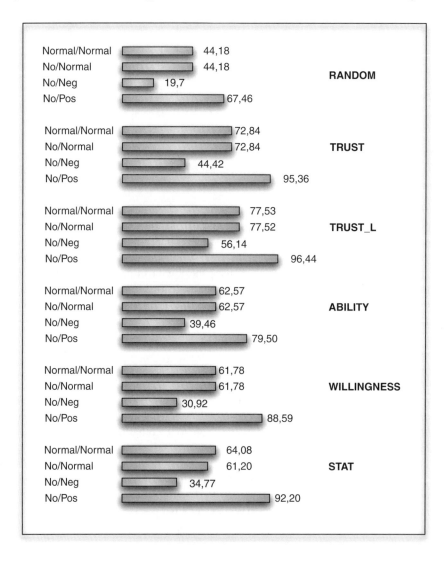

Figure 11.11 Experiments 1, 2, 3 and 4 compared (measuring the success rates). (Reproduced with kind permission of Springer Science+Business Media © 2005)

shows four cases:

1) learning in normal environment, task in normal environment (normal/normal);
2) learning without environment, task in normal environment (no/normal);
3) learning without environment, task in negative environment (no/neg);
4) learning without environment, task in positive environment (no/pos).

11.14.7   Using Partial Knowledge: the Strength of a Cognitive Analysis

The results achieved in the above experiments are quite interesting, but rather predictable. More interesting and with high degree of difficulty of prediction is the experiment in which we try to individuate the level of approximation in the knowledge of a cognitive trustor about both the properties of other agents and of the environment. In other words, we would like give an answer to the questions: *when is it better to perform as a cognitive trustor with respect to a statistical trustor? What level of approximation in the a priori knowledge is necessary in order that this kind of trustor will have the best performance?*

To answer this interesting question we have made some other experiments (as **EXP5**) about errors in evaluation. As already stated, all the values we assume about cognitive features are true values: each agent knows all the real features of the others.

This is an ideal situation that is rarely implemented in the real world/system. In particular, in a multi-agent system environment there can be an evaluation process that is prone to errors. In order to evaluate how much error the cognitive trustor can deal with without suffering from big performance losses, we have compared many cognitive trustors introducing some different levels of 'noise' in their data.

Figure 11.12 shows the data for the *random trustor (RANDOM); the best willingness trustor (WILLINGNESS); the best ability trustor (ABILITY)*; the normal *cognitive trustor (NOERR)*, as well as some other *cognitive trustors (ERR_40, ERR_20, ...)* with 40%, 30%, 20%, 10%, 5%, 2.5% error; the *statistical trustor* (STAT). While all the experiments used a set of six agents, we present aggregated data of different experiments. We have ordered the strategies depending on their performance; it is easy to see that even the worst cognitive *trustor* (40% error) beats the statistical trustor. Under this threshold we have worse performances.

Real Time Experiments

We have performed some real time experiments, too. **EXP6** (see Figure 11.13) involves three cognitive strategies in a normal environment (250 simulations, 500 tasks).

The differences between the cognitive trustor without environment and the two with environment are statistically meaningful; the difference between the two cognitive trustors with environment are not. The results are very close to those that use turns; the differences depend on the limited amount of time we set for performing all the tasks: by augmenting this parameter more quickly, strategies become more performing.

Another experiment (**EXP7**) aims at testing the performance of the first trustworthy trustor (**FIRST**). Here there are two parameters for performance: *Credits* and *Time*. *Time* represents

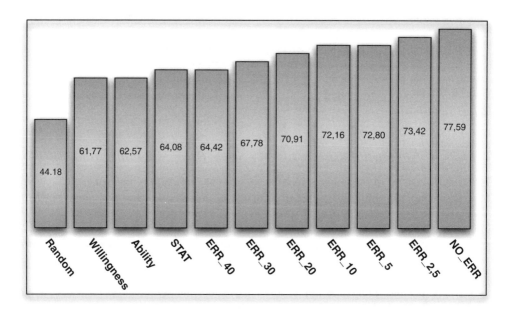

Figure 11.12 Experiment 5: introducing noise (measuring the success rates). (Reproduced with kind permission of Springer Science+Business Media © 2005)

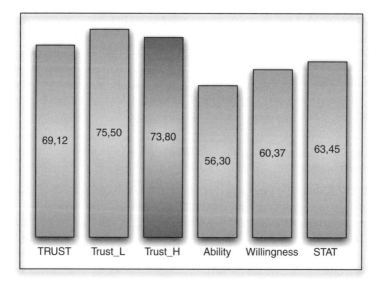

Figure 11.13 Experiment 6: real time (measuring the success rates). (Reproduced with kind permission of Springer Science+Business Media © 2005)

how much time is spent in analyzing offers and delegating, i.e. how much offers an agent collects before choosing.[9]

While in the preceding experiments the agents collected all the offers before deciding, here an agent can delegate when it wants, saving time. This situation is closer to a real MAS situation, where agents act in real time and sometimes do not even know how many agents will offer help. How much time is spent in delegation depends on the strategy and on simulation constraints. The random trustor can always choose the first offer it has, so it results in being the quickest in all cases. If there is a fixed number of agents and the guarantee that all of them will offer, best ability, best willingness, the cognitive trustors and the statistical trustor can build and use an ordered list of the agents: so they have to wait until the offer from the pre-selected agent arrives. In the more interesting MAS scenario, without a fixed number of offering agents, each incoming offer has to be analyzed and compared with the others. In order to avoid waiting ad infinitum, a maximum number of offers (or a maximum time) has to be set.

However, in this scenario there can be other interesting strategies, such as the first trustful trustor, who does not wait until all six offers are collected, but delegates when the first 'good offer' (over a certain threshold) is met; this can lead to more or less time saved, depending on the threshold. Here we present the results of **EXP7** (250 simulations, 100 tasks); in this case all agents wait for exactly six offers (and compare them) before delegating, except for the random trustor (who always delegates to the first one) and the first trustful trustor who delegates to the first one that is over a fixed threshold. Figure 11.14 and Figure 11.15 show the results for credits (as usual) and time spent (analyzed offers).

The first trustworthy trustor still performs better than the statistical trustor, saving a lot of time. Depending on the situation, there can be many ways of aggregating data about credits and time. For example, in a limited time situation agents will privilege quickness over accurateness; on the contrary, in a situation with many agents and no particular constraints over time it would be better to take a larger amount of time before delegating.

Experiments with Costs

In our simulations we assume that the costs (for delegation, for performing tasks, etc.) are always the same; in the future it would be interesting to introduce explicit 'costs' for operations, in order to better model real world situations (e.g. higher costs for more skilled agents).

The costs are introduced as follows. When an agent sends their 'proposal' they quote a price that averages to their ability and willingness (from 0 to 90 Credits).[10] Half the cost is paid on delegation, the second half is paid only if the task is successfully performed. If a task is re-delegated, the delegator has to pay again to delegat. For each successfully performed task the delegator gains 100 credits. There may in the future even be costs for receiving the reports, but at the moment this does not happen.

In order to model an agent who delegates taking into account the gain (and not the number of achieved tasks) we have used a very simple utility function, that simply multiplies trust and (potential) gains minus costs. It has to be noticed that on average a better agent has a

---

[9] There are other possible parameters, such as time spent in reasoning or in performing a task. However, we have chosen only the parameter which is more related to the Delegation Strategy; the other ones are assumed to have fixed values.

[10] We generate randomized values over a bell curve which averages it (ability + willingness)/2.

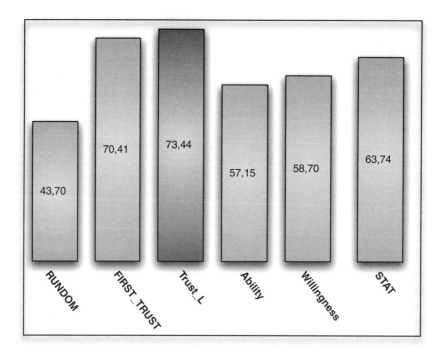

Figure 11.14 Experiment 7: real time, introducing the first trustworthy strategy (measuring Credits). (Reproduced with kind permission of Springer Science+Business Media © 2005)

higher cost; so the agent who maximizes trust is penalized with respect to gain. The agent who maximizes following the utility function chooses agents with less ability and willingness; for this reason it performs fewer tasks than the other agents.[11]

EXP8 (see Figure 11.16) was performed for 200 tasks and 250 simulations. The results refer (1) to the tasks performed;[12] (2) to the gains; (3) to time spent. The policies are: *random*; *cost_trust* (that uses the utility function); *trust_ambient* (that uses trust as usual); *low_cost* (that always delegates to the less expensive); *first_trust* (that delegates to the first over a certain trust threshold), *stat* (that performs statistical learning).

The most important result is about gains; the agent who explicitly maximizes them has a great advantage; the trust strategies perform better than the statistical one, even if they all do not use information about gains but about tasks achieved. It is even interesting to notice that the worst strategy results delegating always to the least expensive: in this case costs are related to performance (even if in an indirect way) and this means that cheap agents are normally bad.

Delegation and Monitoring

In order to model a more complex contract net scenario, we have introduced the possibility of the delegator monitoring the performance of the delegated agents into the intermediate steps

---

[11] We have chosen a little difference between costs and gains in order to keep costs significant: setting lower costs (e.g. 1–10) makes them irrelevant; and the trust agent performs better.

[12] Multiplied * 10 in order to show them more clearly.

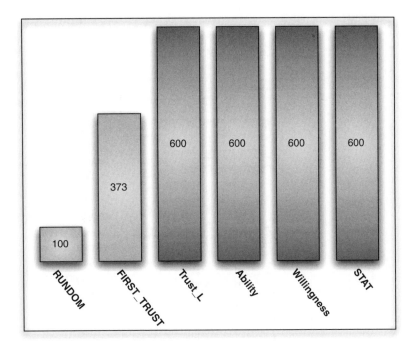

Figure 11.15 Experiment 7: real time, introducing the first trustworthy strategy (measuring time spent). (Reproduced with kind permission of Springer Science+Business Media © 2005)

of the task; he can decide for example to stop the delegation and to change delegee. In this case, the agent has a new attribute: *controllability*. A more complex task resolution scheme and cost model (e.g. the costs of stopping or changing the delegation) is needed, too.

In order to exploit the controllability, we changed the task resolution scheme. In order to be successful, an agent has, as usual, to *score a certain number of hits* (e.g. three). In order to score a hit, instead of performing a number of attempts equal to their willingness, the agent has *exactly 10 attempts*: for each attempt, he first tests his willingness (i.e. if it actually tries to perform it); if this test is successful, it tests his ability (i.e. if the delegee is able to perform the task). Each test is simply 'rolling a random real number in (0,1): if the roll is less than the tested attribute (Willingness or Ability), the test is successful. The environment can interfere with the agent's activity setting the Difficulty of each Ability test. At the end of the ten attempts, if the number of scored hits (e.g. three) is sufficient, the task is performed; otherwise it is not.

In addition to the usual tests of willingness and ability, for each one of the ten attempts each Agent checks if the delegee sends a report (a *controllability* check); the report contains the number of the current attempt, information about the activity in this attempt (tried/not tried; performed/not performed) as well as the number of successes achieved. At the end of the ten attempts, the agent sends a final message that says whether the task was successful or not.

Monitoring consists of receiving and analyzing the reports. Basing on this information, an agent can decide either to confirm or to retire the delegation: this can only be done before the final message arrives. If they retire the delegation, they can repost it (only one more time). Obviously, agents having a higher controllability send more reports and give more opportunities to be controlled and stopped (and in this case the tasks can be reposted).

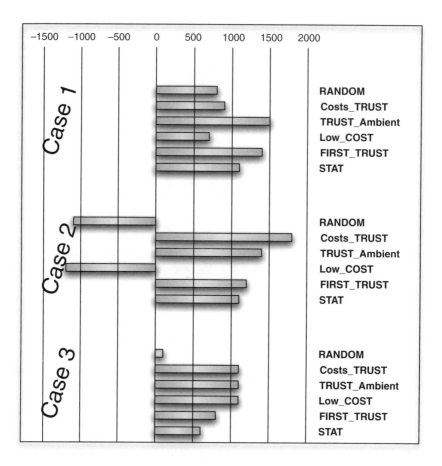

Figure 11.16 Experiment 8: comparing different delegation strategies in different cases: number of tasks performed (success rates, case 1), number of credits (gains, case 2), and number of analyzed offers (time spent, case 3). (Reproduced with kind permission of Springer Science+Business Media © 2005)

In *EXP9* (see Figure 11.17) we use controllability and we offer the possibility of re-posting a task. An agent who uses controllability as an additional parameter for trust (i.e. giving a non null weight to the corresponding edge in its FCM) is compared with agents who do not use it; it is (more or less) biased towards choosing agents who give more reports and so they have more possibilities of re-posting unsuccessful tasks (tasks can be reposted only once). All agents (except *TRUST_BASE*) use the following heuristic in order to decide whether to retire delegation:

$$if \ (10 - current\_essay/current\_essay < 1) \ and \ (success\_number < 3)$$
$$and \ (I\_can\_repost) \ then \ retire\_delegation$$

The experiment was performed for 200 tasks and 250 simulations. *RANDOM* is the baseline; *TRUST* is the agent who uses ability, willingness and environment but without controllability; *TRUST_BASE* is the same agent but the one who never reposts their tasks; *CONTROL_1*

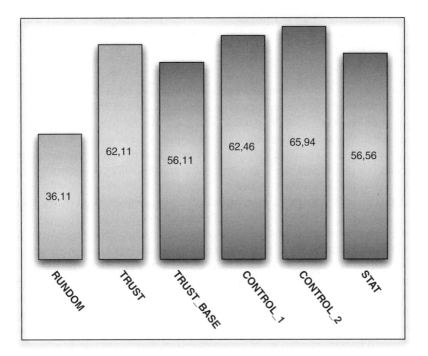

Figure 11.17 Experiment 9: monitoring (measuring the success rates). (Reproduced with kind permission of Springer Science+Business Media © 2005)

introduces a little weight on controllability; ***CONTROL 2*** introduces a significant weight on controllability; ***STAT*** is statistical.

Note that since the task resolution scheme is changed, these results are (on average) lower than in the other experiments; however, it is possible as usual to compare the different strategies. Considering controllability gives a significant advantage over the other strategies.

11.14.8   Results Discussion

In our experiments we have tried to compare different trust strategies to delegate tasks in a contract net. The setting abstracts a simplified real-world interaction, where different agents have different capabilities (mostly represented by ability) and use more or less similar resources (mostly represented by willingness) in order to realize a certain task. The role of the environment is also very relevant because external conditions can make the tasks more or less easy to perform. On the basis of their trust, the delegating agents (with different strategies) decide to whom to assign their tasks.

We analyzed two concepts of trust:

- the first (referred to as the *statistical trustor*), that it is possible to model the trustworthiness of other agents only on the basis of the direct (positive or negative) experience with them; on the fact that there is only one dimension to consider: the number of successes or failures the agents preformed in the previous experiences.

- the second (referred to as the *cognitive trustor*), based on a more complex set of factors that have to be considered before trusting an agent; in particular, both a set of trustor features and environmental features.

In all our experiments the cognitive trustors perform better than the statistical ones, both from the point of view of global successes (number of *Credits*) and stability of behavior (less standard deviation in the simulations data). The cognitive strategy models the agents and environment characteristics more effectively and it allows the resources to be allocated in a highly accurate way. Introducing a changeable environment does not decrease performance, providing that it is considered as a parameter; but even if it is not considered, the results are largely better than with a statistical trustor.

The fact that an algorithm that knows the real processes implemented by the agents when achieving tasks uses a simulation mechanism of these processes for selecting the best performances is quite predictable. For this reason we have made new experiments introducing a significant amount of noise in the cognitive agent knowledge. The results show that the performance of the cognitive agent remains better than the statistical one up to an error of 40%. So, the cognitive trustor is very accurate and stable under many experimental conditions. On the contrary, even with a large amount of data from learning (the training phase), the statistical strategy is not performing well. Moreover, if the learning is done in a different environment, or if the environment is particularly negative, the results are even worse.

With a low number of hits (e.g. three) the task is designed to privilege ability over willingness; however, augmenting the number of hits, the relative relevance changes. A strong environmental influence shifts the equilibrium, too: it modifies the ability scores which become more variable and less reliable. Modifying the relative weight of those parameters (depending on the situation) into the FCM of the cognitive trustor can lead to an even better performance.

In the real time experiments, when time is implicitly introduced as an additional performance measure, a variant of the *cognitive trustor*, the *first trustful trustor*, becomes interesting: it maintains high task performance (measured by credits) with a limited amount of time lost.

Introducing costs into the experiment leads the agents to maximize another parameter, gain, with respect to tasks achieved. It is not always the case that more tasks mean more gain, because many agents who perform well are very costly; in fact the best strategy optimizes gains but not the number of achieved tasks.

Introducing a monitoring strategy, with the possibility of retiring the delegation and reposting the task, introduces an extra possibility for the agents, but also another difficulty, because each re-post is costly. Considering explicitly the controllability as a parameter for trusting an agent gives a significant advantage, because more controllable agents – especially in real time – enable a more precise distribution of the tasks and even a recovery from wrong choices.

Depending on the situation (e.g. with or without environment; with or without the possibility of retiring the delegation) and from the goals (e.g. maximize tasks, time or gains) the possible delegation strategies are many. In all cases, trust involves an explicit and elaborated evaluation of the current scenario and of the involved components – and our results demonstrate that this gives a significant advantage in terms of performance with respect to a mono-dimensional statistical strategy.

11.14.9   Comparison with Other Existing Models and Conclusions

Many existing trust models are focused on reputation, including how trust propagates into recommendation networks (Jonker and Treur, 1999) (Barber and Kim, 2000) (Jøsang and Ismail, 2002). On the contrary, our model evaluates trust in terms of beliefs about the trustee's features (ability, willingness, etc.); reputation is only one kind of source for building those beliefs (other kinds of source are direct experience and reasoning). In the present experimental setting there is not any reputational mechanism (that we could also simulate in the cognitive modeling), so a comparison with these models is not appropriate.

There are some other approaches where trust is analyzed in terms of different parts; they offer a more concrete possibility for comparison. For example, in (Marsh, 1994) trust is split into: *Basic Trust, General Trust* in agents, *Situational Trust* in agents. Basic trust is the general attitude of an agent to trust other agents; it could be related to our model if it is considered as a general attitude to delegate tasks to other agents in the trust relationships; in the experiments already illustrated, we did not consider the possibility of introducing agents with the inclination to delegate to others or to do the task themselves. In any case, the setting can certainly include these possibilities. General trust is more related to a generic attitude towards a certain other agent; the more obvious candidate in our setting is willingness, even if the two concepts overlap only partially. Situational trust is related to some specific circumstances (including costs and utilities, that are not investigated here); there is a partial overlap with the concept of ability, that represents how well an agent behaves with respect to a certain task. So, the model presented in (Marsh, 1994) is, to a certain extent, comparable with our one; however, it lacks any role for the environment (more in general for the external conditions) and it introduces into trust the dimensions of costs and utility that in (Castelfranchi and Falcone, 1998), (Falcone and Castelfranchi, 2001) are a successive step of the delegation process that is presented in a simplified way.

Our experiments show that an accurate socio-cognitive model of trust allows agents in a contract net to delegate their tasks in a successful way. In any case, for better testing the model it is be necessary to realize a set of new experiments in which we even allow the *cognitive trustors* to learn from experience. While the learning of the *statistical trustor* is undifferentiated, the cognitive trustor is able to learn in different ways from different sources. In the model for each trust feature there are (at least) four different sources: direct experience, categorization, reasoning, reputation; each of them contributes in a different way. More, higher level strategies can be acquired: for example, depending on the environment and the task difficulty (number of hits) an optimal weight configuration for the FCMs can be learned.

Other directions of work could be to experiment with agents starting with some a priori knowledge about other agent's stats with a percentage of error, and in which they can refine this percent by analyzing how well they perform in the delegated tasks. This is a kind of statistical, not specific learning. However, in order to learn in a more systematic way, an agent has to discriminate each single stat. In order to do this, they could analyze the incoming reports (e.g. how many times an agent tries a task for willingness; how many times they perform it for ability; how many reports they send for controllability). The controllability stat introduces an upper limit even to how many learning elements an agent can receive, so it becomes even more critical.

Finally, we would like to suggest that a reputation and recommendation mechanism is included, in order to add another trust dimension to the simulations. In this way we could introduce new delegation strategies (based on the existing systems in literature), and study how reputation interacts with the other cognitive features in the cognitive trustor.

References

Barber, S., and Kim, J. (2000) Belief revision process based on trust: agents evaluating reputation of information sources, *Autonomous Agents 2000 Workshop on 'Deception, Fraud and Trust in Agent Societies'*, Barcelona, Spain, June 4, pp. 15–26.

Castelfranchi, C. Reasons: belief support and goal dynamics. *Mathware & Soft Computing*, 3: 233–247, 1996.

Castelfranchi, C. and Falcone, R. Towards a theory of delegation for agent-based systems, *Robotics and Autonomous Systems*, Special issue on Multi-Agent Rationality, Elsevier Editor, 24 (3-4): 141–157, 1998.

Castelfranchi, C., de Rosis, F., Falcone, R., Pizzutilo, S. Personality traits and social attitudes in Multi-Agent Cooperation, *Applied Artificial Intelligence Journal*, special issue on 'Socially Intelligent Agents', 12 (7/8): 649–676, 1998.

Dragoni, A. F. (1992) A model for Belief Revision in a Multi-Agent Environment. In *Decentralized AI - 3*, Y. Demazeau, E. Werner (eds.), pp. 215–231. Amsterdam: Elsevier.

Dubois, D. and Prade, H. (1980) *Fuzzy Sets and Systems: Theory and Applications*, Academic Press, Orlando, FL.

Falcone, R. and Castelfranchi, C. (2001) Social trust: a cognitive approach, in C. Castelfranchi and Y. Tan (eds.), *Trust and Deception in Virtual Societies*, Kluwer Academic Publishers, pp. 55–90.

Falcone, R., Pezzulo, G., Castelfranchi, C. A fuzzy approach to a belief-based trust computation. *Lecture Notes on Artificial Intelligence*, 2631, 73–86, 2003.

Falcone, R., Pezzulo, G., Castelfranchi, C. (2005) A fuzzy approach to a belief-based trust computation, in Falcone, R., Barber, S., Sabater-Mir, J., Singh, M., (eds.), *Trusting Agents for Trusting Electronic Societies*. Lecture Notes on Artificial Intelligence, n°3577, Sprinter (pp. 43–58).

Jonker, C., and Treur, J. (1999) Formal analysis of models for the dynamics of trust based on experiences, Autonomous Agents '99 Workshop on 'Deception, Fraud and Trust in Agent Societies', Seattle, USA, May.

Jøsang, A., and Ismail, R. (2002) *The Beta Reputation System*. In the proceedings of the 15th Bled Conference on Electronic Commerce, Bled, Slovenia, 17–19 June 2002.

Kosko, B. (1986) Fuzzy cognitive maps. *International Journal Man-Machine Studies*, 24: 65–75.

Marsh, S.P. (1994) Formalising trust as a computational concept. PhD thesis, University of Stirling. Available at: http://www.nr.no/abie/papers/TR133.pdf.

Pezzulo, G. and Calvi, G. (2004) AKIRA: a Framework for MABS. *Proc. of MAMABS 2004*.

Schillo, M., Funk, P., and Rovatsos, M. (1999) Who can you trust: Dealing with deception, Autonomous Agents '99 Workshop on 'Deception, Fraud and Trust in Agent Societies', Seattle, USA, May 1. 1: 81–94.

Smith, R.G. (1980) The contract net protocol: High-level communication and control in a distributed problem solver. *IEEE Transactions on Computers*.

12

Trust and Technology

In this book our effort is to model and rationalize the trust notion trying to catch all the different and varying aspects of this broad concept. In fact, there are today many studies, models, simulations and experiments trying to integrate trust in the technological infrastructures: The most advanced disciplines in Human-Computer Interaction (Baecker *et al.*, 1987), (Card *et al.*, 1983) and (Dix *et al.*, 2004), Distributed Artificial Intelligence (Lesser, 1990), (Hewitt and Inman, 1991), (Weiss, 1997), Multi-Agent Systems (Wooldridge, 2002), (Shoham and Leyton-Brown, 2008), and Networked-Computer Systems (Grid, Semantic Web, etc. (Foster and Kesselman, 2003), (Antoniou and van Harmelen, 2008) and (Davies, 2006)) are forced to cope with trust.

But why does trust seem to be so important in the advanced technological contexts? Is it necessary to involve such a complex, fuzzy and human related concept? Is it not sufficient to consider just more technical and simply applicable notions like security?

To give a satisfactory response to these questions we have to evaluate which kind of network infrastructures are taken into consideration in the new communication and interaction scenarios, which kind of peculiar features should have the artificial agents we have to cope with, which kind of computing is going to invade (pervade) the future physical environments?

In fact, trust becomes fundamental in the *open* multi-agent systems where the agents (which could be both human beings and artificial agents owned by other human stakeholders) can (more or less freely) enter and leave the system. The evolution of the interaction and communication technological paradigms toward human style, is, on the one hand, a really difficult task to realize, but, on the other hand, it potentially increases the people accessing to (and fruitful in using) the new technologies. In fact, in the history of their evolution humans have learned to cooperate in many ways and environments; on different tasks; and to achieve different goals. They have intentionally realized (or they were spontaneously emerging) diverse cooperative constructs (purely interactional, technical-legal, organizational, socio-cognitive, etc.) for establishing trust among them.

It is now necessary to remodel the trust concept in the new current and future scenarios (new channels and infrastructures of communication; new artificial entities, new environments) and the efforts in the previously cited scientific fields (HCI, MAS, DAI, NCS) are trying to give positive answers to these main requirements.

Trust Theory   Cristiano Castelfranchi and Rino Falcone
© 2010 John Wiley & Sons, Ltd

Without establishing trustworthy relationships, these new infrastructures and services, these new artificial agents, these new robots, these new pervasive technologies, do not impact with sufficient strength and in fact do not really integrate with the real society.

One of the main features of these new artificial entities (but in fact of this new technological paradigm) is in particular, *autonomy*: having the capacity to realize tasks without direct human control and monitoring, having the capacity to attribute and permit the single (human or artificial) entities to realize their own goals, having the capacity to make decisions on the basis of their own attitudes, beliefs and evaluations (see Chapter 7 for a detailed analysis of this concept and for its relationships with trust). The new environments are increasing the autonomy levels and complexities of these agents offering sophisticated interaction and cooperation. In these environments no agent can know everything, there is no central authority that controls all the agents (due to the features of the environment).

At the same time these complex autonomies (in open environments, broad communities and with indirect interaction) increase human diffidence and risks.

Technology should not only be reliable, safe, secure, but it *should be also perceived* as such, the user must believe that it is reliable, and must feel confident while using it and depending on it. *The unique real answer for coping with others' autonomy is to establish a real trust relationship.* For these reasons, the ability of understand and model the trust concept to transfer its utility in the technological cooperative framework will be in fact the *bottleneck* of the development of the autonomy-based technology that is the technology of the future.

12.1   Main Difference Between Security and Trust

One important thing to underline is the conceptual difference between the *two notions of security and trust*. In general, a secure system should provide mechanisms (Wong and Sycara, 2000) able to contrast (oppose) potential threats and guarantee a set of features:

- *certainty of identification*: in particular techniques of authentication should be able to identify the interacting agents; this identification allows accessibility to defined rights and resources (Grandison and Sloman, 2000);
- *integrity*: the messages and the actions of the agents should not be corrupted by a third party;
- *confidentiality and not intrusivity*: the communication and interaction should remain private if the decision of the agents is so;
- *nonrepudiation*: in specific cases it should be possible to identify unambiguously the author of messages or actions, and they cannot deny this objective identification;
- *secure delegation*: it should be clear who is the delegator of each agent.

There are various research areas (*encryption* (Ellis and Speed, 2001), *cryptography* (Schneier, 1996), (Stallings, 1999), *authentication* (Stallings, 2001), *access control* (Anderson, 2001)) that develop techniques for achieving the above specified features of security.

The objective of automating the procedures of the traditional security systems has viewed and currently views many studies and applications, some of them make explicit reference to trust even if this concept is used in a very reductive and basic sense, oriented toward the strict security rather than to the more complex and general concept of trust. Examples of this use are the so called *Trusted Systems* (Abrams, 1995); the so called *Trusted Computing* (mainly used by industry and regarding the information being processed on a platform with specialized security hardware (see (Josang, 2007) for more details); the so called *Trust Management* ((Blaze *et al.*, 1996)) mainly used for security in the distributed access control; the so called *Trusted Third Parthy* (TTP, (Skevington and Hart, 1995)) describing systems in which the presence of a reputed, disinterested, impartial and responsible entity that is accepted by all the parties is guaranted.

The most interesting, recent works (relatively more oriented towards the very concept of trust) in this area are: *Trust-Serv* (Skogsrud *et al.*, 2003), *PolicyMaker* (Grandison and Sloman, 2000), and *KAoS* (Uszok *et al.*, 2004). In these last systems and approaches the main goal is to provide agents with credentials able to obtain trust from the system on the basis of predefined policies. We have to say that even if these systems are a step towards the real concept of trust, in general the main problem of the multi-agent systems is about how an agent can rely on other agents for achieving its own goals (Huynh *et al.*, 2006).

An interesting distinction in the field is described by Rasmussen and Jansson (Rasmussen and Jansson, 1996) between *hard security* and *soft security*, where hard security is referred to the traditional IT (Information Technology) security mechanisms such as those above defined (access control, authentication, and so on) while soft security is about deceitful and malicious service providers that provide misleading, tricky or false information (Rasmussen and Jansson called this security 'social control mechanisms').

In general we can say that *establishing a true trust relationship is a more complex and different thing with respect to security matter*: the above described techniques cannot guarantee that an interaction partner has the competence he claims or that he is honest about his own intentions.

Trust is more than secure communication, e.g., via public key cryptography techniques: the reliability of information about the status of your trade partner has little to do with secure communication or with its identification. *Maybe perceived security and safety are a precondition and also an aspect of trust, but trust is a more complex and broad phenomenon. Trust must give us tools for acting in a world that is in principle insecure where we have to make the decision to rely on someone in risky situations.*

For this reason the trust challenge is more complex and advanced (and therefore more ambitious) than the one about security, even if there are relationships between them and the solutions to the security problems represent a useful basis for coping with trust problems.

12.2   Trust Models and Technology

In the last fifteen years many studies and researches have been developed in the technological field on trust (for a resume see (Marsh, 1994), (Castelfranchi and Tan, 1999), (Falcone *et al.*, 2001), (Falcone *et al.*, 2003), (Ramchurn *et al.*, 2004), (Falcone *et al.*, 2005), (Huynh *et al.*, 2006), (Cofta, 2007), (Falcone *et al.*, 2008), (Golbeck, 2009)). These works have analyzed different aspects, models, and approaches to trust with the common goal of understanding and introducing trusted relationships within the computational framework.

Let us consider the most relevant approaches to trust in the technological domain: the *logical approaches*, the *computational approaches*, and the *socio-cognitive approaches*. These three approaches often have a varying overlap with each other, but the differences are given by the relevance of the goals they mainly are pursuing. In this part of the book we will omit to introduce the socio-cognitive approaches, which have been discussed at length in the other chapters. In addition you can refer to Chapter 11 for a description of a fuzzy implementation model for the socio-cognitive approach to trust.

12.2.1   Logical Approaches

The *logical approaches* ((Cohen and Levesque, 1990), (Fagin *et al.*, 1994), (Demolombe, 1999), (Jones and Firozabadi, 2001), (Josang, 2001), (Liau, 2003), (Lorini and Demolombe, 2008), (Castelfranchi *et al.*, 2009)) start from models based on mathematical logics for describing, analyzing and implementing the trust relationships. These approaches have the advantage of using powerful methods able to produce inferences and strongly rationalize the conceptual apparatus. The drawbacks are given as the constraints introduced and derived through the same logics that in fact impose their own rules to the conceptualization of thought and action, very often without considering all the elementary criteria of flexibility of the human reasoning and action. In various cases, the approximation of the formalisms to the reality can satisfy specific descriptive purposes of the reality, in other cases this approximation is not appropriate (in the sense that it does not introduce realistic constraints).

A very elegant example of logical approach is given by the Demolombe's analysis (Demolombe, 1999) with respect to the trust in information sources. With a clear (but functional) simplification (maybe too superficial with respect to the social concepts he intends to model), he considers that each information source can have four different properties: *sincerity*, *credibility*, *cooperativity*, and *vigilance*.

- An agent X is *sincere* with respect to another agent Y and a specific content p if X believes p when he is informing Y about p.
- An agent X is *credible* with respect to another agent Y and a content p if X believes p and p is true in the world.
- An agent X is *cooperative* with respect to another agent Y if what he believes is communicated to Y.
- An agent X is *vigilant* with respect to the world if what is true in the world is believed by X.

Demolombe also derives two other concepts: *Validity*, as the conjunction of sincerity and credibility; and *Completeness*, as the conjunction of cooperativity and vigilance.

Using the modal logic (Chellas, 1990) Demolombe formalizes these concepts and defines different kinds of trust: trust with respect to sincerity, credibility, cooperativity, and vigilance. Then he is able to derive consequences from this representation: For example, he is able to derive additional properties like 'if Y trusts X as sincere, then the information p (received by X) let infer to Y that X believes p', and so on.

Jones (Jones and Firozabadi, 2001) represents some interesting aspects of trust (like deception in terms of trust, trust in other's trust) applying the logic of belief together with the deontic logic, and with the logic of 'count as'. We deeply analyze the Jones's model in another part of this book (see Chapters 2 and 8 and in particular Section 2.2.2).

One of the main problems of applying the classical logical framework to the mental attitudes (on which trust is based) is the difficulty of taking into consideration *uncertainty* and *ignorance*: typical features of the beliefs. Different and interesting attempts (in particular the one of

Dempster and Shafer (Shafer, 1976) to cope with these limits are presented in (Motro and Smets, 1997). However, given his direct work on trust and reputation models, it is of interest to cite Josang's approach (Josang, 2001) that introduces the *subjective logic*: an attempt to overcome the limits of the classical logics and also taking into consideration the uncertainty, ignorance and the subjective characteristic of the beliefs.

This approach is strongly influenced by Dempster and Shafer's work but with some specific interesting intuitions. The *opinions* in Josang's approach are belief/trust metrics denoted by:

$$\omega_x^A = (b, d, u, a) \tag{12.1}$$

where ω_x^A expresses the trust of agent A about the truth of the statement of x. b represents the positive A's beliefs, d represents the A's disbeliefs, u represents the uncertainty with:

$$b, d, u \in [0, 1] \qquad b + d + u = 1 \qquad a \in [0, 1]$$

The parameter a is called *relative atomicity* and represents the base rate probability in the absence of evidence: it determines how uncertainty contributes to an opinion's probability expectation value $E(\omega_x^A)$:

$$E\left(\omega_x^A\right) = b + au \tag{12.2}$$

The subjective logics introduce two operators (*discounting* and *consensus*) useful for trust derivation from other opinions.

12.2.2   Computational Approach

The *computational approach* to trust has as a main goal the implementation of a trust model in an automatic system independent from the representational framework. The computational trust models can follow different principles on the basis of the adopted approaches with respect to:

- the *sources* on which the model evaluates the trustee's trustworthiness,
- the kind of *metric* for trust measures.

Different Kinds of Sources

With respect to the trust sources, as shown in Chapter 6, we can distinguish among *direct experience* and *indirect experience*. Direct experience is the more simple and elementary agent's source deriving from its previous direct experiences with other agents and with the world; it strongly depends on the agent's perceptive apparatus: the kind of input it is able to perceive.

Indirect experience sources can be, in their turn, articulated in the so-called *reputation* (others' experience directly communicated or made available (and possibly mediated) by some central or decentral mechanism) and types of *general reasoning and deduction*, like inference over categories, classes of agents, situations (scripts), and so on.

The sources based on the reasoning and their influences on the trust decision are partially shown in Chapter 6 of this book. Reputation has become a very diffused and practiced study field; in fact, on the basis of reputation a set of automatic models and systems for attributing trust were directly studied and built in the last 15 years. In many cases the reputational approach represents the only criterion for defining the trustworthiness of the interactive agents.

Reputation mechanisms are distinguished in *centralized* and *decentralized* mechanisms.

Centralized Reputation Mechanisms

Centralized reputation mechanisms are widespread in electronic commerce: *eBay* [eBay site] (Resnick and Zeckhauser, 2002), *Amazon* (Amazon site], and many others' e-commerce systems manage these kinds of mechanisms in which all the users have their own reputational profile stored in a centralized database. In these systems each user, after an interaction (transaction) with other users, reports on the behavior of the other providing appropriate ratings and giving textual comments. These ratings and comments are public and each user can read them before starting a new interaction/business with a specific agent. In the eBay system the rate scale is from $-1, 0, +1$ (respectively negative, neutral and positive). All the ratings are stored centrally and the global reputation value is the sum of all the ratings in the last six months. The main limit of this approach is given by the extreme simplicity of the model. In fact, just one dimension of the trustworthiness is taken in consideration (that is a more complex entity) and the acritical aggregation of the performances do not give account of the possibility of cheating in few interactions maintaining a good reputation value.

To overcome the limits of the reputation systems shown above, *SPORAS* (Zacharia and Maes, 2000) has been developed and it introduces new methods for aggregating the ratings. In particular, the updating of the ratings follows these principles:

1. New users start with a minimum reputation value and they build up reputation during their activity on the system.
2. The reputation value of a user never falls below the reputation of a new user;
3. After each transaction, the reputation values of the involved users are updated according to the feedback provided by the other parties, which reflect their trustworthiness in the latest transaction;
4. If two users happen to interact more than once the system keeps the most recently submitted rating.
5. Users with a very high reputation value experience much smaller rating changes after each update.
6. Ratings must be discounted over time so that the most recent ratings have more weight in the evaluation of a user's reputation.

The six above principles define a more interesting dynamics of the reputation model with respect to more static ones like *eBay*, or *Amazon*. In addition, *SPORAS* introduces a measure of the reliability of the users' reputations: This reliability is based on the deviation of rating values. In this way this system introduces an indication of the predictive power of the algorithm. High deviations correspond to high degrees of variation (or to insufficient activation in the transactions).

Decentralized Reputation Mechanisms

The main problem with the centralized reputation mechanisms is that they are not suitable for open Multi-Agent Systems given that the MAS nature is intrinsically not referred to a central authority, but prevails in the distribution of the various attitudes to the agents who are possibly considering the emergence of the global phenomenon.

An interesting mechanism, developed by Yu and Singh (Yu and Singh, 2003), take into consideration witnesses as information sources. In this way the system (called *referral system*) is based on the individual agents' knowledge and help, exploiting the agent's contacts for each agent in the system. In addition, agents cooperate with each other with respect to these referrals building in fact recommendations to contact other agents. Using referrals each agent has to store a list of agents it knows (acquaintances) with their expertise and to contact them in case of necessity. It is also possible that an agent who is unable to give answer to a query gives back referrals pointing to agents it believes to have the desired information.

A mixed centralized/decentralized approach is introduced by (Jurca and Faltings, 2003). They start from two main questions: Why should agents report reputation information? And, why should they report it truthfully? They think about a set of broker agents, buying and aggregating other agents' reports and selling back reputation information to the agents when they need it. The proposed payment scheme guarantees that agents reporting incorrect or false information will gradually lose money; on the contrary, the honest agents will not lose money.

A reputation model in which the trust evaluation process is completely decentralized has been introduced by (Sabater and Sierra, 2001) and its name is *Regret* (Sabater, 2003). In this model each agent can evaluate others' performance after each direct interaction has recorded its ratings in a local database. In fact in *Regret* the agents can evaluate trust by themselves, without any reference to a central mechanism. The Regret's authors called *direct trust* the trust value derived from these ratings: They are calculated as the weighted means of all ratings (where the weight depends from the recency). Similarly to *SPORAS*, *Regret* measures the predictive power of each trust value through the two reliability measures: the number of ratings taken into account in producing the value of trust and the deviation of these ratings.

In *Regret* the agents can also share their opinions. To this end the system develops a very interesting and sophisticated witness reputation component (for aggregating witness reports). This component strictly depends on the social network built up by each agent: in fact *Regret* uses social network to find witnesses, to select the witnesses for consulation and for weighting the witnesses' opinions. Finally, *Regret* introduces the concepts of *neighbourhood reputation* and *system reputation*. With *neighbourhood reputation* is meant the reputation of the target's neighbour agents: It is calculated by fuzzy rules. With *system reputation* is meant a mechanism to assign default trust values to the target agent based on its social role in an interaction (e.g. service provider, consumer).

Concluding *Regret* uses various sources of trust information, is decentralized, and as a consequence, satisfies the requirements for modelling trust in multi-agent systems. Its main limit is about the fact that it does not specify how the social networks are to be built.

Another interesting reputation model is *FIRE* (Huynh *et al.*, 2006). It is a decentralized model designed for general applications based on a variety of trust sources. In particular, these sources include: *direct experiences* (coming from the agent's interactions), *witness reports* (coming from third-party references), *role-based rules* (provided by end users encoding beliefs about the environment), and *certified reports* (references obtained by the target agent).

The addition of the fourth source (certified reports) to the traditional first three allows trust to measure in situations in which without this source there is no reference.

The *FIRE*'s authors show a set of experimental results in which the performance of their system is really good and better than other developed systems. In particular they show that:

- Agents using the trust measure provided by *FIRE* are able to select reliable partners for interactions and, thus, obtain better utility gain compared to those using no trust measure.
- Each component of *FIRE* plays an important role in its operation and significantly contributes to its overall performance.
- *FIRE* is able to cope well with the various types of changes in an open MAS and can maintain its properties despite the dynamism possible in an environment.
- Although decentralized, to suit the requirements of a trust model in open MAS, *FIRE* still outperforms or maintains a comparable performance level with *SPORAS*, a centralized trust model.

The main problem and limit with *FIRE* is about its assumption that agents are honest in exchanging information with one another: The authors are aware of this limit and are working to introduce reliability measures for witness ratings and certified ratings.

Different Kinds of Metrics

The computational approaches to trust can be classified also on the basis of the different forms of metrics to rate the trust performances. An analysis about the metrics of the different approaches is particularly relevant because it should give account about the underlying assumptions of these approaches.

Discrete or continuous measures (in general in the numerical range $(-1,+1)$ where -1 means no trust and $+1$ full trust) are the most common. But there are also *discrete verbal statements* like those used in (Abdul-Rahman and Hailes, 2000): *Very Trustworthy, Trustworthy, Untrustworthy, Very Untrustworthy*, that give a more direct representation of the human evaluations: The problem is then to translate these statements in adequate misures for the computational process.

We can cite (Schillo *et al.*, 2000) and (Banerrjee *et al.*, 2000) as examples of the use of bi-stable trust values (good or bad) (Ramchurn *et al.*, 2004); while (Witkowsky *et al.*, 2001) proposed the calculation of trust (a continuous measure) that deal with measurable quantities of bandwidth allocation and bandwidth use (they presented a scenario for telecommunications Intelligent Network in which bandwidth is traded by different agents).

Probabilistic approaches have the advantage of exploiting the consolidated apparatus of the probabilistic methods: in particular they can profit from the different derivation methods (from the probability calculus to the advanced statistical methods) (Josang, 2001), (Krukow and Nielsen, 2006).

An interesting and useful approach is given by the Belief Theory that responds to the limits of the probabilistic approaches with respect to uncertain information. In fact, the aim of *belief theory* is to give a formal representation of the *inaccurate* and *uncertain* aspect of information. In this case the sum of the probabilities over all possible outcomes do not necessarily sum up to 1 (the remaining probability can be interpreted as uncertainty).

We have to say that each of these metrics can present both advantages and problems, but what is important in our view is the fact that trust has an intrinsic multi-factorial nature and this peculiar feature has to be represented and implemented also in the quantitative aspects of this phenomenon (see Chapter 3).

One of the more interesting attempts in this direction is represented by the *REGRET* approach where to overcome the mono-dimensionality of the trust performance some fuzziness over the notion of performance itself is used. *REGRET* introduces a rich semantics for the ratings (called *impressions*) by defining their specific features (for example: *delivery date*, *price*, and so on). On the basis of the fuzzy reasoning techniques, the system is able to compose the different dimensions producing a general and global impression of one agent on another. With respect to the fuzzy approach applied to the trust evaluation we showed in Chapter 11 a specific implementation based on the socio-cognitive approach to trust.

Other Models and Approaches to Trust in the Computational Framework

Recent works on designing models for propagating trust in social networks and for selecting the most appropriate services in the service-oriented environments are of particular interest to analyze. In (Yu and Singh, 2003) the models of trust propagation are distinguished by:

- *Trust in expertise*: ability of providing services; and
- *Trust in sociability*: ability of providing referrals.

These two functions are quite different even if relevant: the former is becoming more important given the increasing use of the Internet by people (Internet users are 'now', 40 years after its birth, 1,663 million) and its circulation of both traditional and innovative services.

With regard to the latter, if we use trust to evaluate people and information we have to compute trust between people who do not know one another and expect to achieve the result that each agent in the network will evaluate the trustworthiness of its potential, often anonymous, partners (Golbeck, 2009).

In *CertProp* model (Hang *et al.*, 2009) i a trust propagation model based on three operators (*aggregation, concatenation* and *selection*) s introduced to efficiently and accurately propagate trust in social networks. In this approach a social network (system of interacting agents) is modeled as a directed graph with weights on the links: a sort of *social graph*, in which the nodes represent the users and the edges represent the relationships between these users (the social graph representation is quite widespread and used, see also (Ziegler, 2009), (Levien, 2009)). The problem of propagating trust in this network using the introduced operators is interesting even if not all the problems connected with the trust propagation are solved: the already presented question about the so-called 'trust transitivity' (see Chapter 6) remains an issue that has not been well addressed.

In any case this work, defining a set of algebric properties for the three operators, determines trust propagation in an efficient and accurate way (even if in a simplified domain). Another interesting question analyzed in this work is the classical mapping between *opinions* and *evidences*: in the networks the weights are subjective opinions and not objective evidence. Then, the authors propose approaches for transforming opinions in evidence: in particular they motivate a new way of this transformation based on Weber_fechner law (that describes the relationship between physical magnitudes of stimuli and the perceived intensity of the stimuli). This transformation also allows the idea that the average opinion yields the lower certainty of transformed trust. It helps to reduce the subjectivity in opinion-based datasets so that the evidence-based approaches like *Cert-Prop* can apply.

The attention to the mathematical properties of the operators, fails sometimes to catch the deeper nature of the trust phenomenon. For example, when I receive different, diverging evaluations (about Y) from two or more agents (say J and Z), not only can I discount the degree of trust in Y on the basis of J or Z's trustworthiness and believability in me (Hang *et al.*, 2009), not only do I have to combine those converging or diverging values with some mathematical 'aggregation', but I have to choose *among different heuristics*, strategies. Not necessarily is the final value of trust a mix of the various values. I may, for example, be a very suspicious and prudent guy (or adopt a prudent strategy), and, although J and Z say that Y is sure and good, since W says that Y is not good (or sure) I adopt W's view, and put aside J and Z's evaluations. Or I might have an optimistic attitude and adopt always the best, more favorable estimation. Or I might have a strong esteem of Z (as evaluator) and trust him very much; although J and W have different opinions I do not care about them, I adopt Z's opinion (trust) without discounting it by combining it with the other evaluations. In sum, there are different possible heuristics in combining (or not) various evaluations; there is not a unique ('rational') equation.

In (Richardson *et al.*, 2003) the trust propagation model allows each user to maintain trust in a small number of other users. This method first enumerates all paths between the user and every other user who has a local belief in a given statement. Then, the belief associated with each path (concatenation operator) is calculated, and combined with the beliefs associated with all paths (aggregation operator). The aggregation operator is the same as the *Cert-Prop's* one while the concatenation operator is different.

Trust metrics[1] compute quantitative estimates of how much trust an agent X should have in Y, taking into account trust ratings from other agents on the network.

Two main important applications of trust metrics are: *Advogate* (Levien, 2009) and *Appleseed* (Ziegler, 2009). Both these metrics can be classified as *local group trust metrics*. *Local* is intended versus *Global*: where *Global* take into account all peers in the network and the links connecting them; while *Local* trust metrics take into account personal bias. They operate on partial trust graph information (the web of trust for an agent X is the set of relationships emanating from X and passing through nodes X (directly or indirectly) trusts.

Advogate computes a set of accepted nodes in three steps. First, it is assigned a capacity to every node as a function of the shortest path distance from the seed to that node. Second, there is a transformation of the graph, adding extra edges from each node to a special node (called supersink). Third, it is computs the maximum network flow for the new graph: the accepted nodes are those that have a flow across the special node (supersink).

In contrast to *Advogate*, *Appleseed* uses spreading activation (Quilian, 1968). It spreads energy across the graph, and when propagated through a node, divides energy among sucessors based on the edge weights. The main idea in *Appleseed* is to simulate the spectral decomposition and it requires several iterations to converge towards the set of acceptable nodes.

---

[1] Here considered with a different meaning with respect to the previous section 'Different kinds of metrics in this chapter.

Another interesting work about trust propagation in social networks is focused on a different aspect of trust inference: in particular, the change in trust values in the network and the impact of that change on the results of the existing algorithms. In their work (Golbeck and Kuter, 2009), Golbeck and Kuter show an experimental study to understand the behavior of different trust inference algorithms with respect to the changes occuring in the social network. Their contribution is on different relevant items that they define with the following questions:

'How far does a single change propagate through the network? How large is the impact of that change? How does this relate to the type of inference algorithm?' ((Golbeck and Kuter, 2009), p. 170). They show the relevance of the chosen algorithm in all three questions.

As we have claimed above the problem of selecting trustworthy services on the web is becoming really relevant given the variety and quantity of the offer. In particular, two main problems should be taken in consideration:

a) The trust evaluation of a service should take into consideration the fact that very often a service is a *composed* service (with different providers, functions and responsibilities).
b) The *dynamism* of service providers and consumers (both the needs of the consumers and the providers' quality of service, continuously change).

Singh (Singh, 2003), has coined the term '*service-oriented computing*' for meaning 'the emergence of independent services that can be put together dynamically at run time and possibly across administrative domains' (p.39). In fact, while works exist about modeling trustworthiness of individual service providers, very few results were reached in modeling groups of providers working to a specific composed service. This approach is also important because – like in our model – consider the 'quality' or the 'competence' intrinsic part of trustwortiness and of trust; differently from many models (in game theory, in philosophy, etc.) that want to eliminate this component/dimension of trust to reduce it only to honesty, reliability, morality, etc.

Hafizoglu and Yolum (Hafizoglu and Yolum, 2009), propose a group trust model to understand the behavior of such teams that realize a composed service. Their work is in turn based on the service graphs model (Yolum and Singh, 2004) where graphs are helpful for reasoning about services that are related to each other. In (Hafizoglu and Yolum, 2009), the authors have to cope with the problem that 'the behavior of an agent in teamwork environment may differ from its behavior in single service environment'. In fact, collaboration may have some influence on the agents' performances. So the individual features of the agents in providing specific services are not so useful for selecting the right (best) team of the composed service. The authors analyze a set of possible tendencies for the agents (*ideal behavior, group antipathy, group motivation, colleague effect, task effect, familiarity effect*) that have influence on the agents' collaborative performances.

Another interesting work on this problem is (Hang and Singh, 2009) in which is proposed a trust-aware service selection model based on a Bayessan network. The model evaluates the service trustworthiness on both direct and indirect (from referrals) experience. The method models causal relationships between services with Bayesan networks.

The main characteristic of this model is that it can deal with incomplete observation (in such a way taking into account the possibility that underlying services may not be exposed to the consumer) for this introduces a specific parameter representing the percentage of missing data.

12.3 Concluding Remarks

It is now clear – even in engineering – that when we build a new technology, for direct human use, in fact we are building a new 'socio-tecnical system' and even a new 'cognition' dealing with and incorporating that mental, pragmatic, and social 'extension'. This is true with mechanical engineering (factories, cars, tractors, etc,) but is much more important with cognitive and social technologies: like computers, web, and their mediation and support of the entire human individual and social activity, from study and learning to work, from friendship and communities to political participation, from market and business to smart learning environments.

We have to design hand in hand with technology the cognitive, interactive, and collective dimensions. More precisely, we have to design technology with those incorporated dimensions. But in order to do this one should have the appropriate understanding of those dimensions and some theoretical abstraction of them and some possible modeling of them. Otherwise we proceed just in an empirical, haphazard (trials and errors) way.

This is why we believe that a deep and complete model of trust (including the cognitive, emotional, decisional, social, institutional) dimension be not just useful but necessary.

In particular, we believe that to support this kind of human computer, human ambient, human robot interaction, and computer-mediated/supported interaction, organization, work, etc. a technology able to deal with typical human cognitive and social features and phenomena (like expectation, intentions, preferences, like emotions, trust, etc. like norms, roles, like institutions, collectives, etc.) must be designed. A technology endowed with autonomous learning, decentralization, acquisition of local and timely information; able to reason and solve problems; endowed with some proactivity and a real collaborative (not just executive) attitude. We think that autonomous computational 'agents' will play a significant role. But if this is true this will make even more central the role of trust and delegation and of their modelling.

References

Abdul-Rahman, A. and Hailes, S. (2000) Supporting Trust in Virtual Communities. In Proceedings of the Hawaii International Conference on System Sciences, Maui, Hawaii, 4–7 January 2000.

Abrams, M.D. (1995) Trusted system concepts. *Computers and Security*, 14 (1): 45–56.

Amazon site, http://www.amazon.com, world wide web.

Anderson, R. (2001) *Security Engineering: A Guide to Building Dependable Distributed Systems*, John Wiley & Sons Ltd.

Antoniou, G. and van Harmelen, F. (2008) *A Semantic Web Primer*, 2nd edition. The MIT Press.

Baecker, R.M. and Buxton, W.A.S. (eds.) (1987) Readings in Human-Computer Interaction. A multidisciplinary approach. Los Altos, CA: Morgan-Kaufmann Publishers.

Barber, S., and Kim, J. (2000) Belief Revision Process based on trust: agents evaluating reputation of information sources, *Autonomous Agents 2000 Workshop on 'Deception, Fraud and Trust in Agent Societies'*, Barcelona, Spain, June 4, pp. 15–26.

Bishop, M. (2005) *Introduction to Computer Security*. Reading, MA: Addison-Wesley.

Blaze, M., Feigenbaum, J., Lacy, J. (1996) Decentralized trust management. In Proceedings of the 1996 IEEE Conferente on Security and Privacy, Oakland, CA.

Card, S.K., Moran, T.P. and Newell, A. (1983) The Psychology of Human-Computer Interaction. Hillsdale, NJ: Lawrence Erlbaum Associates.

Castelfranchi, C. (1996) Reasons: belief support and goal dynamics. *Mathware & Soft Computing*, 3: 233–247.

Castelfranchi, C., de Rosis, F., Falcone, R., Pizzutilo, S. (1998) Personality traits and social attitudes in Multi-Agent Cooperation, *Applied Artificial Intelligence Journal*, special issue on 'Socially Intelligent Agents', 12 (7/8): 649–676.

Castelfranchi, C., Falcone, R., Lorini, E. (2009) A non-reductionist approach to trust. In Golbeck, J. (ed.) *Computing with Social Trust. Human Computer Interaction Series*, Springer.

Castelfranchi, C. and Falcone, R. (1998) Towards a theory of delegation for agent-based systems, *Robotics and Autonomous Systems*, Special issue on Multi-Agent Rationality, Elsevier Editor, 24 (3-4): 141–157.

Castelfranchi, C. and Falcone, R. (1998) Principles of trust for MAS: cognitive anatomy, social importance, and quantification, *Proceedings of the International Conference on Multi-Agent Systems (ICMAS'98)*, Paris, July, pp. 72–79.

Castelfranchi, C. and Tan, Y.H. (eds.) (1999) *Trust and Deception in Virtual Societies*. Kluwer, Dordrecht.

Chellas, B.F. (1990) *Modal Logic: an introduction*. Cambridge University Press. Cambridge.

Cofta, P. (2007) *Trust Complexity and Control*. John Wiley & Sons Ltd.

Cohen, P.R. and Levesque, H.J. (1990) Intention is choice with commitment. *Artificial Intelligence*, 42, 213–261.

Davies, J. (2006) *Semantic Web Technologies: trends and research in ontology-based systems*. John Wiley & Sons Ltd.

Demolombe, R. (1999) To trust information sources: A proposal for a modal logic framework. In Castelfranchi, C., Tan, Y.H. (eds.) *Trust and Deception in Virtual Societies*. Kluwer, Dordrecht.

Dix, A. Flay, J., Abowd, G. and Beale, R. (2004) *Human-Computer Interaction*. Third edition. Prentice Hall.

Dragoni, A. F. (1992) A model for belief revision in a multi-agent environment. In *Decentralized AI - 3*, Y. Demazeau, E. Werner (eds.), 215–231. Amsterdam: Elsevier.

Dubois, D. and Prade, H. (1980) *Fuzzy Sets and Systems: Theory and Applications*, Academic Press, Orlando, FL.

Dubois, D. and Prade, H. (1980) *Fuzzy Sets and Systems: Theory and Applications*, Academic Press, Orlando, FL.

eBay site, http://www.ebay.com, world wide web.

Ellis, J. and Speed, T. (2001) *The Internet Security Guidebook*, Academic Press.

Fagin, R. and Halpern, Y. (1994) Reasoning about Knowledge and probability. *Journal of the Association for Computing Machinery*, 41 (2): 340–367.

Falcone, R., Singh, M., Tan, Y.H. (eds.) (2001) Trust in Cyber-societies. Lecture Notes on Artificial Intelligence, n°2246, Springer.

Falcone, R., Barber, S., Korba, L., Singh, M. (eds.) (2003) Trust Reputation, and Security: Theories and Practice. Lecture Notes on Artificial Intelligence, n°2631, Springer.

Falcone, R., Pezzulo, G., Castelfranchi, C. (2003) A fuzzy approach to a belief-based trust computation. *Lecture Notes on Artificial Intelligence*, 2631, 73–86.

Falcone, R., Barber, S., Sabater-Mir, J., Singh, M. (eds.) (2005) Trusting Agents for Trusting Electronic Societies. Lecture Notes on Artificial Intelligence, n°3577, Springer.

Falcone, R., Barber, S., Sabater-Mir, J., Singh, M. (eds.) (2008) Trust in Agent Societies. Lecture Notes on Artificial Intelligence, n°5396, Springer.

Falcone, R. and Castelfranchi, C. (2001) Social trust: a cognitive approach, in C. Castelfranchi and Y. Tan (eds.), *Trust and Deception in Virtual Societies*, Kluwer Academic Publishers, pp. 55–90.

Foster, I. and Kesselman, C. (2003) *The Grid 2: Blueprint for a new computing*. Morgan Kaufmann Publishers Inc. San Francisco, CA.

Golbeck, J. (ed.) (2009) Computing with Social Trust. Human Computer Interaction Series, Springer.

Golbeck, J. and Kuter, U. (2009) The ripple effect: change in trust and its impact over a social network. In Golbeck, J., (Ed.), *Computing with Social Trust. Human Computer Interaction Series*, Springer.

Grandison, T., and Sloman, M. (2000) A survey of trust in internet application. IEEE Communication Surveys & Tutorials, 3 (4).

Hang, C.W., Wang, Y., Singh, M.P. (2009) Operators for Propagating Trust and their Evaluation in Socila Networks. In Proceedings of the 8th International Joint Conference on Autonomous Agents and Multi-agent Systems (AAMAS).

Hewitt, C. and Inman, J. DAI betwixt and between: from 'intelligent agents' to open systems science, *IEEE Transactions on Systems, Man, and Cybernetics*. Nov./Dec. 1991.

Huynh, T. D., Jennings, N. R., and Shadbolt, N. R., (2006) An integrated trust and reputation model for open multi-agent systems. *Autonomous Agents and Multi-Agent Systems Journal*, 13, 119–154.

Jones, A.J.I., Firozabadi, B.S. (2001) On the characterization of a trusting agent: Aspects of a formal approach. In Castelfranchi, C., Tan, Y.H. (eds.) *Trust and Deception in Virtual Societies*, pp. 55–90. Kluwer, Dordrecht.

Jonker, C., and Treur, J. (1999) Formal Analysis of Models for the Dynamics of Trust based on Experiences, Autonomous Agents '99 Workshop on 'Deception, Fraud and Trust in Agent Societies', Seattle, USA, May 1, pp. 81–94.

Josang, A. A logic for uncertain probabilities. *International Journal of Uncertainty, Fuzziness and Knowledge-based Systems* 9 (3): 279–311, 2001.

Josang, A. (2007) Trust and Reputation Systems, in Aldini, A., Gorrieri, R. (eds.), Foundations of Security Analysis and Design IV, FOSAD 2006/2007 Tutorial Lectures. Springer LNCS 4677.

Jøsang, A., and Ismail, R. (2002) *The Beta Reputation System*. In the proceedings of the 15th Bled Conference on Electronic Commerce, Bled, Slovenia, 17-19 June 2002.

Jurca, R. and Faltings, B. (2003) Towards incentive-compatible reputation management, in Falcone, R. et al. (eds.), *Trust, Reputation and Security: Theories and Practice*, LNAI Vol. 2631 (pp. 138–147), Sprinter-Verlag.

Kosko, B. (1986) Fuzzy cognitive maps. *International Journal Man-Machine Studies*, 24: 65–75.

Krukow, K. and Nielsen, M. (2006) From Simulations to Theorems: A position paper on research in the field of computational trust. In Proceedings of the Workshop of Formal Aspects of Security and Trust (FAST 2006), Ontario, Canada, August 2006.

Lesser, V.R. An overview of DAI: viewing distributed AI as distributed search. *Journal of Japanese Society for Artificial Intelligence.* Special issue on Distributed Artificial Intelligence, 5 (4): 392–400, 1990.

Levien, R. (2009) Attack-resistant trust metrics. In Golbeck, J., (Ed.), *Computing with Social Trust. Human Computer Interaction Series*, Springer.

Liau, C.J. (2003) Belief, information acquisition, and trust in multi-agent systems: a modal logic formulation. *Artificial Intelligence*, 149: 31–60.

Lorini, E. and Demolombe, R. (2008) From binary trust to graded trust in information sources: A logical perspective, pp. 205–225. In Falcone, R., Barber, S., Sabater-(Mir and Singh, 2008) Mir, J. and Singh, M. (eds.) (2008) Trust in Agent Societies. Lecture Notes on Artificial Intelligence, n°5396, Springer.

Marsh, S. P. (1994) Formalising Trust as a computational concept. PhD thesis, University of Stirling. Available at: http://www.nr.no/abie/papers/TR133.pdf.

Motro, A. and Smets, Ph. (1997) *Uncertainty Management in Information Systems: from needs to solutions.* Klewer, Boston.

Pezzulo, G. and Calvi, G. (2004) AKIRA: a Framework for MABS. *Proc. of MAMABS 2004.*

Quilian, R. (1968) Semantic memory, In Minsky, M., (editor), *Semantic Information Processing*, MIT Press, Boston, MA, USA, pp. 227–270.

Ramchurn, S., Huynh, T. D., Jennings, N. R. (2004) Trust in multi-agent systems. *The Knowledge Engineering Review*, Cambridge University Press, 19 (1): 1–25.

Rasmussen, L. and Janssen, S. (1996) Simulated Social Control for Secure Internet Commerce. In Meadows, C. (Editor), Proceedings of the 1996 New Security Paradigm Workshop. ACM.

Resnick P. and Zeckhauser, R. (2002) Trust among strangers in internet transactions: Empirical analysis of eBay's reputation systems. In Baye, R. (Editor), The economics of the internet and e-commerce. Vol. 11 of *Advances in Applied Microneconomics*. Elsevier Science.

Richardson, M., Agrawal, R., and Domingos, P. (2003) Trust Management for the Sematic Web, in the Second ISWC, LNCS 2870, pp. 351–368.

Sabater, J. (2003) Trust and Reputation for Agent Societies, PhD thesis, Universitat Autonoma de Barcelona.

Sabater, J. and Sierra, C. (2001) Regret: a reputation model for gregarious societies, In 4th Workshop on Deception and Fraud in Agent Societies, (pp. 61–70). Montreal, Canada.

Schillo, M., Funk, P., and Rovatsos M. (1999) Who can you trust: Dealing with deception, Autonomous Agents '99 Workshop on 'Deception, Fraud and Trust in Agent Societies', Seattle, USA, May 1.

Schneier, B. (1996) *Applied Cryptography*, 2nd edn, Wiley.

Shafter, G. (1976) *A Mathematical Theory of Evidence*. Princeton University Press: New Jersey.

Shoham, Y. and Leyton-Brown, K. (2008) *Multi-agent Systems: Algorithmic, game-theoretic, and logical foundations.* Cambridge University Press.

Singh, M.P. (2003) Trustworthy service composition: challenger and research questions. In Falcone, R., Barber, S., Korba, L., Singh, M., (Eds.), Trust Reputation, and Security: Theories and Practice. Lecture Notes on Artificial Intelligence, n°2631, Springer.

Skevington, P. and Hart, T. Trusted third parties in electronic commerce, *BT Technology Journal*, 15 (2), 1997.

Skogsrud, H., Benatallah, B., and Casati, F. (2003) Model-driven trust negotiation for web services. *IEEE Internet Computing*, 7 (6): 45–52.

Smith, R. G. (1980) The contract net protocol: High-level communication and control in a distributed problem solver. *IEEE Transactions on Computers*.

Stallings, W. (1999) *Cryptography and Network Security*, Prentice Hall.

Stallings, W. (2001) *SNMP, SNMPv2, SNMPv3, and RMON 1 and 2*, 3rd edn, Addison Wesley.

Uszok, A., Bradshow, J. M., and Jeffers, R. (2004) KAoS: A policy and domain services framework for grid computing and semantic web services. In Jensen, C., Poslad, S., and Dimitrakos, T. (eds.), *Trust Management: Second International conference*, iTrust 2004, Oxford, UK, March 29–April 1, 2004. Lecture Notes in Computer Sciente n°2995, (pp. 16-26), Sprinter-Verlag, Berlin, Heidelberg.

Weiss, G. Distributed artificial intelligence meets machine learning: learning in multi-agent environments, Berlin: Springer-Verlag, Lecture notes in *Computer Science*, vol. 1237, 1997.

Wong, H.C., Sycara, K. Adding security and trust to multi-agent systems, Proceedings of Autonomous Agents '99. (Workshop on Deception, Fraud and Trust in Agent Societies), 2000.

Wooldridge, M. (2002) *An Introduction to Multi-agent Systems*, John Wiley & Sons Ltd.

Yu, B. and Singh, M.P. (2003) Searching social networks. In Proceedings of the second international joint conference on autonomous agents and multi-agent systems (AAMAS). Pp. 65-72. ACM Press.

Zacharia, G. and Maes, P. Trust management through reputation mechanisms. *Applied Artificial Intelligence Journal*, 14 (9): 881–908, October 2000.

Ziegler, C.N. (2009) On propagation interpersonal trust in social network. In

13

Concluding Remarks and Pointers

Let us conclude this quite complex book, on the one hand, with some brief considerations, on the other hand, with two short paragraphs pointing to two very relevant and very topical developments: the new frontiers of neuro-science of trust; political dimensions of trust and of its crisis.

13.1 Against Reductionism

As for the general remarks, we hope that our readers have realized that trust is not at all a simple object and notion, but that it is possible and necessary (or at least very useful) to recollect its different aspects, functions, and components in a unitary, integrated, principled, and well-structured frame. This has been our attempt; against a lot of very reductive treatments of trust, due to very limited and *ad hoc* domains or disciplinary interests, or to very rigid formal tools.

We have systematically argued against not just abstraction, simplification, or normativity (very necessary in science), but against deformation and reductionism: arguing, for example, against the reduction of trust to a mere mental attitude without a behavioral or relational component, or to a mere vague feeling without specific thoughts (evaluations, predictions, and so on), or to a mere measure of the frequency of successes, or to mere subjective probability of a favorable event, or just to behavioral reliance, or just to exchange situations and expectation of reciprocation, and so on.

Moreover, it is not at all a question of taking into account all the components and functions, just adding one to the other, but of showing how one property is based on the other, one component is integrated with the others in a precise structure.

Finally, the structural nature of trust should not just be its 'statics', but it should also find and justify its 'dynamics': how trust increases or decreases, how we build it, how trust propagates, or how it is generalized, instantiated, transferred from one agent or one task/good to another, and so on.

No doubt our solution is still unsatisfactory, but we hope that our readers will be persuaded that this is the right objective: *an explicit anatomy of trust, an integrated and justified model of its ingredients, of their integration, and of how it works.* We do not need just domain specific

Trust Theory   Cristiano Castelfranchi and Rino Falcone
© 2010 John Wiley & Sons, Ltd

'measures' of some aspect of trust, or some *ad hoc* technical notion of very limited use; we need a real *theory* of trust. This is the (hard) mission of science.

13.2 Neuro-Trust and the Need for a Theoretical Model

In the last few years a set of studies on the neurobiological evidence of trust have been developed. One of the main claims from Kosfeld *et al.*'s neurobiological approach (Kosfeld *et al.*, 2005) about trust is that trust 'cannot be captured by beliefs about people's trustworthiness and risk preference alone, but that social preferences play a key role in trusting behaviour' (Fehr, 2008). In other words, on the basis of their studies (Kosfeld *et al.*, 2005) there is an important and significant distinction between risk constituted by asocial factors and that based on interpersonal interactions. The thesis would be: by analyzing the neurobiological bases of trust it is shown that a trustor does not decrease the general sensitivity to risk, but on the contrary, this sensitivity seems to be decreased in social interaction.

However, how did they determine the neurobiological bases? They experimented with the trust game (involving real monetary exchanges between two actors playing the roles of trustor and trustee) with two groups of students, one that had received neuropeptide oxytocin and the other that had received an inert placebo. As Fehr writes: 'the rationale for the experiment originates in evidence indicating that oxytocin plays a key role in certain prosocial approach behaviours in non-human mammals. (. . .) Based on the animal literature, Kosfeld *et al.* (2005), hypothesized that oxytocin might cause humans to exhibit more behavioural trust as measured in the trust game.' (Fehr, 2008). In these experiments they also show how oxytocin has a specific effect on social behaviour because it impacts differently on the trustor and the trustee (only in the first case is there a positive influence). In addition, it is also shown that the trustor does not reduce the sensitivity to risk as a general behaviour but as a consequence of the partner nature (human versus not-human).

Following this paradigm we could say that when some social framework is given and perceived by the subject, oxytocin is released in a certain quantity, so modifying the activity of precise regions of the brain and consequently producing a more or less trusting behavior. But, what are the features of that social framework? Is this framework just based on the presence of other humans? And what is the role played by past experiences (for modulating the oxytocin release)? How are the unconscious and spontaneous bias (characterizing the emotional, hot, not rationale trust) related to the conscious deliberative reasoning (characterizing the rational and planned trust)?

The results of this discovery are without doubt quite relevant. But they must be complemented and interlinked with a general cognitive theory of trust. Without this link, without a mediation of more complex structures and functions, we risk that an articulate and multi-dependent concept is trivially translated only in a chemical activation of a specific brain area cutting out all the real complexity of the phenomenon (unable to explain, for example, why, before taking a very difficult decision we think on this for hours, sometimes for days. And how the contribution of the different beliefs impacts on that choice). The fact of individuating and establishing some precise and well defined basic neuro-mechanisms can be considered an important advancement for the brain studies and also for founding the cognitive models of the behaviour. In any case, in our view, they cannot be considered as the only description of the external behaviour (based on more sophisticated, articulated and complex notions than a mapping of chemical activations), unable of really playing a role in a detailed and 'usable' prediction and description of the phenomenon. Without this mediation *(psycological interpretation) localization or biochemistry say nothing.*

For example, in addition to the clear general mechanism established by the study of Kosfeld *et al.* about a social predisposition of trusting behavior based on the oxytocin released, it also seems clear how the differences of the emotional part of trust among human individuals, in the same situation, can be explained by the different physiological functions of these release mechanisms. In our cognitive theory of trust these facts can be taken into account by the role played from the *different thresholds of the model* (one of the few parts that in our model includes the emotional component of trust; see Chapters 3 and 5), where the rationality describes only the macro attitudes (the general trends) and the fine-grained differences are due to the various subjectivities.

At the same time, the role of influence and guide of the rational, deliberative, belief-based process on the final trusting behaviour, also interacting and sometimes overlapping on the release mechanism of oxytocin should be also analyzed. The necessity of a socio-cognitive model of trust for a strict interaction, comparison and guidance (in both directions from the experimental results to the model and viceversa) would be really interesting.

It could be possible to experiment with the different trustor's beliefs on the trustee's basic features, evaluating how these beliefs (about individual willingness, individual competence or class, category, group membership, and so on) influence the trusting behavior of the trustor. The same holds for the dynamical aspect of trust.

In sum, trust cannot be reduced to a simple, vague, unitary notion and activation: it is a complex structure of representations, related feelings, dispositions, decisions, and actions. The analytic (compound) model of this phenomenon should *guide* the brain research, which otherwise looks rather blind, reductive, and merely suggestive.

13.3   Trust, Institutions, Politics (Some Pills of Reflection)

During these years of very serious financial, economic, and social crises 'trust' is every day in the newspaper headlines or in the speeches of political leaders, as the crucial issue of the crisis.

It seems in fact that trust is the engine of socio-economic development: there is a clear awareness from the public institutions, from the social and economic authorities, that trust is the glue of society (see Chapter 9). That if trust decreases to below a particular level, it could compromise some relevant social function (for example, those based on implicit reciprocal trust (see Chapter 4)). So many (public and private) actions are designed, planned and realized to increase it, to elicite individual and collective trust in society.

A particularly interesting case is trust in politics and institutions. As we have said (Chapter 2) trust is structurally based on the achievement of some goal, interest, need, desire, objective, etc. Without this motivational element there can not be trust. In the case of representative and institutional entities, these goals are (or better, should be) relative to collective achievements, public interests, ideal attainments. So the frustration of these achievements over a long duration can have different solutions: all of them explainable through our socio-cognitive model of trust. Let us show this very briefly and in a simplistic way.

A first solution is to attribute the direct responsibility of these failures to a specific political party (or to individual leaders). The result is the change of vote or preference. This is the more likely scenario.

A second solution is to attribute the responsibility of these failures to the external conditions in which the representatives of the people work (the world (global) causes for internal crisis are more often considered as unavoidable causes). So it is necessary to have a redefinition of the goals as achievable in these new contexts.

A third solution is to consider the goal in general as not achievable independently from the specific political party or leader. This kind of attribution can be realized in particular in the cases in which relevant changes and advances are expected with respect to the current social, cultural, economic situation. So there could be a disappointment with respect to the possibility of these changes. There needs to be backing of the social goals and an advancement of the individual ones and a departure from politics and institutions and a re-evaluation of private and individual objectives. The result is a weakening of collective goals, values, principles, of social attainments, with clear consequences on the cooperative and interactive climate. At the same time, given the social intrinsic nature of the environment in which humans live, there will emerge, on the one hand, a disorientation with respect to the identity, the ends both of the single individual and society; on the other hand the need will emerge for alternative social infrastructures and entities, persuing collective goals (even of small communities).

A concrete example of the attribution of the responsibilities to the context is given in the Italian political situation: the good public-opinion performance of Berlusconi's government in these years of very serious economical and social crisis. One of the reasons for this endurance is precisely the fact that, despite the continuous very bad trends and results, people ascribe them to external circumstances and international factors. The government (with the help of its quasi-monopolistic control of media and thanks to its 'activism' on the media) is able to convey to the public the image of 'doing as much as possible!'. This is actually false (while reading for example the data about the strong increase in tax evasion, or comparing investments in Italy and Germany, or listening to trade unions, Confidustria, or Bank Italia analyses), but what matters obviously is the information and the image: 'We are doing our best! Circumstances are adverse').

On the contrary, the previous government (Prodi) was seriously affected by a trust crisis; not only for having too high expectations, but also because there was a systematic *internal attribution* of all their difficulties and partial successes. This was very much due to the continuous media-communication of disagreements and fights inside government or its majority, a weak decision style, etc. This gave the impression that they (the government and those parties) were inadequate, that any partial success was obvious and not so much due to their ability or commitment, while any retard or failure was definitely due to their internal limits and fragility.

13.3.1   For Italy (All'Italia)

In a recent editorial of an Italian newspaper a journalist was synthesizing the core problem of Italy and of its crisis, with these words: ' . . . . *It is a collective disaster, the greater tragedy: we are losing trust, will to fight, hope'.*

Why is losing trust – especially for a country – a big tragedy? And what is the link between will to fight and hope? Maybe, because trust is, on the one hand, the social glue of any collective and collaborative aspect and action? Of any institution not overwhelmed by fear? Maybe because trust is the oxygen of the individual enterprise?

We hope we have explained this and/or provided the fundamental tools to explain it. We tried to build explicit and operational models to explain the nature of the phenomenon, its subtle relationships, to understand the (mental and interpersonal) backwater 'processes' and 'mechanisms', which are not easily observable.

What – for example – is the link between trust and '*will to fight*', that is the felt motivation to fight for achieving our own (individual or collective) objective against perceived obstacles and difficulties? The link is rich; on the one hand, I cannot feel sure and be convinced of fighting without some self-trust, some positive expectation of possible success, but also some positive evaluation of me, of my capacity and persistence, some sense of control, the feeling that this result also depends on me (on us) (not on providence or some generous powerful man). Moreover, if this should be a collective effort and result I have to trust the others, their motivation, conviction, self-esteem and their trust in me. There must be reciprocation and collaboration, no betrayal and cowardice. Moreover, I have to trust the fair rules of the social competition, my institutions, legal conditions, of our possibility and willingness to change them. I have to trust possible organizations in this fight, and their leaders. Even some form of trust in the adversaries is needed (except in some form of real war). And so on. Moreover, those trusts are not simply independent and additional; they are circular: the one influencing the other; growing together or crowding down together. Moral dimensions influence the institutional one, and collective atmosphere influences the individual motivation and attitude, and so on. This is the Italian 'tragedy'.

And what is the link between the crisis of trust and the 'perdita' (loss) of hope and of a sense of the future? As we have explained, trust (especially trust as decision, counting on, action) entails some 'hope', necessarily, as a subcomponent: the positive expectation that those results are possible, will hopefully become true. This is the core of hope in its broadest sense.

References

Fehr E., On the Economics and Biology of Trust, Technical Report of the Institute for the Study of Labor, no 3895, December 2008.

Kosfeld, M., Heinrichs, M., Zak, P.J., Fischbacher, U. & Fehr, E., Nature 435: 673–676, 2005.

Index